Encyclopedia of Environmental Protection and Management

Volume I

Encyclopedia of Environmental Protection and Management
Volume I

Edited by **Chuck Lancaster**

R CALLISTO REFERENCE

New York

Published by Callisto Reference,
106 Park Avenue, Suite 200,
New York, NY 10016, USA
www.callistoreference.com

Encyclopedia of Environmental Protection and Management
Volume I
Edited by Chuck Lancaster

International Standard Book Number: 978-1-63239-243-5 (Hardback)

Printed in the United States of America.

Contents

Preface

The term 'environmental protection' can be traced to the year 1184, when during the German occupation of Tanzania, conservation laws were enacted for the protection of game and forests. As the name suggests in itself, the term involves all activities which attempt to conserve the environment. Environmental protection operates at individual, organizational and governmental levels.

Degradation of environment is happening in all spheres of life. From air pollution, to global warming, to soil and water pollution, the number of pollutants in the environment is only increasing. Thus, in the present scenario, environmental protection is the need of the hour as it will benefit both the natural environment and human beings.

Governments have begun placing restraints on activities that cause environmental degradation. Conservation of environment however, requires a combined effort of both the individual and the society. There's a need to sensitize and create awareness about the significance of the environment and relevance of environmental protection. And this awareness needs to start at the grass-root level. The consequences of constant environment degradation need to be brought to the forefront and emphasized.

I'd like to thank all the contributors who have shared their valuable studies and researches on environmental protection with us to make this book a valuable one. I would also like to thank my family for supporting me at every step.

Editor

Sustainable Management of Algae in Eutrophic Ecosystems

William W. McNeary[1], Larry E. Erickson[2]

[1]Department of Chemical Engineering, University of Missouri-Columbia, Columbia, USA; [2]Department of Chemical Engineering, Kansas State University, Manhattan, USA.

ABSTRACT

The accelerated eutrophication of the world's freshwater and marine ecosystems is a complex problem that results in decreased productivity, loss of biodiversity, and various economic woes. Controlling algae populations in a eutrophic water body has values in mitigating some of these negative effects. This paper reviews a number of strategies for algae management, with a focus on sustainable practices that have minimal environmental impact. The information in the literature is then used to propose a design for an integrated algae-aquaculture system to be used for the dual purposes of nutrient assimilation and production of fish and algal biomass. Effectiveness of the proposed system and possible revenue streams to offset capital costs are examined; other solutions that utilize the techniques in the literature are also explored.

Keywords: Algae; Nutrients; Eutrophication; Ecosystem; Dead Zone

1. Introduction

Anthropogenic eutrophication of the world's aquatic ecosystems, both marine and freshwater, is a serious environmental, social, and economic concern. As human society advances, many of the agricultural and industrial processes necessary to maintain our quality of life inadvertently release nutrients (primarily nitrogen and phosphorus) into nearby waterways. Increased nutrient loading into lakes, reservoirs, and coastal areas promotes sudden biomass growth, which inevitably results in large amounts of dead plant and animal matter in the water; the aerobic bacteria that consume the detritus create a low-oxygen (hypoxic) environment that damages other aquatic life and can disrupt delicate ecosystems [1]. Algae and cyanobacteria blooms resulting from the eutrophication process often compound the problem by causing taste and odor problems and producing harmful toxins that can endanger the health of those who utilize the water body for recreation or drinking water [2]. In the Gulf of Mexico, where stratification between the incoming freshwater and the saline Gulf water exacerbates oxygen deficiency, the "Dead Zone" that appears annually averages over 5000 square miles every year, and is severely detrimental to both the tourism and fishing industries—the NOAA estimates losses for both to be on the order of $82 million per year [3]. Freshwater eutrophication in the United States results in value losses in recreational water usage, waterfront real estate, threatened/endangered species recovery, and drinking water, all of which incur annual costs of nearly $2.2 billion [4]. Obviously, effective and sustainable solutions are required to address a problem of this magnitude.

A key to any eutrophication management strategy is the control of the nutrient sources affecting the water body. Point sources, such as industrial effluents, and non-point sources, like agricultural fields, require different methods of control. Point source nutrient output is commonly influenced by governmental policies, such as phosphate bans or discharge limits; non-point sources require collaboration between landowners and governing bodies across watersheds to implement best management practices for fertilizer application and erosion control [2]. In order to increase the effectiveness of these tactics and minimize recovery time, other mitigation strategies in and around the eutrophic water body are often used in tandem with source reduction.

This paper will conduct a review of established mitigation technologies, as well as newly-developed methods that hold promise in combating eutrophication. All strategies discussed will have greatest utility when used as a supplement to source reduction; on their own, their effects may be marginal. In addition, a brief overview of integrated algae-aquaculture systems will be conducted, followed by a proposal for a sustainable aquaculture system in the Mississippi delta that is meant to address the Gulf of Mexico dead zone and produce marketable products.

2. Mitigation Strategies

2.1. Biomanipulation

The effects of increased nutrient loading are felt throughout the ecosystem of a body of water experiencing eutrophication. The trophic cascade model (see **Figure 1**) is often used to visualize the relationships between organisms involved in the pelagic food web, which is the first system affected by nutrient influx. It is widely understood that phytoplankton populations explode when limiting nutrients are delivered in excess; fish biomass also increases with them. However, zooplanktivorous fish species often come to dominate in the new nutrient rich environment [5], which results in a "top down" effect on the food web in which the diminished zooplankton population is no longer able to hold the booming phytoplankton in check, thereby worsening the algae bloom issues.

Ecosystem restructuring that utilizes the trophic cascade model as a tool to increase water quality and decrease phytoplankton population is known as biomanipulation [6]. Typically, biomanipulation schemes involve restructuring the eutrophic food web by supplementing the population of piscivorous fish in order to control zooplanktivorous fish and alleviate pressure on the algae-grazing zooplankton populations [6,7]. Addition of filter-feeding fish such as tilapia has also been attempted in hopes to directly control the algae population [8]. Biomanipulation is an attractive option for combating the effects of cultural eutrophication due to its comparatively low implementation costs and utilization of existing natural systems, but potential drawbacks do exist. Strategies proven successful in temperate climate water bodies do not always find the same level of success when implemented in tropical environments—indeed, some instances of further water quality degradation due to invasive species and nutrient excretion levels of certain fish species have been recorded [5,7,8]. It is hypothesized that unintended consequences such as these result from the more diverse trophic structure of tropical water bodies. The long-term effectiveness of biomanipulation also remains in doubt; therefore, it is recommended that all such strategies be evaluated on a case-by-case basis and undertaken only by parties with a firm working knowl-

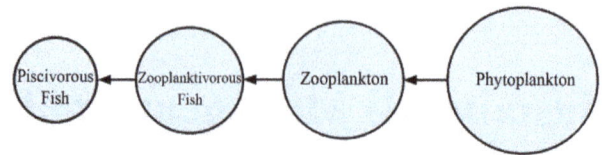

Figure 1. Trophic cascade model (adapted from [6]).

edge of the food webs at work in the ecosystem of the water body in question [6].

2.2. Artificial Circulation

Artificially-induced mixing of hypoxic water bodies holds value in eutrophication control. Typical circulation techniques can be broken down into two categories: destratification, in which water or air is forced into the bottom of the water column, thereby encouraging mixing and destroying the temperature gradient, and hypolimnetic aeration, where water is removed, aerated and returned to its original depth; the latter aims to actually inhibit mixing and preserve the natural biochemical cycling of the water body [1]. A summary of the beneficial effects of destratification can be seen in **Figure 2**.

Destratification has been proven as an efficient method of cyanobacteria bloom prevention; however, experimental evidence suggests that it may promote diatom growth and does little to reduce total algae populations [9]. Hypolimnetic aeration technologies are well-established and effective at alleviating hypoxic conditions [10], but have been found to be of little benefit in reducing internal phosphorus loading from P stored in lakebed sediments [11]. As many of these setups use pneumatic technologies, the associated capital costs and energy consumption rates are quite high [12], which limits the applicability of artificial circulation to smaller inland water bodies, such as lakes and reservoirs.

2.3. Removal of Harmful Species

Direct removal of the algae biomass in a eutrophic water body provides immediate environmental improvement by addressing both the cause of the algae bloom and its effects on surrounding aquatic life. Nutrients synthesized in the growth of the algae will be removed along with the biomass, thereby providing a pathway for permanent nutrient reduction and addressing the cause of the bloom; the reduction is further enhanced since harvesting a lower-trophic organism such as algae eliminates nutrient loss incurred through upper-level food chain inefficiencies [1]. Algae harvesting will also limit the growth of the heterotrophic bacteria that are the direct cause of hypoxia by removing organic matter [13]. Conventional algae harvesting methods include trawling with plankton nets and traveling screens; a pilot-scale experiment focused on harvesting cyanobacteria in the Baltic Sea was

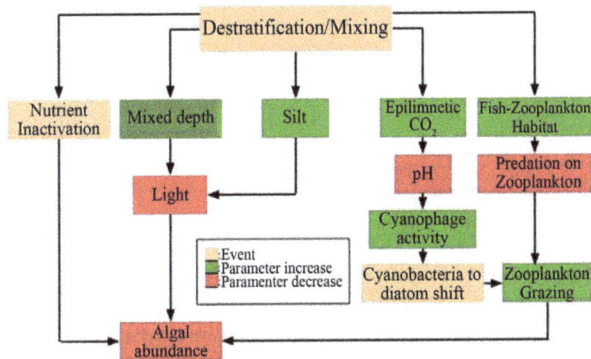

Figure 2. Effects of destratification on aquatic ecosystems (adapted from [1]).

able to scrub 5.5 ha/hr of sea surface using an oil boom modified with polyester weave forming fabric, a material commonly used in the paper industry [14]. Obvious downsides to algae harvesting are fuel, equipment, and labor costs, particularly when the afflicted area is substantial, as is often the case in cultural eutrophication. However, it has been suggested that the marketability of various algae as protein supplements or bioenergy feedstock may be able to offset some of those costs. The Harvested Algae Biofuel Energy Recovery model (HABER) developed by Kuo [13] found that a harvesting operation in the Gulf of Mexico using plankton net trawling could conceivably break even at standard eutrophic chlorophyll concentrations if the harvested biomass was converted into crude oil by a hydrothermal liquefaction process or into methane via anaerobic digestion. Unfortunately, further environmental analysis also found that the expected biomass harvest amount for this technology would have minimal impact on the overall size of the Gulf hypoxic zone, thus it was suggested that algae harvesting operations may hold more utility for smaller-scale areas, such as lakes and reservoirs experiencing eutrophication.

Copper-based algaecides and chemical flocculants have been used to control algae blooms in many instances; however, these practices may create more environmental problems than they solve [2]. A recent interest in more sustainable treatment methods has led to the exploration of using clay minerals as flocculation agents. Dense clay particles attach to the algae cells and promote conglomeration and sinking, despite the buoyancy of the algae cells [15]. Field tests in Japan and South Korea found that clay sediments were effective in controlling harmful algae blooms and restoring threatened mariculture operations [16]. Concerns regarding the cost of clay transport and the effectiveness of clay flocculation in low-salinity environments have inspired investigation into alternatives; recent work has found that sand modified with biodegradable *Moringa oleifera* seed extract and chitosan has great flocculation potential in both fresh and saltwater environments [17]. The modified sand

flocculant was able to remove over 90% of three different species of algae in suspension, but its effectiveness has yet to be proven on a large scale.

2.4. Nutrient Assimilation

Whereas the previous topics have focused on addressing and mitigating the effects of eutrophication within a water body, nutrient assimilation strategies are focused on "treating" influent water before it is deposited into the water body. The following management techniques are used as preventative, rather than curative, methods in eutrophication and algae bloom control.

2.4.1. Wetlands

Wetlands naturally provide a variety of ecological services; foremost among them is nutrient removal. Mechanisms for the removal of nitrogen and phosphorus in a wetland involve either burial in sediment layers or uptake and storage by plants. Riparian vegetation is very effective at removing these nutrients from runoff water, and, if the plants are harvested, their assimilation of the nutrients can be considered a permanent loss pathway. Wetland microbes also facilitate the denitrification process in which NO_3 is converted into N_2 gas. The natural efficacy of these processes has already been put to use in the Mississippi River delta as a method of combating eutrophication and hypoxia [18].

A hydrodynamic model of nutrient loading in the Maurepus forested wetland in Louisiana was used to assess the nutrient removal potential of a proposed Mississippi River diversion into the wetland [19]. It was estimated that the diversion would result in a 90% - 95% reduction of introduced nitrate by the wetlands. Given the anticipated water quality improvements, the study suggested that plans for the diversion move forward. An assessment of the denitrification efficacy of the Breton Sound estuary, a wetland already receiving diverted Mississippi River water, found that average removal rates were between 21 and 32 g $N/m^2/yr$, which indicated that the estuary was capable of processing significant quantities of nitrate from the Mississippi [20]. Wetlands are so effective at nutrient removal that they are sometimes used as methods of tertiary water treatment, receiving effluent directly from wastewater treatment plants. Conventional tertiary treatments methods are often very expensive; wetland assimilation can provide an effective, low-cost alternative if geographically feasible. Day Jr. *et al.* [18] compiled a list of different wetland treatment projects undertaken in coastal Louisiana using data compiled from multiple studies of the individual sites. A selection of site-specific data on nutrient removal, as well as estimated cost savings over a 20-year period (compared to conventional treatment methods) to the communities can be seen in **Table 1**.

Table 1. Nutrient removal and cost savings data for two LA wetlands used in tertiary wastewater treatment [18].

Site	Treatment Basin (ha)	N loading ($g/m^2/yr$)	P loading ($g/m^2/yr$)	%N reduction	%P reduction	Cost Savings ($)
Breaux Bridge	1475	1.87	0.94	100	80	2,636,000
Thibodaux	231	3.1	0.6	69	66	500,000

2.4.2. High-Rate Algal Ponds

Microalgae are often hailed as an important "third generation" biofuel source due to their rapid growth rate (often doubling biomass within 24 hours), as well as their high oil content [21]. There are also well-established markets for some species as animal feeds and nutritional supplements. Wastewater treatment utilizing the nutrient uptake abilities of microalgae was first proposed by Oswald and Golueke [22] in the form of a high-rate algae pond (HRAP). HRAPs are shallow, open raceway-style ponds that are typically mixed with a paddlewheel and can cover up to 5000 m^2 in large-scale applications [23]. Though perhaps not as cost-effective as natural wetlands (due to capital and construction costs), high-rate algae ponds have potential as a combined source of water treatment and marketable products.

A recently published study examined the possibilities for the inclusion of HRAPs at the Western Treatment Plant in Melbourne, Australia in wastewater treatment and biofuel production [24]. The authors posit that biofuel production from algae can be profitable when wastewater treatment is the primary goal. A spreadsheet economic model found that algae oil of biodiesel quality could be produced for less than $1/L (US dollars) when conventional oil extraction techniques were used on a portion of the algal biomass, and the rest was converted into electrical energy by anaerobic digestion. This production cost was said to be possible due to the low-cost water and nutrients readily supplied by the wastewater treatment plant, as well as "free" supplemental carbon dioxide provided by on-site diesel generators. In Castilla y Leon, Spain, two 464 L HRAPs (mixed cultures) were studied for a period of nine months in order to examine their long-term nutrient removal ability when fed with effluent from a pig farm [23]. Nitrogen removed in the biomass ranged from 21% - 48%. Phosphorus removal efficiencies remained low (<10%) for the duration of the study due to the low pH of the effluent; regardless, the authors claimed that HRAPs hold great promise in the field of waste-to-value pollution control. The potential of microalgae *Chlorella vulgaris* in removing nitrate-nitrogen from tilapia pond effluent, as well as possibilities for separation of algal biomass (sedimentation or filtration) from the effluent have been examined [25]. The algae were found to have a substrate utilization rate (a measure of nitrogen removed per dry biomass per unit time) of 22.3 mg N/g DW/day. The small cell size of *C. vulgaris*

(2.5 - 12 μm) lowered the effectiveness of sedimentation and filtration via sieve trays, which led to the conclusion that a more energy-intensive process, such as membrane filtration, may be required to concentrate the biomass.

3. Integrated Algae-Aquaculture Systems

As the world population increases, demands on food production will rise. Aquaculture is expected to be the primary source of seafood by 2050, and it is essential that sustainable methods of production be devised and implemented so that land, water, and nutrients can be used efficiently [26]. Extensive aquaculture systems that utilize algae as a method of capturing excess nutrients and a self-sustaining process input (*i.e.* fish food) represent an innovative path towards this goal.

In Mbour, Senegal, a prototype aquaculture system was constructed for tilapia production; both algae (*Chlorella* sp.) and zooplankton (*Brachionus plicatilis*) were cultivated in wastewater ponds outside the intensive fish tanks, and then distributed to the juvenile tilapia as additional food [27]. A primary feature of this freshwater system was that, with the exception of evaporation losses, water was completely recycled for the duration of this study. The system was able to reach a global productivity 1.85 $kg/m^2/yr$ in tilapia biomass, and a nitrogen balance found that N was assimilated at a rate of 1.90 mg/L/day. The authors found these results to be comparable to others in the literature for similar systems.

A pilot-scale system focused on reducing water demand via recirculation and fish food production using algae grown in a bioremediation pond was developed and tested [28]. The first units in the system were tanks for the intensive pellet feeding of catfish; water from these tanks flowed into an algae pond, wherein the algae utilized the excess nutrients for growth. Continuous mixing was applied to ensure maximum light exposure. The wet algal biomass was then pumped into a second pond stocked with carp and tilapia, which were not pellet-fed and used the algae as their food source. The system was closed and water was emptied only at the end of the twelve-week study. As expected, the catfish in the intensive tanks had a high yield of 20.25 kg/m^3, while the carp and tilapia in the pond produced 0.12 and 0.38 kg/m^3, respectively. Nutrient retentions of the system during the 12 weeks for nitrogen and phosphorus were 1074 kg N/ha and 327 kg P/ha, which represented 57.1% of the N load and 77.3% of the P load. These numbers were significantly higher than those reported for other fishponds; the authors posited that nutrient reduction efficiency of their system was greatly improved by the addition of treatment organisms (algae and secondary feeder fish) in two separate ponds.

4. Design of a Novel System

4.1. Overview

In 2008, the Mississippi River/Gulf of Mexico Watershed Nutrient Task Force, a consortium made up of federal organizations such as the EPA, USGS, USDA, and others, stated a goal of reducing the five-year running average area of the Gulf dead zone to less than 1930 square miles by the year 2015. It was specified that nitrogen and phosphorus nutrient loads would each need to be reduced by about 45% with respect to 1980-1996 levels in order to reach this goal. The Task Force aims to accomplish this through a combination of source reduction and remediation techniques spearheaded by their various constituent organizations [29]. The design presented here was devised with the intent to assist in the achievement of the aforementioned goal by providing a nutrient assimilation service, while also providing marketable products in the form of algal biomass and tilapia.

This concept for this design was inspired by the efficiency of both wetlands and water treatment by microalgae in nutrient load reduction projects, and it combines certain aspects of both remediation technologies with an aquaculture operation. In a manner similar to a wetland project, all water for this system would be sourced directly from the Mississippi River via a diversion. A series of screen filters would remove debris from the incoming water, which would then enter into a fish pond for extensive rearing of tilapia. Effluent from the fish pond, still containing nutrients from the Mississippi, would be emptied into a large raceway pond system, similar to the HRAPs described by Park et al. [30]. In order to minimize costs and energy usage, pure carbon dioxide will not be piped into the pond; the paddlewheel mixing system will be counted upon to facilitate mass transfer of CO_2 into the water. The HRAP will be inoculated by a smaller separate vessel housing a pure culture of Chlorella microalgae. Following the pond system, algae biomass will be physically separated from the water using a cross-flow filtration system. A portion of wet biomass will be sent back to the fish ponds as feed, while the residual biomass will be sent to further processing and converted into a marketable product. Effluent water, now with reduced nutrient loads, will be sent back into the Mississippi. **Figure 3** shows a basic process flow diagram (PFD) of the system; unit operations are discussed in detail in the following sections of the paper.

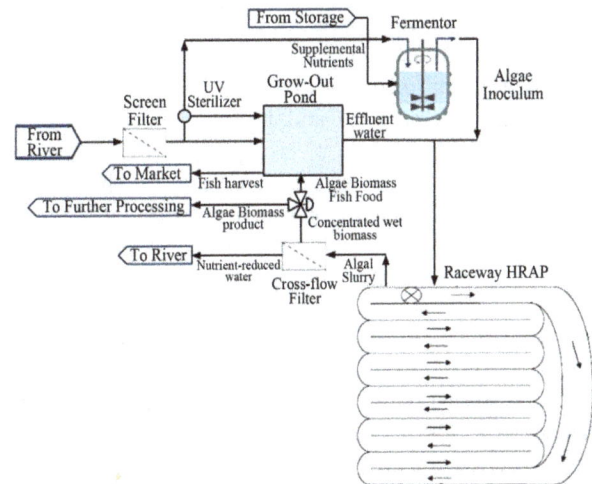

Figure 3. PFD of integrated algae-aquaculture system (HRAP image credit: [31]).

4.2. Fish Ponds

In extensive aquaculture, fish are cultivated without supplemental feeding. As the diet of tilapia naturally consists of algae, they are a convenient choice for extensive algae-aquaculture systems like the one developed by Gal et al. [28]. They are also a marketable commodity, having recently become the most popular farmed fish in the United States [32]. In the proposed system, tilapia will be stocked in an extensive fish pond where their only food source is the algal biomass harvested from the HRAP. Apportionment of the concentrated wet biomass stream will be determined based on the nutrient requirements of the fish.

Extensive feeding is reported to yield best results for tilapia ponds stocked with densities at or below 4,000/acre [33]. Assuming a pond depth of 1 meter, an area of 1 acre, and an initial stocking weight of 29 g/fish, it was determined that biomass density at stocking (B_i) should be roughly 0.029 kg/m^3. Based on this stocking value in a 1 m-deep pond, a model was developed to test the effect of different pond areas on fish production and algae consumption over time.

Total tilapia biomass at a given time in kg/m^3 (B_f) can be represented by Equation (1), where n is the culture period in days and i is the daily rate of increase of biomass (DRIB, measured in percent) divided by 100 [34]. Coche [34] lists multiple studies of extensive tilapia cultures that have DRIBs ranging from 1.7% - 4.8%; for the purposes of this estimation, a DRIB of 3% was assumed.

$$B_f = B_i \left(1+i\right)^n \tag{1}$$

B_f and the pond volume can be used to calculate the total kilograms of fish present in the pond, and, assuming uniform mass and negligible spawning, the mass per fish.

Understandably, as the fish mature, they will require greater amounts of food; a study on wild tilapia [35] yielded a linear relationship between the weight of a single fish in grams (x) and its daily algae intake in milligrams (y), seen below in Equation (2). The daily algae intake was assumed to be the same for all fish and multiplied by the total number of fish to obtain the total algae biomass requirement for the fish pond in kg/d.

$$y = 271 + 13.3x \qquad (2)$$

A culture period of 100 days was assumed based on harvesting recommendations in [33]. At harvest time, the pond will be partially drained and seine nets with a 1-inch mesh will be used to capture the fish. The amount of labor required for this task is dependent on the optimal pond area. In addition, the authors advocate a pond size between 1 - 10 acres for extensive operations; for ease of management, a 5-acre pond was decided upon for this design. The tilapia-algae relationship for such a pond over the recommended culture period can be observed in **Figure 4**.

4.3. Pure Culture Inoculum Reactor

One method of algae ecosystem management is to seed the system with a desirable species of algae that can be produced in pure culture. Continuous seeding may be sufficient to control concentrations of algae or cyanobacteria that produce toxins.

Photobioreactors are commonly used for algae cultivation on a small scale; however, though many innovative designs are being tested [36], photobioreactors often must be highly specialized for the task at hand and can prove costly to operate [37]. If an algae can grow in heterotrophic conditions (which *Chlorella* can), it is often considerably more economical and efficient to produce the algae in a fermentor; reasons for this include both the elimination of photosynthetic inefficiencies as well as a large pre-existing knowledge base regarding fermentor operation [38]. Heterotrophic growth has also been proven to be an effective method to produce inoculum for large-scale phototrophic algae ponds due to its high productivity [39,40]. Taking this information into consideration, it was determined that a fermentor would be the best choice for the pure culture inoculum reactor.

Kinetics parameters in the fermentor were modeled after the results of experiments in [39] on the heterotrophic growth of *Chlorella* using molasses as a carbon source and yeast extract as a nitrogen source. The highest maximum biomass concentration on a dry weight basis (X) and specific growth rate (μ) obtained in the study occurred for a molasses concentration of 20 g/L and a yeast extract concentration of 6 g/L. These parameters were taken as the basis for calculation, since the goal of inoculation is to add as high a concentration of biomass

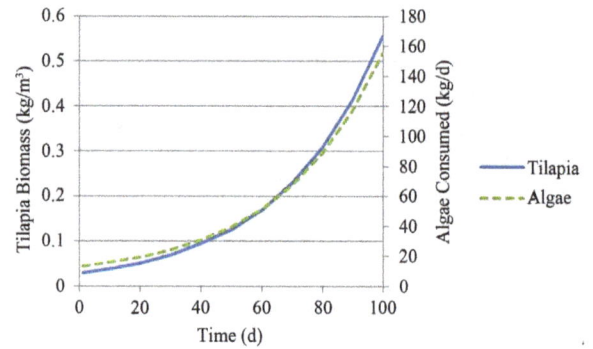

Figure 4. Relationship between tilapia biomass and algae consumption requirements in a 5-acre extensive fishpond.

as possible to the raceway pond so that the *Chlorella* growth will not be overtaken by less desirable species of algae. However, assuming the yeast extract was approximately 10% nitrogen [41] and the mean total N concentration in the Mississippi River is 2.26 mg/L [42], we would need to increase nitrogen concentration in the incoming water substantially (~26,450%) in order to match their experimental conditions. Thus, supplemental N will have to be added along with the molasses. Seeding of the raceway is intended to be continuous, so the fermentor will be at steady-state; thus the specific growth rate is equal to the dilution rate (D). This value, along with an assumed medium flow rate (F) can be used to calculate volume of the reactor (V_F) and the mean residence time (τ). All results can be seen in **Table 2**.

As contamination is a primary concern when preparing a pure culture, an ultraviolet sterilizer will treat the river water before it enters the fermentor.

4.4. Raceway High-Rate Algae Pond

The raceway pond is where the system will perform its primary function of nutrient assimilation. Unless a low-cost source of CO_2, such as a coal-fired power plant, is located near the pond, it will be assumed that paddle-wheel mixing and turbulence will be the sole facilitators of mass transfer for the system. When performing design calculations for the HRAP, multiple equations found in a reference text [43] were used. Equation (3) is Manning's equation, where V is the mean velocity (m/s), R is the mean hydraulic radius (m), S is the rate of energy loss per unit length (dimensionless; can be split into $\Delta d/L$), and n is Manning's friction coefficient (s/m$^{1/3}$). Equation (4) is an equation developed from empirical observation of large-scale open pond algae cultures where C_C is the light-limited algae concentration on a dry weight basis (mg/L) and d is the pond depth in centimeters. Equation (5) is obtained from a differential mass balance on the pond where C_0 is entering concentration, μ is specific growth rate, and t_r is retention time. Equation (6) is the standard dilution equation, adapted for the transition of

Table 2. Inoculum fermentor kinetics parameters with molasses concentration of 20 g/L and total N concentration of 0.60 g/L.

μ (1/hr)	D (1/hr)	F (L/hr)	V_F (L)	τ (hr)	X (mg/L)
0.023	0.023	2.3	100	44	7180

algal biomass between the fermentor and the HRAP where V_F is volume of the fermentor, V_P is volume of the pond, and X is the concentration exiting the fermentor.

$$V = \frac{1}{n} R^{2/3} S^{1/2} \qquad (3)$$

$$C_C = 9000/d \qquad (4)$$

$$\ln \frac{C_C}{C_0} = \mu t_r \qquad (5)$$

$$XV_F = C_0 V_P \qquad (6)$$

In order to perform design calculations, a number of parameters had to be estimated or set. Assuming an asphalt lining for the pond, n was set to 0.015 s/m$^{1/3}$. The Δd component was assumed to be 0.5 d. Pond dimensions were set at 3 m wide and 30 cm deep, which made $R = 0.25$ m. Using that depth, Equation 4 yields a maximum light-limited concentration of 300 mg/L. Specific growth rate in the pond was estimated at 0.6/day [25]. Inoculum concentration C_0 (X diluted to V_P) was unknown, since the pond length was also undetermined. Iterative calculations to solve for unknowns were performed in the following manner (results follow):

- Value guessed for C_0 and pond retention time solved for using Equation (5) (τ doubled to account for the fact that Attasat *et al.*'s growth rate assumed constant illumination).
- Equation (3) algebraically manipulated (assuming $V = L/t_r$) to solve for pond length L, which was then used to calculate the required pond area.
- L calculated separately by Equation (6) using the width and depth of the pond.
- Residual between L values minimized by new value for C_0.

4.5. Biomass Separation

There are many different methods for dewatering and concentrating algal biomass, and they vary greatly in both cost of implementation and energy consumption [44]. For the purposes of this initial design, preference was given to processes that would minimize the use of chemical coagulants and electricity (even if higher cost was incurred); as one of the primary goals of this project is sustainable water treatment for the lower Mississippi, it seemed counterintuitive to suggest processes that involve adding more chemicals to the water right before

discharging it back into the river. Under these constraints, it was postulated that cross-flow filtration may be the best option for this process. A diagram of a cross-flow filtration apparatus can be seen in **Figure 5**.

Cross-flow filtration is a conventional purification method in which a pressure gradient forces a fluid mixture flowing across a membrane to separate; it has been successfully utilized for separation of microalgae slurry [46], and an economic study of its utility in an industrial-scale algal dewatering process estimated costs at around $0.75/kg dry biomass [47]. Cheaper options exist when chemical pretreatment of the algae slurry or dynamic filtration using a rotor are considered; however, those practices may not support the project's goal of minimal environmental impact.

Once separated, the filtered water will be discharged back into the Mississippi. The concentrated biomass will now be split into two streams: one for further processing, and one to return to the fish pond as food. Since the tilapia require greater amounts of food as they mature, the portion of algal biomass returning to the extensive pond will be adjusted daily according to the nutrient requirements calculated by the model.

4.6. Further Processing

There are a number of options for processing the residual algal biomass into a marketable product. Markets exist for the algae biomass to be directly sold as an agricultural feed [48,49]; however, increased transportation costs would have to be taken into consideration if this avenue was to be pursued. Also, quality assurance procedures would have to be implemented in order to ensure food safety [50].

Many algae-based operations aim to produce bio-energy in their processes; three different methods of algal biomass conversion were considered for this system (see **Table 3**).

Dewatering and drying are typically the most costly aspects of any algae-to-energy process—drying alone usually accounts for 70% - 75% of the entire production cost [44]. Dewatering is already built into the proposed design through the use of cross-flow filters; regardless, in keeping with the sustainability goals of this project, it

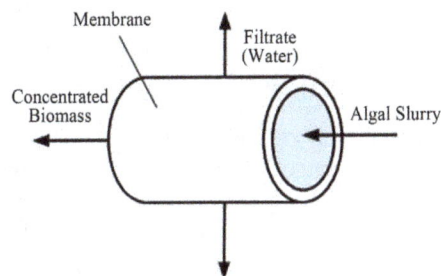

Figure 5. Cross-flow filtration unit (adapted from [45]).

Table 3. Analysis of algal bio-energy conversion methods.

Method	Product	Drying Required?	Net Energy Yield	Relative Cost
Transesterification	Biodiesel	Yes	Low	High
Anaerobic Digestion	Biogas	Yes	High	Low
Hydrothermal Liquefaction	Bio-crude oil	No	High	Low

Sources: [13,51,52].

would be prudent to consider only those processes which minimize energy input and cost. From that standpoint, anaerobic digestion or hydrothermal liquefaction may be the best choices if energy is determined to be the highest-value product attainable from the algal biomass. Further research is required to determine the cost of implementation for said processes, as well as the market price of their products.

4.7. Expected Productivity

The theoretical maximum algae biomass productivity in $g/m^2/day$ (P_{max}), assuming ideal conditions, can be estimated using Equation (4), where I_0 is the average solar radiation in $MJ/m^2/day$, η_{max} is the maximum efficiency of photosynthetic conversion of solar energy, and H is the energy value of the algal biomass in kJ/g [30].

$$P_{max} = \frac{I_0 \eta_{max}}{H} \times 1000 \qquad (7)$$

Average solar radiation for the Mississippi delta area is approximately 16.2 $MJ/m^2/day$ [53]; using this value along with $\eta_{max} = 2.4\%$ and $H = 21$ kJ/g [30], it was estimated that the maximum areal productivity for a HRAP system in that region would be about 18.5 $g/m^2/day$. Taking the previously calculated area into account, this would mean that the proposed system would have a maximum biomass productivity of 5487 kg/day.

As previously mentioned, the fish will be harvested every 100 days. According to the tilapia growth model, total mass of fish at harvest time will be (under ideal conditions) approximately 11,277 kg. Assuming a tilapia market price at \$3/kg [54], this will translate to an income of almost \$123,500 per year from the aquaculture aspect of the project.

4.8. Processing Capacity and Expected Nutrient Load Reduction

The average discharge rate of the Mississippi River is estimated to be around 17,000 m^3/s [55]. The volumetric flow rate in a channel (Q) can be determined using Equation (8) [43]. Upon comparison of the flow in the HRAP and the discharge rate of the river, it appears that the proposed design would have the capacity to process ap-

proximately 0.000173% of the Mississippi River discharge.

$$Q = wdV \qquad (8)$$

Nutrient retentions of 1074 kg N/ha and 327 kg P/ha from the algae-aquaculture system described in [28] were used as the basis for estimation regarding nutrient load reduction potential of the proposed system. The retention rates were converted into yearly rates per hectare assuming the twelve-week study could be extrapolated out to one year. The area of the proposed HRAP was used to calculate the yearly nutrient removal potential. In order to determine annual nutrient removal percentages, these rates were then compared to the mean nutrient fluxes delivered from the Mississippi to the Gulf of Mexico for the period of 1980-1996: 1,567,900 metric tons/yr of nitrogen and 136,500 metric tons/yr of phosphorus [42]. Results can be seen in **Table 4**.

There are many locations along the Mississippi where integrated algae-aquaculture systems may be designed and operated to remove nutrients and harvest fish. River water can be diverted through constructed lakes with flow controlled by gates that restrict that amount of water that enters the lake. The nutrient removal percentages calculated here may not represent valid predictions for systems implemented further upstream, since mean nutrient fluxes vary along the length of the river.

5. Discussion

Though these removal percentages are a far cry from the 45% reductions recommended by the Mississippi River/Gulf of Mexico Watershed Nutrient Task Force, it is worth noting that the proposed design will be more efficient per unit area at nutrient removal than the wetland projects discussed earlier. The added value products of tilapia and algal biomass may be able to offset capital and maintenance costs of the facility, and perhaps even allow for a profitable enterprise. Further research, particularly into the comparative costs of algae-to-energy processes, is needed to investigate the economic viability of the system proposed in this paper. This system may have utility in enclosed freshwater environments as well; the nutrient loads entering lakes and reservoirs will not be as massive as those entering the Gulf, therefore, the system will provide greater net reductions of nitrogen and phosphorus when used to treat influent to smaller water bodies.

A scaled-down alternative to the algae-aquaculture system may also be the construction and improved maintenance of farm ponds and recreational lakes throughout the watershed of a eutrophic ecosystem. If these man-made water bodies are built to cultivate algae and tilapia in a single environment, they will be able to provide nu-

Table 4. Raceway pond nutrient load reduction.

N removal rate (kg/yr)	P removal rate (kg/yr)	%N removed annually	%P removed annually
138,415	42,143	0.00883	0.0309

trient removal services and marketable products, as well as catchment areas for flooding and storm water management. Biomanipulation, artificial circulation, and algae harvesting may prove to be useful tools in smaller, less complex aquatic ecosystems such as these. Productive reservoir/lake/pond ecosystems involving fish and algae represent permanent nutrient loss pathways, since the organisms that take up the nutrients are being harvested and removed from the system; case studies often attest to the natural nutrient uptake abilities of these ecosystems [56]. If enough of these water bodies are constructed and maintained, their cumulative effect may be substantial. Preliminary calculations indicate that approximately 1,850,000 hectares of such productive ecosystems would be needed to reduce the annual Mississippi River phosphorus load by 50%. Further research is required to gauge the cost-effectiveness and feasibility of constructing and maintaining pond/lake ecosystems compared to algae-aquaculture systems similar to the one proposed herein.

One of the significant aspects of ecosystem management is that of finding methods to allocate costs and pay for each project. The reduction of nitrogen and phosphorus associated with an algae-aquaculture system has value to society, but methods to convert this value to project funding may require government action.

6. Conclusion

The proposed integrated algae-aquaculture system was developed from knowledge of existing technologies and techniques for sustainable eutrophication management. The mitigation strategies discussed in this paper will be most effective when coupled with a comprehensive nutrient control strategy that addresses the sources of anthropogenic nutrient loading in the watershed of a eutrophic water body. Benefits of improved eutrophication management include more productive ecosystems, increased real estate values, fewer taste and odor issues in reservoirs, and greater recreational usage of water bodies. It is our hope that the ideas presented in this paper can be utilized and expanded upon in order to create synergistic strategies that allow eutrophication management to become a successful, cost-effective, and sustainable enterprise.

7. Acknowledgements

Financial support for this project was provided by the National Science Foundation through award EEC-1156549. We would like to thank Dr. Walter K. Dodds for his assistance with the project. The Process Flow Diagram drawing template from EngineeringToolBox.com was used in developing **Figure 3**.

REFERENCES

[1] B. Henderson-Sellers and H. R. Markland, "Decaying Lakes: The Origins and Control of Cultural Eutrophication," John Wiley & Sons Ltd., Hoboken, 1987.

[2] W. K. Dodds, "Chapter 17: Trophic State and Eutrophication" In: Thorp, J.H., Ed., *Freshwater Ecology: Concepts and Environmental Applications*, Academic Press, San Diego, 2002, pp. 337-365,391-410.

[3] National Oceanic and Atmospheric Administration, "Economic Impacts of Harmful Algal Blooms," 2008. http://www.cop.noaa.gov/stressors/extremeevents/hab/current/econimpact_08.pdf

[4] W. K. Dodds, W. W. Bouska, J. L. Eitzmann, T. J. Pilger, K. L. Pitts, A. J. Riley, J. T. Schloesser and D. J. Thornbrugh, "Eutrophication of U.S. Freshwaters: Analysis of Potential Economic Damages," *Environmental Science & Technology*, Vol. 43, No. 1, 2009, pp. 12-19.

[5] E. Jeppesen, M. Søndergaard, N. Mazzeo, M. Meerhoff, C. C. Branco, V. Huszar and F. Scasso, "Lake Restoration and Biomanipulation in Temperate Lakes: Relevance for Subtropical and Tropical Lakes," In: V. Reddy, Ed., *Restoration and Management of Tropical Eutrophic Lakes*, Science Publishers, Enfield, 2005, pp. 331-349.

[6] G. Closs, B. Downes and A. Boulton, "Freshwater Ecology: A Scientific Introduction," Blackwell Science Ltd., Malden, 2004.

[7] E. Jeppesen, M. Meerhoff, B. A. Jacobsen, R. S. Hansen, M. Søndergaard, J. P. Jensen, T. L. Lauridsen, N. Mazzeo and C. W. C. Branco, "Restoration of Shallow Lakes by Nutrient Control and Biomanipuation—The Successful Strategy Varies with Lake Size and Climate," *Hydrobiologia*, Vol. 581, No. 1, 2007, pp. 269-285.

[8] C. C. Figueredo and A. Giani, "Ecological Interactions between Nile Tilapia (*Oreochromis niloticus*, L.) and the Phytoplanktonic Community of the Furnas Reservoir (Brazil)," *Freshwater Biology*, Vol. 50, No. 8, 2005, pp. 1391-1403.

[9] W.-M. Heo and B. Kim, "The Effect of Artificial Destratification on Phytoplankton in a Reservoir," *Hydrobiologia*, Vol. 524, No. 1, 2004, pp. 229-239.

[10] A. W. Fast and M. W. Lorenzen, "Synoptic Survey of Hypolimnetic Aeration," *Journal of the Environmental Engineering Division*, Vol. 102, No. 6, 1976, pp. 1161-1173.

[11] R. Gächter and B. Wehrli, "Ten Years of Artificial Mixing and Oxygenation: No Effect on the Internal Phos-

phorus Loading of Two Eutrophic Lakes," *Environmental Science & Technology*, Vol. 32, No. 23, 1998, pp. 3659-3665.

[12] A. W. Fast, M. W. Lorenzen and J. H. Glenn, "Comparative Study with Costs of Hypolimnetic Aeration," *Journal of the Environmental Engineering Division*, Vol. 102, No. 6, 1976, pp. 1175-1187.

[13] C.-T. Kuo, "Harvesting Natural Algal Blooms for Concurrent Biofuel Production and Hypoxia Mitigation," M.S. Thesis, University of Illinois at Urbana-Champagne, Urbana, 2010.

[14] F. Gröndahl, "Removal of Surface Blooms of the Cyanobacteria *Nodularia spumigena*: A Pilot Project Conducted in the Baltic Sea," *Ambio*, Vol. 38, No. 2, 2009, pp. 79-84.

[15] M. Sengco and D. M. Anderson, "Controlling Harmful Algal Blooms through Clay Flocculation," *Journal of Eukaryote Microbiology*, Vol. 51, No. 2, 2004, pp. 169-172.

[16] R. H. Pierce, M. S. Henry, C. J. Higham, P. Blum, M. R. Sengco and D. M. Anderson, "Removal of Harmful Algal Cells (*Karenia brevis*) and Toxins from Seawater Culture by Clay Flocculation," *Harmful Algae*, Vol. 3, No. 2, 2004, pp. 141-148.

[17] L. Li and G. Pan, "A Universal Method for Flocculating Harmful Algal Blooms in Marine and Fresh Waters Using Modified Sand," *Environmental Science & Technology*, Vol. 47, No. 9, 2013, pp. 4555-4562.

[18] J. W. Day Jr., J.-Y. Ko, J. Rybczyk, D. Sabins, R. Bean, G. Berthelot, C. Brantley, L. Cardoch, W. Conner, J. N. Day, A. J. Englande, S. Feagley, E. Hyfield, R. Lane, J. Lindsey, J. Mistich, E. Reyes and R. Twilley, "The Use of Wetlands in the Mississippi Delta for Wastewater Assimilation: A Review," *Ocean and Coastal Management*, Vol. 47, 2004, pp. 671-691.

[19] R. R. Lane, H. S. Mashriqui, G. P. Kemp, J. W. Day, J. N. Day and A. Hamilton, "Potential Nitrate Removal from a River Diversion into a Mississippi Delta Forested Wetland," *Ecological Engineering*, Vol. 20, No. 3, 2003, pp. 237-249.

[20] R. D. Delaune and A. Jugsujinda, "Denitrification Potential in a Louisiana Wetland Receiving Diverted Mississippi River Water," *Chemistry and Ecology*, Vol. 19, No. 6, 2003, pp. 411-418.

[21] Y. Chisti, "Biodiesel from Microalgae," *Biotechnology Advances*, Vol. 25, No. 3, 2007, pp. 294-306.

[22] W. J. Oswald and C. G. Golueke, "Biological Transformation of Solar Energy," *Advances in Applied Microbiology*, Vol. 2, 1960, pp. 223-262.

[23] I. de Godos, S. Blanco, P. A. Garcia-Encina, E. Becares and R. Munoz, "Long-term Operation of High Rate Algal Ponds for the Bioremediation of Piggery Wastewaters at High Loading Rates," *Bioresource Technology*, Vol. 100, No. 19, 2009, pp. 4332-4339.

[24] D. Batten, T. Beer, G. Freischmidt, T. Grant, K. Liffman, D. Paterson, T. Priestley, L. Rye and G. Threlfall, "Using Wastewater and High-Rate Algal Ponds for Nutrient Removal and the Production of Bioenergy and Biofuels," *Water Science & Technology*, Vol. 67, No. 4, 2013, pp. 915-924.

[25] S. Attasat, P. Wanichpongpan and W. Ruenglertpanyakul, "Cultivation of Microalgae (*Oscillatoria okeni* and *Chlorella vulgaris*) using Tilapia-Pond Effluent and a Comparison of their Biomass Removal Efficiency," *Water Science & Technology*, Vol. 67, No. 2, 2013, pp. 271-277.

[26] J. S. Diana, H. S. Egna, T. Chopin, M. S. Peterson, L. Cao, R. Pomeroy, M. Verdegem, W. T. Slack, M. G. Bondad-Reantaso and F. Cabello, "Responsible Aquaculture in 2050: Valuing Local Conditions and Human Innovations will be Key to Success," *BioScience*, Vol. 63, No. 4, 2013, pp. 255-262.

[27] S. Gilles, L. Fargier, X. Lazzaro, E. Baras, N. De Wilde, C. Drakides, C. Amiel, B. Rispal and J.-P. Blancheton, "An Integrated Fish-Plankton Aquaculture System in Brackish Water," *Animal*, Vol. 7, No. 2, 2013, pp. 322-329.

[28] D. Gal, F. Pekar, E. Kerepeczki and L. Varadi, "Experiments on the Operation of a Combined Aquaculture-Algae System," *Aquaculture International*, Vol. 15, No. 3-4, 2007, pp. 173-180.

[29] Mississippi River/Gulf of Mexico Watershed Nutrient Task Force, "Gulf Hypoxia Action Plan 2008 for Reducing, Mitigating, and Controlling Hypoxia in the Northern Gulf of Mexico and Improving Water Quality in the Mississippi River Basin," Washington, DC, 2008.

[30] J. B. K. Park, R. J. Craggs and A. N. Shilton, "Wastewater Treatment High Rate Algal Ponds for Biofuel Production," *Bioresource Technology*, Vol. 102, No. 1, 2011, pp. 35-42.

[31] I. Rawat, R. Ranjith Kumar, T. Mutanda and F. Bux, "Dual Role of Microalgae: Phycoremediation of Domestic Wastewater and Biomass Production for Sustainable Biofuels Production," *Applied Energy*, Vol. 88, No. 10, 2011, pp. 3411-3424.

[32] E. Rosenthal, "Another Side of Tilapia, the Perfect Factory Fish," *The New York Times*, 2 May 2011, p. A6.

[33] J. E. Rakocy and A. S. McGinty, "Pond Culture of Tilapia," 1989. https://srac.tamu.edu/index.cfm/event/getFactSheet/which factsheet/50/

[34] A. G. Coche, "Cage Culture of Tilapias," In: R. S. V. Pullin and R. H. Lowe-McConnell, Eds., *The Biology and Culture of Tilapias*: ICLARM Conference Proceedings 7, *International Center for Living Aquatic Resources*

Management, Manila, 1982, pp. 205-246.

[35] C. M. Moriarty and D. J. W. Moriarty, "Quantitative Estimation of the Daily Ingestion of Phytoplankton by *Tilapia nilotica* and *Haplochromis nigripinnis* in Lake George, Uganda," *Journal of Zoology*, Vol. 171, No. 1, 1973, pp. 15-23.

[36] D. Briassoulis, P. Panagakis, M. Chionidis, D. Tzenos, A. Lalos, C. Tsinos, K. Berberidis and A. Jacobsen, "An Experimental Helical-Tubular Photobioreactor for Continuous Production of *Nannochloropsis* sp.," *Bioresource Technology*, Vol. 101, No. 17, 2010, pp. 6768-6777.

[37] K. K. Vasumathi, M. Premalatha and P. Subramanian, "Parameters Influencing the Design of Photobioreactor for the Growth of Microalgae," *Renewable and Sustainable Energy Reviews*, Vol. 16, No. 7, 2012, pp. 5443-5450.

[38] P. W. Behrens, "Chapter 13: Photobioreactors and Fermentors: The Light and Dark Sides of Growing Algae," In: R. A. Anderson, Ed., *Algal Culturing Techniques*, Elsevier Academic Press, Burlington, 2005, pp. 189-203.

[39] R. Leesing and S. Kookkhunthod, "Heterotrophic Growth of *Chlorella* sp. KKU-S2 for Lipid Production using Molasses as a Carbon Substrate," *International Proceedings of Chemical, Biological, and Environmental Engineering*, Vol. 9, 2011, pp. 87-91.

[40] Y. Zheng, Z. Chi, B. Lucker and S. Chen, "Two-Stage Heterotrophic and Phototrophic Culture Strategy for Algal Biomass and Lipid Production," *Bioresource Technology*, Vol. 103, No. 1, 2012, pp. 484-488.

[41] Acumedia Manufacturers, "Yeast Extract 7184," 2011. http://www.neogen.com/Acumedia/pdf/ProdInfo/7184_PI .pdf

[42] D. A. Goolsby, W. A. Battaglin, G. B. Lawrence, R. S. Artz, B. T. Aulenbach, R. P. Hooper, D. R. Keeney and G. J. Stensland, "Flux and Sources of Nutrients in the Mississippi-Atchafalaya River Basin: Topic 3 Report for the Integrated Assessment on Hypoxia in the Gulf of Mexico," NOAA Coastal Ocean Program Decision Analysis Series 17, NOAA Coastal Ocean Program, Silver Spring, Maryland, 1999.

[43] M. A. Borowitzka, "Chapter 14: Culturing Microalgae in Outdoor Ponds," In: Anderson, R.A., Ed., *Algal Culturing Techniques*, Elsevier Academic Press, Burlington, 2005, pp. 205-18.

[44] K. Y. Show, D. J. Lee and J. S. Chang, "Algal Biomass Dehydration," *Bioresource Technology*, Vol. 135, 2013, 720-729.

[45] S. Wronsky, E. Molga and L. Rudniak, "Dynamic Filtration in Biotechnology," *Bioprocess Engineering*, Vol.

4, No. 3, 1989, pp. 99-104.

[46] X. Zhang, Q. Hu, M. Sommerfeld, E. Puruhito and Y. Chen, "Harvesting Algal Biomass for Biofuels Using Ultrafiltration Membranes," *Bioresource Technology*, Vol. 101, No. 14, 2010, pp. 5297-5304.

[47] S. D. Rios, J. Salvado, X. Farriol and C. Torras, "Antifouling Microfiltration Strategies to Harvest Microalgae for Biofuel," *Bioresource Technology*, Vol. 119, 2012, pp. 406-418.

[48] P. Spolaore, C. Joannis-Cassan, E. Duan and A. Isambert, "Review: Commercial Applications of Microalgae," *Journal of Bioscience and Bioengineeering*, Vol. 101, No. 2, 2006, pp. 87-96.

[49] J. A. Stamey, D. M. Shepherd, M. J. de Veth and B. A. Corl, "Use of Algae or Algal Oil Rich in n-3 Fatty Acids as a Feed Supplement for Dairy Cattle," *Journal of Dairy Science*, Vol. 95, No. 9, 2012, pp. 5269-5275.

[50] A. G. Day, D. Brinkmann, S. Franklin, K. Espina, G. Rudenko, A. Roberts and K. S. Howse, "Safety Evaluation of a High-Lipid Algal Biomass from *Chlorella prototheocides*," *Regulatory Toicology and Pharmacology*, Vol. 55, No. 2, 2009, pp. 166-180.

[51] P. E. Wiley, J. E. Campbell and B. McKuin, "Production of Biodiesel and Biogas from Algae: A Review of Process Train Options," *Water Environment Research*, Vol. 83, No. 4, 2011, 326-338.

[52] G. Yu, Y. Zhang, L. Schideman, T. L. Funk and Z. Wang, "Hydrothermal Liquefaction of Low Lipid Content Microalgae into Bio-Crude Oil," *Transactions of the American Society of Agricultural and Biological Engineers*, Vol. 54, No. 1, 2011, pp. 239-246.

[53] National Renewable Energy Laboratory, "Photovoltaic Solar Resource of the United States [graphic]," 2012. http://www.nrel.gov/gis/images/eere_pv/national_photov oltaic_2012-01.jpg

[54] K. Fitzsimmons, "Marketing of Tilapia in the USA," 2000. http://ag.arizona.edu/azaqua/tilapia/Thailand/paper.htm

[55] J. C. Kammerer, "Largest Rivers in the United States," United States Geological Survey Open-File Report 87-242, United States Geological Survey, Reston, 1990.

[56] S. H. Wang, D. G. Huggins, F. deNoyelles, J. O. Meyer and J. T. Lennon, "Assessment of Clinton Lake and its Watershed," Kansas Biological Survey Report 96, Kansas Biological Survey, University of Kansas, Lawrence, 2000.

2

Long-Term Release of Iron-Cyanide Complexes from the Soils of a Manufactured Gas Plant Site

Magdalena Sut[1*]**, Thomas Fischer**[2]**, Frank Repmann**[3]**, Thomas Raab**[1]

[1]Department of Geopedology and Landscape Development, Brandenburg University of Technology Cottbus-Senftenberg, Cottbus, Germany; [2]Central Analytical Laboratory, Brandenburg University of Technology Cottbus-Senftenberg, Cottbus, Germany; [3]Department of Soil Protection and Recultivation, Brandenburg University of Technology Cottbus-Senftenberg, Cottbus, Germany.

ABSTRACT

Iron-cyanide (Fe-CN) complexes have been detected at Manufactured Gas Plant sites (MGP) worldwide. The risk of groundwater contamination depends mainly on the dissolution of ferric ferrocyanide. In order to design effective remediation strategies, it is relevant to understand the contaminant's fate and transport in soil, and to quantify and mathematically model a release rate. The release of iron-cyanide complexes from four contaminated soils, originating from the former MGP in Cottbus, has been studied by using a column experiment. Results indicated that long-term cyanide (CN) release is governed by two phases: one readily dissolved and one strongly fixed. Different isotherm and kinetic equations were used to investigate the driving mechanisms for the ferric ferrocyanide release. Applying the isotherm equations assumed an approach by which two phases were separate in time, whereas the multiple first order equation considered simultaneous occurrence of both cyanide pools. Results indicated varying CN release rates according to the phase and soil. According to isotherm and kinetic models, the long-term iron cyanide release from the MGP soils is a complex phenomenon driven by various mechanisms parallely involving desorption, diffusion and transport processes. Phase I (rapid release) is presumably mainly constrained by the transport process of readily dissolved iron-cyanide complexes combined with desorption of CN bound to reactive heterogeneous surfaces that are in direct contact with the aqueous phase (outer-sphere complexation). Phase II (limited rate) is presumably driven by the diffusion controlled processes involving dissolution of precipitated ferric ferrocyanide from the mineral or inner-sphere complexation of ferricyanides. CN release rates in phase I and II were mainly influenced by the pH, organic matter (OM) and the total CN content. The cyanide release rates increased with increasing pH, decreased with low initial CN concentration and were retarded by the increase in OM content.

Keywords: Prussian Blue; Elovich Equation; Freundlich Equation; Parabolic Diffusion Equation; Multiple First Order Equation

1. Introduction

Cyanide in the form of iron-cyanide (Fe-CN) complexes is a potentially toxic compound that once exposed to UV or visible light radiation, in solution, can be broken down to free cyanide (CN^- and HCN) [1]. Anthropogenic activities, like the process of gas purification after coal gasification in Manufactured Gas Plants (MGPs), yielded side products in the form of ferric ferrocyanide (Prussian Blue), leading possibly to the contamination of soil and groundwater. The manufactured gas was conducted through wood shavings, impregnated with hydrated iron

oxide, in order to remove hydrogen sulfide (H_2S) and hydrogen cyanide (HCN). When the iron oxide lost its absorbing capacity it was often deposited in the vicinities of MGP, which generated a potential environmental pollution due to high amounts of sulfur, tar and various complex iron-cyanides.

Knowledge concerning the behavior, particularly dissolution and desorption, of contaminants can help in reducing the extent of cleanup technologies. In order to design effective remediation strategies, it is relevant to understand contaminant fate and transport in soil, and to quantify and mathematically model the release rate [2]. The mobility of iron-cyanide complexes in soil is mainly

---

[*]Corresponding author.

governed by the characteristics of the soil solution (pH, pE), the presence of complexing cations (K^+, Mn^{2+}, Fe^{2+}, etc.), the presence of UV light as well as the substrate composition and stratigraphy (e.g. clay content, hydrological barriers) of the site. Fe-cyanide complexes are negatively charged and can form inner-sphere complexes on positively charged surfaces, which makes adsorption on the soil particles a possible Fe-CN retention mechanism [3]. With decreasing pH the adsorption of iron-cyanide complexes on iron and aluminum oxides surfaces, which are positively charged under acidic conditions, increases. Hence, neutral and alkaline soils sorb CN anions to a lower extent than acidic soils. Depending on the pH, Fe-CN complexes can be adsorbed on the soil organic matter (SOM). According to Mansfeldt [4], the adsorption takes place through hydrogen bonds under acidic conditions and through charge transfers complexes under neutral to caustic conditions. Fuller [5] stated that the sorption of ferricyanide in soil is driven by the pH, iron-oxides and clay mineral content. According to Ohno [6] sorption of ferrocyanide was increased, when the pH of the soil decreased. Rennert and Mansfeldt [3] found that ferrocyanides adsorb on goethite surfaces rather than ferricyanides. Rennert and Mansfeldt [7] predicted that ferricyanide forms outer-sphere and weak inner-sphere surface complexes on goethite. According to them, ferrocyanide was sorbed inner-spherically and by precipitation of a Prussian Blue-like phase. Cheng and Huang [8] found that the adsorption of either ferrocyanide or ferricyanide complexes onto aluminum oxide is achieved through outer-sphere complexation. Ghosh *et al.* [9] carried out a column experiment, where both ferricyanide and ferrocyanide were not restrained by the sandy aquifer material.

Sorption of iron-cyanide complexes by soils, as shown above, is a subject that is studied in soil chemistry, but the reverse process (release/desorption) should be of an equal environmental interest, due to its practical importance. Column studies can provide key information concerning the mechanism of the iron-cyanide complexes dissolution or desorption. Release rate parameters can be estimated from the isotherms of the time dependent data using various mathematical models. The aim of this study was to use different isotherm and kinetic equations to investigate the phenomena of iron-cyanide complexes release from the MGP soils. Applying various models to the column experimental data, was believed to provide the knowledge whether the contaminant discharge is driven by the kinetics of desorption from the heterogeneous substrates (Elovich, Freundlich), the diffusion-controlled phenomena (Parabolic Diffusion Equation) or by the transport following Multiple First Order Equation. Additionally the influence of soils parameters such as pH, texture, OM content, initial CN concentrations on the iron-cyanide complexes release rate was studied.

2. Materials and Methods

2.1. Field Data

Field data on total and dissolved cyanide content of the soils, pH, EC, OM were obtained by sampling the site of aformer MGP in Cottbus (51°45,161'N; 14°18,529'E). The investigation field covers an area of 2500 m² and is relatively flat. The annual average temperature is 8.8°C and average annual precipitation sum is 589 mm [10]. The local climate is characterized as humid continental [11]. Own pre-studies show that the groundwater table is situated at a depth about 7 m below the surface and the soil pH varies between 3.2 and 7.7 [12]. The top soil layer is composed of varying fractions of sand, coal, slag and organic matter (up to 0.5 m deep). The deeper soil (0.5 - 2.0 m) has a sandy texture (texture classes according to German classification system).

Gas works produced a variety of largely hazardous waste products (like iron-cyanide complexes) that were used as a filling material contaminating the surrounding field. Soils (labeled A, B, C and D) used in the column experiment originate from the former MGP site in Cottbus. Soils A, C and D are the top soils (up to 0.5 m deep), whereas soil B was the lower sandy layer (0.5 - 1.5 m deep) of soil A. Selected chemical and physical properties of the investigated soils are presented in **Table 1**.

Grain size analysis was performed by sieving (>20 μm) and X-ray granulometry (XRG) using the SediGraph 5120™ particle-size analyzer [13]. Organic matter was determined with the Loss on Ignition method (LOI). pH and EC were studied with a bench pH/mV meter MultiLab 540 (WTW). Total and water soluble cyanide (**Table 1**) determination was performed according to the micro dist procedure US QuickChem Method 10-204-00-1-X [15]. After distillation cyanide was determined with the flow injection analyzer (FIA Compact, MLE) [16]. The detection limit for both (total and water soluble CN) extractions is 0.02 mg·l^{-1} of cyanide in analyte.

2.2. Column Experiment (Dissolution/ Desorption)

The release of iron-cyanide complexes from the MGP soils (A, B, C and D) was studied by conducting laboratory column experiments at constant flow rates under unsaturated conditions [14]. Eight percolation columns (two replicates for each soil) were constructed from Plexiglas® (ID 5.4 cm, height 30 cm) and positioned perpendicular toeach other. A peristaltic pump fed distilled water to each column, in the beginning of the experiment at a flow rate of 20 ml·h^{-1} once per day. Introduced soil was homogenized by hand and each column was loaded with ≈ 700 g of field fresh soil. The system was daily percolated with distilled water and the obtained

Table 1. Selected chemical and physical properties of the investigated soils [14].

Soil	Soil characteristic	OM	Water content	Tot. CN conc.	Tot. water soluble CN conc.	pH	EC	Clay	Silt	Sand
		(%)	(%)	$(mg \cdot kg^{-1})$	$(mg \cdot kg^{-1})$		$(\mu S \cdot cm^{-1})$	(%)	(%)	(%)
A	top soil	3.4	12.6	875	148	7.6	1455	9.0	14.1	76.9
B	0.5 - 1.5 m deep	1.2	6.4	401	26	5.9	2041	11.8	17.6	70.6
C	top soil	3.1	12.9	1718	21	5.0	2253	7.4	15.2	77.4
D	top soil	4.2	10.6	40	0.6	7.7	780	8.0	14.1	77.9

leachate was subsequently analyzed with the FIA. The experimental set up is shown in **Figure 1**.

2.3. Isotherm Equations

Three isotherm models were applied to the CN experimental data in order to better understand the release process of iron-cyanide complexes from the MGP soils with the varying pH, OM content, CN concentration and soil texture. The gathered data were computed according to the following equations that often describe time-dependent data sufficiently [17].

2.3.1. Elovich Equation [17]
The Elovich equation is generally considered an empirical equation. It has been used in the soil chemistry to describe the kinetics of sorption and desorption of various inorganic materials on the soil [18], and the soil chemical reaction rates [19,20].

$$q = (1/\alpha) \cdot \ln(a \cdot \alpha) + (1/\alpha) \cdot \ln t \quad (1)$$

where:
q—the amount of released CN in time t $(mg \cdot CN \cdot kg^{-1})$;
α—a release constant $(mg \cdot CN \cdot kg^{-1} \cdot day^{-1})$; and
a—a constant related to the initial velocity of the reaction $(mg \cdot CN \cdot kg^{-1})$.

Plot of "q" vs. "$\ln t$" gives a linear relationship with the slope of $(1/\alpha)$ and the intercept of $(1/\alpha)$, $\ln(a \cdot \alpha)$.

2.3.2. Parabolic Diffusion Equation [21]
The parabolic diffusion equation is often used to indicate that diffusion-controlled phenomena are rate limited. The diffusion models have been developed to predict the dynamic character of release and have been successfully used to describe for example metal reactions on soil and soil constituents [22].

$$q = a + K_d \cdot t^{1/2} \quad (2)$$

where:
q—the amount of released CN in time t $(mg \cdot CN \cdot kg^{-1})$;
a—constant $(mg \cdot CN \cdot kg^{-1})$; and
K_d—apparent diffusion rate constant $(mg \cdot CN \cdot kg^{-1} \cdot day^{-1/2})$.

Plot of "q" vs. "$t^{1/2}$" gives linear relationship if the reaction confirms the parabolic diffusion law. The "a" and

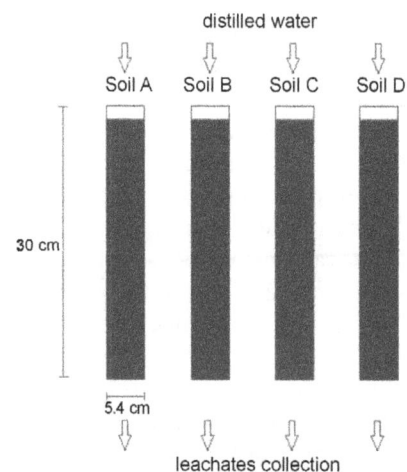

Figure 1. Scheme of the column experiment set-up.

"K_d" parameters are determined from the intercept and the slope of the function respectively.

2.3.3. Freundlich Equation [23]
Freundlich equation is generally considered an empirical relationship describing the adsorption of solutes from a liquid to a solid surface, and have been widely applied to experimental data. Elkhatib et al. [24] used a modified Freundlich equation to describe the kinetics of lead and copper desorption [25] from soils.

$$q = k \cdot t^{v} \quad (3)$$

where:
q—the amount of released CN in time t $(mg \cdot CN \cdot kg^{-1})$;
k—release rate coefficient (day^{-1});
t—reaction time (day); and
v—a constant.

The Freundlich isotherm is a power function, where "k" and "v" are constants that can be determined from the coefficient and the exponent respectively.

2.4. Kinetic Equation

Transport models assuming chemically controlled non-equilibrium, which describes the kinetic of a release or dissociation reactions is often defined as a first order reaction [26]. The heterogeneity of a system as well as the controlling mechanism of the release process (such as mass transfer or chemical reaction) determines the rate constants that are required to describe the experimental data.

First Order Equation [27]

Release kinetics based on the first order equation, where the total released amount (q) within a certain time (t), is expressed by the following equation:

$$q = q_0 \cdot \left(1 - e^{-kt}\right) \tag{4}$$

where:

q—the amount of released CN in time t (mg·CN·kg^{-1});
q_0—the amount of CN released at equilibrium (mg·CN·kg^{-1}); and
k—apparent release rate coefficient (day^{-1}).

Assuming that CN release is constrained by more than one pool, total released CN amount should be expressed as:

$$q_{tot} = q_1 + q_2 \tag{5}$$

where:

q_{tot}—is the total amount of released CN in time t (mg·CN·kg^{-1});
q_1—is the fast releasing CN pool (mg·CN·kg^{-1}); and
q_2—is the stronger fixed CN pool (mg·CN·kg^{-1}).

The release kinetics for two pools concept (one readily and one slowly liberating) can be expressed using the multiple first order rate equation, where each pool has its capacity and rate constant:

$$q = q_1 \cdot \left(1 - e^{-k_1 t}\right) + q_2 \cdot \left(1 - e^{-k_2 t}\right) \tag{6}$$

Numerical parameters (k) fit was based on least sum of squares.

2.5. Statistical Analysis

The predictive performances of the developed models were assessed by adjusted correlation coefficient (R^2), standard error (SE) and the probability value (p), using the analysis of variance (ANOVA).

3. Results

3.1. Column Experiment

Release of cyanide from MGP soils (A, B, C and D), was investigated with the soil column experiment. The CN release rate was studied for four soils with different pH's and textures (**Table 1**). According to Bodenkundliche Kartieranleitung [28], soil A and B are medium loamy sandsoils (SI3), whereas soils C and D are characterized as weak loamy sand soils (SI2). **Figure 2** represents the relation of the released cyanide, plotted cumulatively, vs. release time. **Figure 2** indicates that long-term CN release from soil can be described using two separate cyanide pools: one available and one strongly fixed. The amount of released cyanide representing each pool was visually obtained from the graph (**Figure 2**).

It is assumed that amount of cyanide in the column leachates is influenced by mobilization of readily soluble

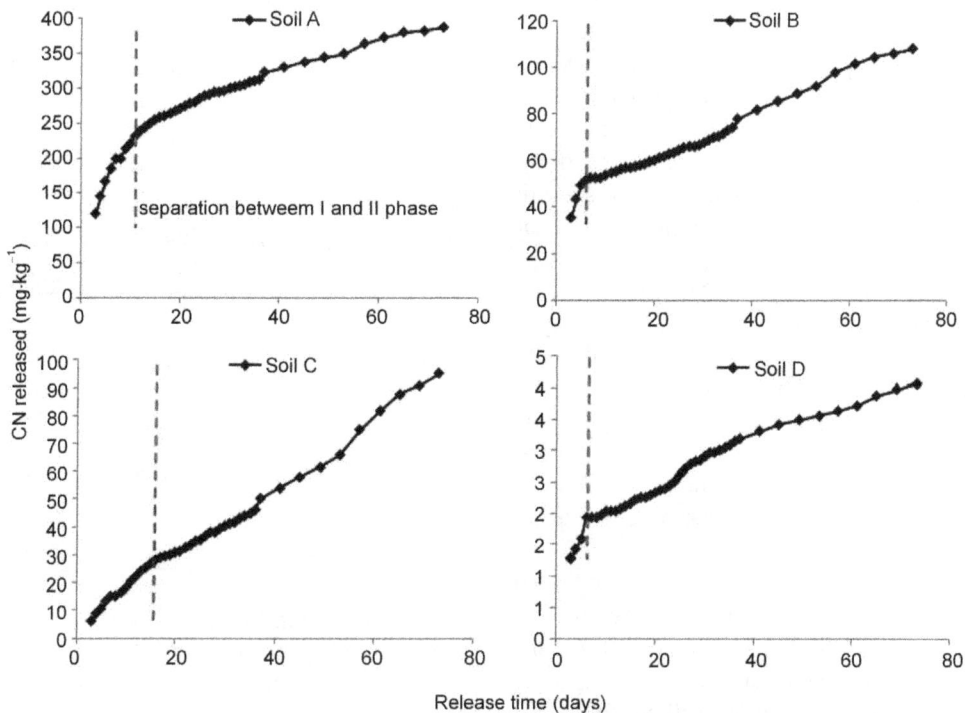

Figure 2. Cumulative CN release curves for the four investigated soils [14].

hexacyanoferrats (phase I) and slow dissolution of ferric ferrocyanide (phase II) [29]. The kinetics of CN release will be based on deriving a constant release rate for each phase, based on the continuously measured CN rerelease as a function of time.

3.2. Isotherm Models

Modeling of the CN release experimental data using isotherm equations assumes that the above mentioned two phase approach is separate in time and that phase I precedes phase II. Treating the processes separately, intent to derive the cyanide release rates for each phase.

3.2.1. Elovich Equation

The empirical equation [19] was used to describe the CN release rate from the MGP soils (A, B, C and D) in the column experiment. **Figure 3** demonstrates the Elovich equation plots of released CN vs. ln of reaction time obtained for phase I and phase II. In **Figure 3** it can be noticed that a linear relationship exists between the released CN "q" and ln of release time "ln (release time)" for both phases in all investigated soils.

The Elovich equation parameters, determined from the slope and intercept of the linear plots, are given in **Tables 2** and **3**. In the Elovich equation a decrease in "α" values and increase in "a" values would increase the reaction rates [22]; [30]. Regression analysis (**Table 2**) indicated significant (<0.01) correlation in all investigated soil.

The Elovich equation parameters for phase II are listed in **Table 3**. Regression analysis (**Table 3**) indicated significant (<0.01) correlation in all investigated soils. As indicated by the regression analysis, the Elovich equation resulted to be adequate for describing the kinetics of CN release from contaminated soils in a column experiment. Moreover, the Elovich equation provides a very good fit ($R^2 > 0.95$) for phase I and a good fit ($R^2 > 0.84$) for phase II of CN release.

3.2.2. Parabolic Diffusion

The parabolic diffusion equation was subsequently used to describe the CN release from the contaminated soils (A, B, C and D) in the column experiment. A parabolic diffusion plot of CN release vs. $t^{1/2}$ is shown in **Figure 4**. The parabolic diffusion equation parameters, determined from the slope and intercept of the linear plots, are given in **Tables 4** and **5**.

Regression analysis for phase I (**Table 4**) indicated significant (<0.01) correlation and high correlation coefficient (>0.91) in all investigated soil. In phase II (**Table 5**), regression analysis demonstrates significant (<0.01) correlation in all investigated soil, as well as high correlation coefficient (>0.97) and low SE.

Table 2. The Elovich equation parameters and correlation coefficients for phase I CN release in the MGP soils.

| Soil | Phase I | | | | |
| | α | a | R^2 | SE | p |
	mg·CN·kg^{-1}·day^{-1}	mg·CN·kg^{-1}			
A	0.01	125.00	0.99	3.29	<0.01
B	0.05	39.92	0.96	1.54	<0.01
C	0.07	5.82	0.98	1.37	<0.01
D	2.02	2.21	0.98	0.03	<0.01

Table 3. The Elovich equation parameters and correlation coefficients for phase II CN release in the MGP soils.

| Soil | Phase II | | | | |
| | α | a | R^2 | SE | p |
	mg·CN·kg^{-1}·day^{-1}	mg·CN·kg^{-1}			
A	0.01	88.99	0.98	4.46	<0.01
B	0.04	15.77	0.85	5.72	<0.01
C	0.02	3.79	0.96	3.81	<0.01
D	1.27	0.58	0.93	0.08	<0.01

Table 4. The parabolic diffusion equation parameters and correlation coefficients for phase I CN release in the MGP soils.

| Soil | Phase I | | | | |
| | K_d | a | R^2 | SE | p |
	mg·CN·kg^{-1}·day$^{-1/2}$	mg·CN·kg^{-1}			
A	83.00	69.63	0.94	4.96	<0.01
B	32.18	35.81	0.99	0.81	<0.01
C	10.03	5.57	0.99	0.77	<0.01
D	0.54	0.02	0.92	0.11	<0.01

Table 5. The parabolic diffusion equation parameters and correlation coefficients for phase II CN release in the MGP soils.

| Soil | Phase II | | | | |
| | K_d | a | R^2 | SE | p |
	mg·CN·kg^{-1}·day$^{-1/2}$	mg·CN·kg^{-1}			
A	29.30	139.92	0.99	2.26	<0.01
B	10.29	15.98	0.94	3.60	<0.01
C	17.12	53.96	0.98	2.60	<0.01
D	0.26	0.95	0.97	0.05	<0.01

3.2.3. Freundlich Equation

The Freundlich equation was also used to describe the CN release from the MGP soils in a column experiment. The Freundlich isotherm is a power function, where "v" and "k" are constants. Isotherms of this form have been

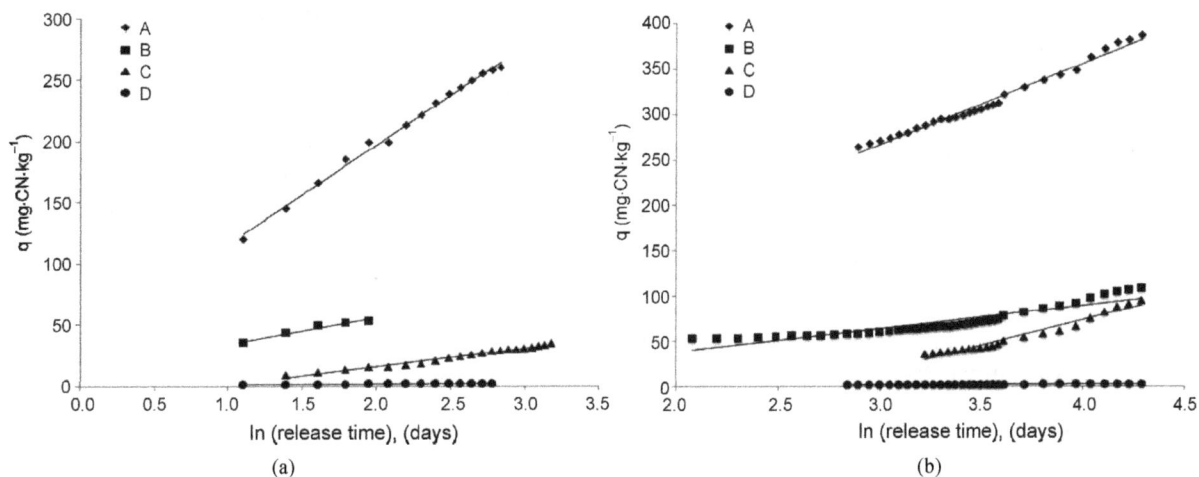

Figure 3. The Elovich equation plots for CN release from the MGP soils in (a) phase I and (b) phase II.

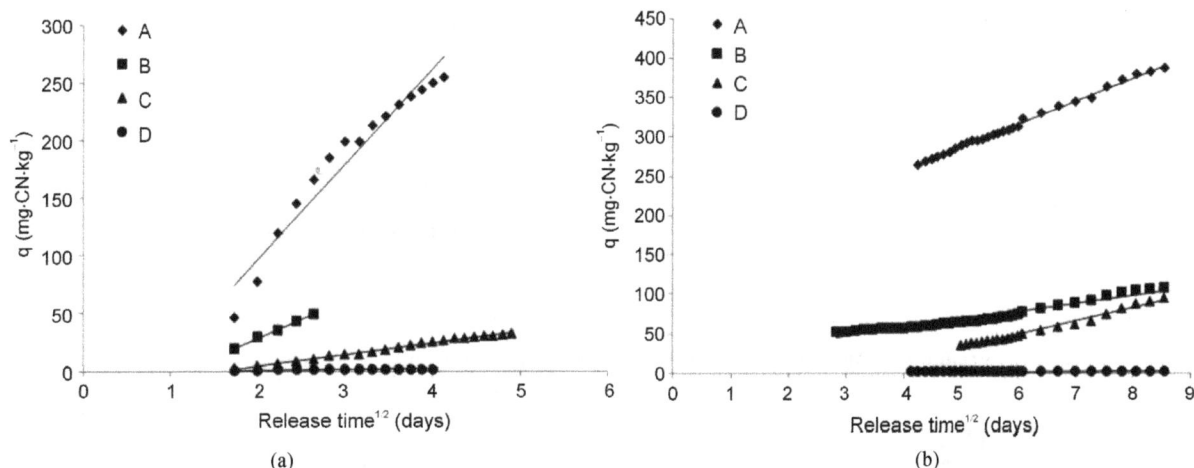

Figure 4. The parabolic diffusion equation plots for CN release from the MGP soils in (a) phase I and (b) phase II.

observed for a wide range of heterogeneous surfaces, including activated carbon, silica, clays, metals, and polymers [31]. The release of CN in phase I and II was well modeled by the Freundlich equation (**Figure 5**).

In both phases (I and II), a power function was able to fit the data with a high degree of correlation: $R^2 > 0.93$ and $R^2 > 0.90$ respectively. Regression analysis (**Tables 6** and **7**) indicated significant (<0.01) correlation in all investigated soil. The Freundlich equation proved to be successful in describing both phases in CN release from the MGP soils.

3.3. Kinetic Model

Another consideration assumes that release of iron-cyanide complexes is constrained by two phases that occur simultaneously, which would suggest non-equilibrium liberation. In this approach transport phenomena of phase I is not considered separately from the slow chemical reaction of phase II. For this approach, a modified first order equation was used. The total released CN amount

Table 6. The Freundlich equation parameters and correlation coefficients for phase I CN release in the MGP soils.

Soil	Phase I				
	$v \times 10^{-3}$	$\dfrac{k}{day^{-1}}$	R^2	SE	p
A	427.60	81.40	0.97	7.30	<0.01
B	477.30	21.77	0.94	2.00	<0.01
C	596.50	5.58	0.98	0.70	<0.01
D	295.40	0.94	0.97	0.04	<0.01

Table 7. Freundlich equation parameters and correlation coefficients for phase II CN release in the MGP soils.

Soil	Phase II				
	$v \times 10^{-3}$	$\dfrac{k}{day^{-1}}$	R^2	SE	p
A	278.40	117.00	0.99	3.60	<0.01
B	345.90	22.37	0.91	3.80	<0.01
C	948.70	1.60	0.99	2.70	<0.01
D	300.30	0.87	0.95	0.06	<0.01

was determined as a sum of phase I and phase II (Equation 6), where each phase had its capacity and rate constant.

Multiple First Order Equation

A multiple, two-component first-order equation was used to describe the CN release form the MGP soils in a column experiment. **Figure 6** represents fitted release curves for the investigated soils, the measured CN and the released quantities form both phases. **Figure 6** shows that the two-component first order model provides a good fit of the experimental data for soil A, B and D. The multiple first order equation parameters and correlation coefficients are listed in **Table 8**. Applying this kinetic approach, it was assumed that each phase has its release rate (k), which is proportional to the amount present in a specific pool.

Regression analysis (**Table 8**) indicated significant (<0.01) correlation in all investigated soil. According to the correlation coefficient and standard error, a modified two-component first-order equation was successful in describing the experimental data from soil A, B and D. Slightly worse correlation was obtained for soil C (R^2 = 0.89; SE = 7.42).

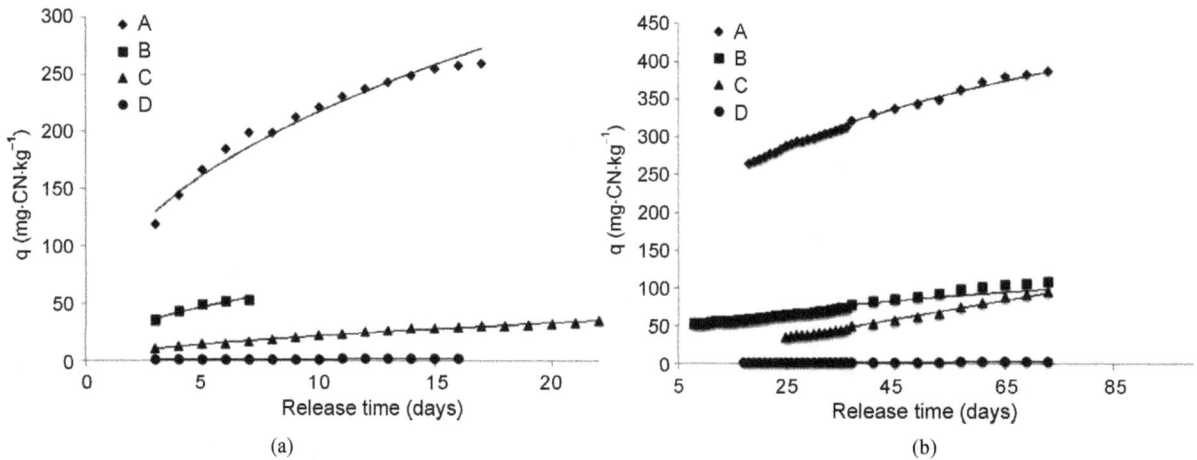

Figure 5. The Freundlich equation plots for CN release from the MGP soils in (a) phase I and (b) phase II.

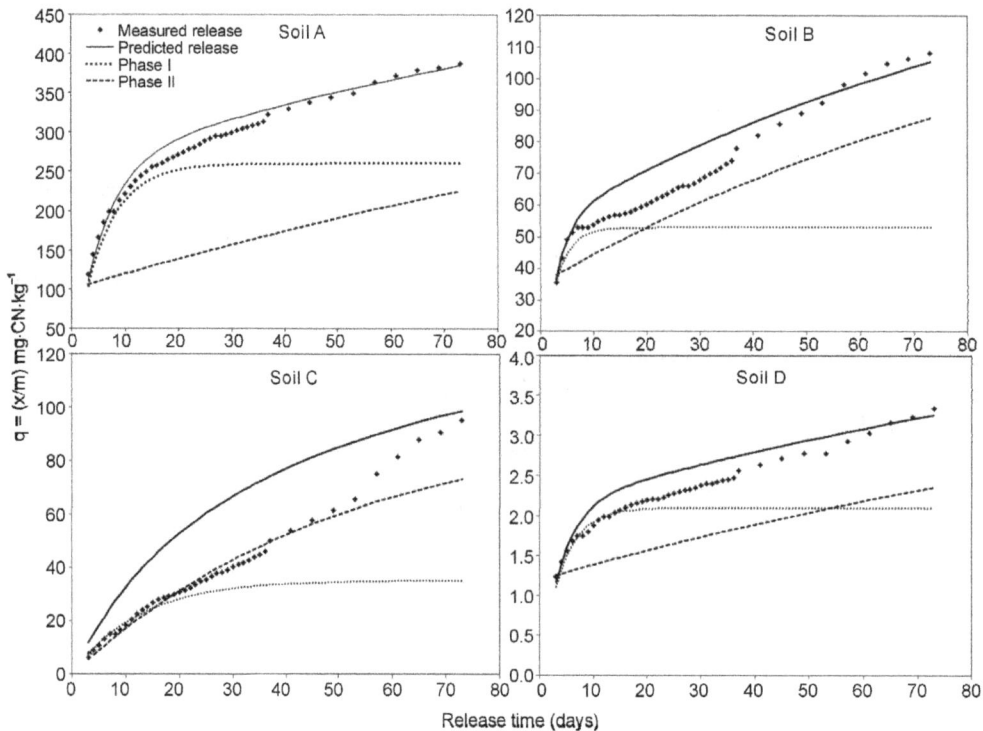

Figure 6. Cumulative measured CN release plots with predicted CN release, phase I and phase II, using the modified first order equation, for the MGP soils.

Table 8. The multiple first order equation parameters and correlation coefficients for CN release in the MGP soils.

Soil	Phase I	Phase II	R^2	SE	p
	k_1	k_2			
	day^{-1}	day^{-1}			
A	0.17	0.53×10^{-2}	0.97	9.28	<0.01
B	0.37	0.01	0.92	4.14	<0.01
C	0.08	0.01	0.89	7.42	<0.01
D	0.5×10^{-2}	0.25	0.94	0.10	<0.01

4. Discussion

In contaminated soils, on the sites of former MGPs, the mobility of iron-cyanide complexes is mainly governed by the dissolution and precipitation of ferric ferrocyanide and adsorption on soil minerals [32]. The purpose of our work was to gain better knowledge concerning the iron-cyanide complexes release from the MGP soils by applying various isotherm and kinetic equations. The column experiment simulated the experimental conditions relevant to anthropogenically altered soils. The approach presented here assumes that the release of iron-cyanide complexes is constrained by two phases. According to Aharoni et al. [23], the rate of release is rapid, when it is governed by the transport process taking place in the liquid phase, or diffusion in the bulk of the liquid, at the film adjacent to the solid particle, in liquid-filled pores, etc. Theis et al. [33] attributed quick and complete desorption of ferricyanide from goethite to outer-sphere complexation.

If the release rate is slow, it is probably limited by the process taking place in the solid phase. It can be constrained by the constant dissolution of the ferric ferrocyanide like precipitate, which according to Mansfeldt and Dohrmann [34] may originate from coprecipitation on the soil surface or from precipitation of iron-cyanide complexes with alkali and alkaline earth cations. Rennert and Mansfeldt [7]; [35] proposed that slow and incomeplete desorption of ferrocyanide was attributed to inner-sphere surface complexation, which occurs through the formation of direct chemical bonds with the mineral surface, (typically with surface oxygen atoms), and by precipitation of a Prussian Blue-like phase on the goethite surface.

Pursuant to the results obtained in the column experiment (**Figure 2**), it is believed that the release of iron-cyanide complexes from the contaminated soils can be described using two separate cyanide pools: one available (like transport of readily dissolved hexacyanoferrats or desorption of weak outer-sphere complexation) and one strongly fixed (like dissolution of precipitate in from of ferric ferrocyanide or desorption of inner-sphere complexation). Applying the isotherm models to the column

experimental data required handling the phases separately in order to derive the release rate constants. Implementing the equations to the complete data set (**Figure 2**) resulted in very low correlations, which proves the hypothesis that the release of iron-cyanide complexes from the MGP soils is constrained by two phases.

The Elovich equation has been frequently used to study the chemical release processes and is suitable for systems with heterogeneous adsorbing surfaces [36]. The kinetic behavior of inorganic materials like metals (Pb and Cu) has been successfully described by the Elovich equation [22]; [23]. Mathematical analysis of the CN release data indicated that the Elovich equation is suitable to describe the kinetic behavior of iron-cyanide complexes in the MGP soils (**Figure 3**). In the Elovich equation, a decrease in "α" increases the reaction rate. In phase I (**Table 2**), the CN release rate increases with the increasing soil pH (**Table 1**), except for the soil D, where release rate is very low according to soil pH (7.7). The low release rate of soil D is most probably caused by the low total CN concentration (**Table 1**) hence, the cyanide release rate in phase I is influenced by the soil pH as well as the initial CN concentration in soil. Generally, in the desorption processes, one of the most important parameter is the initial pH value of the solution, which influences both the contaminant surface binding sites and the contaminant chemistry in water. Our findings are consistent with the study made by Ohno [6], who investigated sorption of ferrocyanide by five soils, where increasing sorption was observed with decreasing soil pH. In phase II, the relation of "α" and "a" values are analogous with the ones obtained for phase I, where the release rates for phase I are higher than for phase II. Low pH of the soil C (pH = 5) most probably reduces the amount of readily dissolved iron-cyanide complexes in the phase I, which results in only slightly higher "α" values for the phase II. Lower, but still significant, was the correlation in soil B ($R^2 = 0.85$), which may be affected by the low OM content (**Table 1**). On the other hand, depending on pH, the overall charge of SOM is either neutral or negative, hence the anion adsorption cannot be expected. However, Rennert and Mansfeltd [3] state that SOM promotes the sorption, hence the content of C_{org} is possibly enhancing the sorption of iron-cyanide complexes on soils. The suggested reaction for this process was the charge transfer complexes, formed by cyanide ion (CN^-), via cyanide-N with quinone groups of humic acids [37].

Simultaneously, the parabolic diffusion equation was used to describe the CN release from the MGP soils. This model has been used by many scientists to characterize the diffusion-controlled phenomena in soil constituents and the release of ion in soil and soil minerals [38]. It assumes that described CN release is determined by the sum of various diffusion processes with different diffusion coefficients and various particle sizes. Linear relationship visible in phase II indicates (**Figure 4**) that the parabolic equation adequately describes the CN release process, suggesting that phase II is driven by the diffusion of CN out of the mineral matrix. On the other hand, in phase I, the regression line for the soil A doesn't pass through the origin, suggesting that the diffusion is not the main driving mechanisms. Additionally, in phase I, the "a" value was determined from the y-intercept ($t = 0$). The intercept is most probably affected by the rapid CN release, which would be much slower if not influenced by the transport of already dissolved phase. The apparent CN diffusion rate coefficient "K_d" in the parabolic diffusion law is considered the measure of the relative rate of CN release [23]. The difference between the "K_d" values indicates that the release power of the soils is different. In phase I (**Table 4**), the "K_d" values for the studied soils were increasing with the increasing soil pH, except for the soil D, which despite of the alkaline character, indicated low CN release, most probably induced by low initial CN concentration. Soil B, despite of acidic character, indicated comparably high CN release rate, which can be attributed to the low OM content. The diffusion coefficient "K_d" is higher in sandy soils with lower organic matter. More heterogenic soils are more likely to have an increase in transport-limited process [23]. In phase II, the relative rate of CN release (K_d) seems to be affected by the CN concentration. Major decrease in "K_d" value can be noticed in soil A, despite of basic soil pH. Release rate in soil B also decreased, whereas in soil C, continuous release, comparable to the one obtained in phase I, can be observed (**Table 5**). Based on correlation coefficient it can be stated that the parabolic diffusion law effectively describes the phase II of CN release from the MGP soils. For the phase I, the results revealed ("a" value determined from the y-intercept) that the diffusion phenomena is not the driving mechanisms, however it doesn't imply that CN release does not include a slow diffusion reaction. It may rather indicate that the kinetics of this process shouldn't be considered separately from the transport phenomena. More study need to be done to determine whether CN release is driven by intraparticle diffusion, external-film diffusion or internal-pore diffusion.

Subsequently, the release of CN from the MGP soil was described using the Freundlich equation. This power function exhibits increasing release rate with increasing time, but decreasing positive slope as time increases (**Figure 5**). The Freundlich equation is often used for heterogeneous surfaces and describes desorption from solid to the solutes in liquid and assumes that different sites with several adsorption energies are involved. Many organics and inorganics follow this type of behavior [19]; [23]. According to mathematical analysis (**Tables 6** and

7), the Freundlich reaction based model was successful in describing, both phase I and phase II, CN release from the MGP soils. The exception is phase I in soil A, where the regression line doesn't pass through the origin (**Figure 5**), suggesting that desorption is not the driving mechanism. Soil A is alkaline (pH = 7.6) and has high CN content, which would explain high amount of dissolved cyanide in the pore water and imply that the CN release in phase I is mainly constrained by the transport of readily dissolved compounds rather than desorption. The values of release rate coefficient "k", in phase I and II, decrease with the decreasing soil pH, except for the soil D, where low "k" value might be a result of low CN concentration in soil.

Applying the isotherm equations to the column data was aimed at better understanding the mechanisms of the CN release that, prior to the kinetic study, was divided in two phases. This modeling approach assumed that phase I and phase II are separated in time. Results revealed various release rates in both phases, implying that the driving mechanisms are different. The column experimental data for phase II showed good correlation with the Elovich, Freundlich and Parabolic Diffusion Equations leading to inconclusive results about the driving mechanisms of the CN release. For the phase I, poor fitting of the regression line (Freundlich) and the negative intercept values (Parabolic diffusion), implied the transport of dissolved iron-cyanide complexes as the main process.

The First order equation was previously used by many researchers to describe time-dependent data [26]; [39]. This modeling approach assumes that both CN release phases occur simultaneously. The modified first order model well described the CN release data (**Table 8**), which is supported by the graphical test presented in **Figure 6**. This result suggests that the release of CN from the MGP soils followed the multiple first order kinetics. Worst graphical and regressional correlation was obtained for soil C. The release of CN from soil C is almost linear, most probably due to low soil pH, constrained mainly by one strongly fixed pool. According to Meeussen *et al.* [40] the mobility of cyanide in the soil largely depends on pH. Under acidic conditions, solid iron-cyanide complexes in the form of precipitated Prussian Blue are likely to be expected. It could explain why the two-component approach didn't manage to describe the kinetics of CN release from the soil C. Due to the low pH, the amount of dissolved iron-cyanide complexes is relatively low, so the difference in the release rates for phase I and phase II is relatively small (**Table 8**).

Rate constants for each soil vary (**Table 8**), indicating the highest release rate in soil B for phase I and in soil D for phase II. The low initial release rate in phase I for soil C is consistent with the study made by Meeussen *et al.*

[40]. They stated that acidic character of soil will considerably decrease the CN concentration in groundwater and reduce the mobility of iron-cyanide complexes in such soils. High initial release rate (k_1) in soil B can be constrained by the low OM content, despite of a slightly acidic character of the soil. Using the multiple first order kinetic equation for modeling of the long-term cyanide release probably closer reflects the phenomena that occur in the MGP soils. It is more probable that the release of phase I and phase II appear simultaneously rather than completely separate in time.

5. Conclusions

The study of iron-cyanide complex release, in a column experiment, was conducted to investigate the long-term desorption or dissolution mechanisms. The research revealed that the cyanide liberation from the investigated MGP soils is driven by two phases. From the kinetic studies, it was observed that the cyanide release was initially rapid (phase I) followed by a much slower release rate (phase II). Most probably, one more fraction exists (an amount that is not released), but our experimental time scale didn't allow for that observation.

Modeling with isotherm equations, assuming that both phases are separate in time, delivered inconclusive results concerning the driving mechanisms for the cyanide release in phase II. The Elovich equation was in good agreement to describe the CN release in phase I and II, suggesting desorption from the heterogeneous surfaces to the liquid. Analogously good correlation was obtained by using the Freundlich equation, except for phase I in soil A, where too high CN content and alkaline pH imply transport of readily dissolved cyanide as a main driving release mechanism. The parabolic diffusion adequately describes the rate-limiting CN release (phase II), implying that it's driven by the diffusion of CN out of the mineral matrix. For phase I, obtained results imply that transport of dissolved cyanide is the main mechanisms. Indefinite results for phase II, obtained from applying the isotherm equations, most probably indicate that the long-term iron cyanide release from the MGP soils is a complex phenomenon driven by various mechanisms parallely involving desorption, diffusion and dissolution processes.

The multiple first order equation assumed simultaneous occurrence of both phases and adequately described the CN release from soil A, B and D, except for the soil C, where due to it's acidic character, the CN mobility is most probably constrained by one strongly fixed pool. This non-equilibrium approach is considered to closer reflect the probable cyanide release mechanisms from the MGP soils.

Based on conducted isotherm and kinetic modeling, we attribute the fast release rate (phase I) to the transport process of readily dissolved iron-cyanide complexes (hexacyanoferrats) that is taking place in the liquid phase combined with the desorption of CN bound to heterogeneous surfaces that are in direct contact with aqueous phase (outer-sphere complexation).

Mobility governed on the low release level (phase II) is probably controlled by the desorption, dissolution or diffusion processes, like the dissolution of precipitated ferric ferrocyanide or of inner-sphere complexed ferricyanides.

The iron-cyanide release rates for phase I and II, obtained in the kinetic modeling, revealed that the CN mobility is mainly influenced by the pH (which affects both the contaminant surface binding sites and the contaminant chemistry in water), by the initial CN concentration and by the possible sorption on soil organic matter. The cyanide release rates increased with the increasing pH, decreased with low initial CN concentration and was retarded by the increase in OM content.

6. Acknowledgements

This work was supported by the Brandenburg Ministry of Science, Research and Culture (MWFK) as part of the International Graduate School (IGS) at Brandenburg University of Technology (BTU). This study was partially funded by Deutsche Bahn AG within the project "Stabilisierung des DB AG-Standortes ehem. Leuchtgasanstalt Cottbus durch Verfahren der Bioremediation (Phytoremediation)".

REFERENCES

[1] P. Kjeldsen, "Behavior of Cyanides in Soil and Groundwater: A Review," *Water, Air & Soil Pollution*, Vol. 115, No. 1-4, 1998, pp. 279-307.

[2] C. M. Saffron, J. H. Park, B. E. Dale and T. C. Voice, "Kinetics of Contaminant Desorption Form Soil: Comparison of Model Formulations Using the Akaika Information Criterion," *Environmental Science Technology*, Vol. 40, 2006, pp. 7662-7667.

[3] T. Rennert and T. Mansfeldt, "Sorption of Iron-Cyanide Complexes in Soil," *Soils Science Society of American Journal*, Vol. 66, No. 2, 2002, pp. 437-444.

[4] T. Mansfeldt, "Mobilität und Mobilisierbarkeit von Eisenkomplexierten Cyaniden. Materialien zur Altlastensanierung und zum Bodenschutz (MALBO)," *Landesumweltamt Nordrhein-Westfalen (LUA NRW)*, Vol. 16, 2003, pp. 17-44.

[5] W. H. Fuller, "Cyanides in the Environment with Particular Attention to the Soil," In: D. Van Zyl, Ed., *Cyanide and the Environment*, Colorado State University, Fort Collins, 1985, pp. 19-44.

[6] T. Ohno, "Levels of Total Cyanide and NaCl in Surface

Waters Adjacent to Road Salt Storage Facilities," *Environmental Pollution*, Vol. 67, No. 2, 1990, pp. 123-132.

[7]   T. Rennert and T. Mansfeldt, "Sorption of Iron-Cyanide Complexes on Goethite," *European Journal of Soil Science*, Vol. 52, No. 1, 2001, pp. 121-128.

[8]   W. P. Cheng and C. Huang, "Adsorption Characteristics of Iron-Cyanide Complex on γ-Al_2O_3," *Journal of Colloid Interface Science*, Vol. 181, No. 2, 1996, pp. 627-637.

[9]   R. S. Ghosh, D. A. Dzombak, R. G. Luthy and D. V. Nakles, "Subsurface Fate and Transport of Cyanide Species at a Manufactured Gas Plant Site," *Water Environment Research*, Vol. 71, No. 6, 1999, pp. 1205-1216.

[10] M. Linder, H. Bugmann, P. Lasch, M. Fleschig and W. Cramer, "Regional Impacts of Climatic Change on Forests in the State of Brandenburg, Germany," *Agricultural and Forest Meteorology*, Vol. 84, No. 1-2, 1997, pp. 123-135.

[11] M. C. Peel, B. L. Finlayson and T. A. McMahon, "Updated World Map of the Köppen Geiger Climate Classification," *Hydrology and Earth System Sciences*, Vol. 11, 2007, pp. 1633-1644.

[12] M. Sut, T. Fischer, F. Repmann, T. Raab and T. Dimitrova, "Feasibility of Field Portable Near Infrared (NIR) Spectroscopy to Determine Cyanide Concentrations in Soil," *Water, Air & Soil Pollution*, Vol. 223, No. 8, 2012, pp. 5495-5504.

[13] H. W. Müller, R. Dohrmann, D. Klosa, S. Rehder and W. Eckelmann, "Comparison of Two Procedures for Particle-Size Analysis: Köhn Pipette and X-Ray Granulometry," *Journal of Plant Nutrition and Soil science*, Vol. 172, No. 2, 2009, pp. 172-179.

[14] M. Sut, F. Repmann and T. Raab, "Stability of Prussian Blue in Soils of a Former Manufactured Gas Plant Site," *Soil and Sediments Contamination an International Journal*.

[15] USEPA, "Method 10-204-00-1-X, Lachat US EPA Approved and Equivalent Method," Revision 3, 2008.

[16] DIN EN ISO 14 403, "Bestimmung von Gesamt Cyanid und Freiem Cyanid mit Derkontinuerlichen Fließanalytik-Teil D," 2002.

[17] S. H. Chien, W. R. Clayton and G. H. McClellan, "Kinetics of Dissolution of Phosphate Rocks in Soil," *Soils Science Society of American Journal*, Vol. 44, No. 2, 1980, pp. 260-264.

[18] R. J. Atkinson, F. J. Hingston, A. M. Posner and J. P. Quirk, "Elovich Equation for the Kinetics of Isotope Exchange Reaction at Soild-Liquid Interfaces," *Nature*, Vol. 226, 1970, pp. 148-149.

[19] S. H. Chien and W. R. Clayton, "Application of Elovich Equation to the Kinetics of Phosphate Release and Sorption in Soils," *Soils Science Society of American Journal*, Vol. 44, No. 2, 1980, pp. 265-268.

[20] J. Torrent, "Rapid and Slow Phosphate Sorption by Mediterranean Soils. Effect of Iron Oxides," *Soils Science Society of American Journal*, Vol. 51, 1987, pp. 78-82.

[21] K. J. Laidler, "Chemical Kinetics," McGraw-Hill, New York, 1965.

[22] P. M. Jardine and D. L. Sparks, "Potassium-Calcium Exchange in Multireactive Soil System. I. Kinetics," *Soils Science Society of American Journal*, Vol. 48, No. 1, 1984, pp. 39-45.

[23] C. Aharoni, D. L. Sparks, S. Levinson and I. Ravina, "Kinetics of Soil Chemical Reactions: Relationships between Empirical Equations and Diffusion Models," *Soils Science Society of American Journal*, Vol. 55, 1991, pp. 1307-1312.

[24] E. A. Elkhatib, G. M. ElShebiny and A. M. Balba, "Comparison of Four Equations to Describe the Kinetics of Lead Desorption from Soils," *Zeitschrift für Pflanzenernährung und Bodenkunde*, Vol. 155, 1992, pp. 285-291.

[25] E. A. Elkhatib, A. M. Mahdy, M. E. Saleh and N. H. Barakat, "Kinetics of Copper Desorption from Soils as Affected by Different Organic Ligands," *International Journal of Envirnomental Science Technology*, Vol. 4, No. 3, 2007, pp. 331-338.

[26] M. M. Nederlof, W. H. Van Riemsdijk and L. K. Koopal, "Analysis of the Rate of Dissociation of Ligand Complexes," *Environmental Science and Technology*, Vol. 28, 1994, pp. 1048-1053.

[27] S. E. A. T. M. Van der Zee and W. H. Van Riemsdijk, "Model for Long-Term Phosphate Reaction Kinetics in Soil," *Journal of Environmental Quality*, Vol. 17, No. 1, 1998, pp. 35-41.

[28] Bodenkundliche Kartieranleitung, AG Boden, 5. Aufl., 438 S., 41 Abb., 103 Tab., 31 Listen, Hannover, 2005.

[29] F. J. Hingston, A. M. Posner and J. P. Quirk, "Anion Adsorption by Goethite and Gibbsite. II. Desorption of Anions from Hydrous Oxide Surfaces," *Journal of Soil Sciences*, Vol. 25, No. 1, 1974, pp. 16-26.

[30] M. J. D. Low, "Kinetics of Chemisorption of Gases on Solids," *Chemical Reviews*, Vol. 60, No. 3, 1960, pp. 267-312.

[31] R. J. Umpleby II, S. C. Baxter, M. Bode, J. K. Berch Jr., R. N. Shah and K. D. Shimizu, "Application of the Freundlich Adsorption Isotherm in the Characterization of Molecularly Imprinted Polymers," *Analitica Chimica Acta*,

Vol. 435, No. 1, 2001, pp. 35-42.

[32] J. L. Meeussen, M. G. Keizer, W. H. Van Riemsdijk and F. A. M. de Haan, "Dissolution Behavior of Iron Cyanide (Prussian Blue) in Contaminated Soils," *Environmental Science & Technology*, Vol. 26, No. 9, 1992, pp. 1832-1838.

[33] T. L. Theis and M. L. West, "Effects of Cyanide Complexation on the Adsorption of Trace Metals at the Surface of Goethite," *Environmental Technology Letters*, Vol. 7, No. 1-12, 1986, pp. 309-318.

[34] T. Mansfeldt and R. Dohrmann, "Identification of a Crystalline Cyanide-Containing Compound in Blast Furnace Sludge Deposits," *Journal of Environmental Quality*, Vol. 30, No. 6, 2001, pp. 1927-1932.

[35] T. Rennert and T. Mansfeldt, "Sorption of Iron-Cyanide Complexes on Goethite in the Presence of Sulfate and Desorption with Phosphate and Chloride," *Journal of Environmental Quality*, Vol. 31, No. 3, 2002, pp. 745-751.

[36] F. C. Wu, R. L. Tseng and R. S. Juang, "Characteristics of Elovich Equation Used for the Analysis of Adsorption Kinetics in Dye-Chitosan Systems," *Chemical Engineering Journal*, Vol. 150, 2009, pp. 366-373.

[37] B. Schenk and B. M. Wilke, "Cyanidadsorption an Sesquioxiden, Tonmineralen und Huminstoffen," *Zeitschrift für Pflanzenernährung und Bodenkunde*, Vol. 147, No. 6, 1984, pp. 669-679.

[38] R. L. Evans and J. J. Jurinak, "Kinetics of Phosphate Release from a Desert Soil," *Soil Science*, Vol. 121, No. 4, 1976, pp. 205-211.

[39] D. Freese, "Criteria and Methods for the Assessment of Long-Term Phosphate Sorption and Desorption in Soils," *Habilitationsschrift, Landwirtschaftlich-Gärtnerischen Fakultät*, Humboldt-Universität zu Berlin, 1994.

[40] J. C. Meeussen, G. Meindert, W. H. Van Riemsdijk and F. A. M. de Haan, "Solubility of Cyanide in Contaminated Soils," *Journal of Environmental Quality*, Vol. 23, No. 4, 1994, pp. 785-787.

3

Short-Term Effects of Ozone and PM$_{2.5}$ on Mortality in 12 Canadian Cities[*]

Nawal Farhat[1#], Tim Ramsay[1], Michael Jerrett[2], Daniel Krewski[1,3]

[1]Department of Epidemiology and Community Medicine, University of Ottawa, Ottawa, Canada; [2]School of Public Health, University of California at Berkeley, Berkeley, USA; [3]McLaughlin Center for Population Health Risk Assessment, University of Ottawa, Ottawa, Canada.

ABSTRACT

Numerous recent epidemiological studies have linked health effects with short-term exposure to air pollution levels commonly found in North America. The association between two key pollutants—ozone and fine particulate matter—and mortality in 12 Canadian cities was explored in a time-series study. City-specific estimates were obtained using Poisson regression models, adjusting for the effects of seasonality and temperature. Estimates were then pooled across cities using the inverse variance method. For a 10 ppb increase in 1-hr daily maximum ozone levels, significant associations were in the range of 0.56% - 2.47% increase in mortality. For a 10 µg/m^3 increase in the 24-hr average PM$_{2.5}$ concentration of, significant associations varied between 0.91% and 3.17% increase in mortality. Generally, stronger associations were found among the elderly. Effects estimates were robust to adjustment for seasonality, but were sensitive to lag structures. There was no evidence for effect modification of the mortality-exposure association by city-level ecologic covariates.

Keywords: Air Pollution; Ozone; Particulate Matter; Mortality; Canada

1. Introduction

Health effects of air pollution have become a major public health concern in North America, Europe and other developed regions in the past several years. The World Health Organization estimated 1.34 million premature deaths (2.4% of total deaths) were attributable to outdoor air pollution in 2008 [1]. Further, using satellite imaging data to predict tropospheric PM$_{2.5}$ concentrations globally, Evans *et al.* [2] recently estimated that 7% of global mortality may be attributable to particulate air pollution.

In Canada, the Canadian Medical Association's (CMA) report No Breathing Room—National Illness Costs of Air Pollution published in 2008, stated that air pollution results in considerable health and economic damages that will only increase over time. It was estimated that 21,000 deaths and 92,000 emergency department visits in Canada could be attributed to short- and long-term exposure to air pollution in year 2008. Associated economic costs for the year, including worker absenteeism, higher health care costs, loss of life, and other factors were expected to exceed $10 billion [3].

Numerous studies have shown positive and significant associations between adverse health effects and short-term exposure to ozone (O$_3$) and particulate matter (PM), both of which are major components of smog in Canada [4-16]. Both pollutants have been linked to various health effects including premature mortality, deaths, and hospital admissions due to respiratory or cardiovascular diseases. Other effects reported include decreases in lung

[*]Declaration of interest: Funding was provided through the McLaughlin Center for Population Health Risk Assessment. D. Krewski holds a Natural Sciences and Engineering Research Council Industrial Research Chair in Risk Science at the University of Ottawa, a peer-reviewed Canadian university-industry partnership program. He also serves as Chief Risk Scientist for Risk Sciences International, a Canadian company established in partnership with the University of Ottawa (www.risksciences.com) which has conducted air pollution risk assessments for public and private sector clients.

[#]Corresponding author.

function and the exacerbation of existing chronic respiratory and cardiovascular diseases. The elderly [6,17-20] and children [21-25] have been reported to be at greater risk than the general population. Exposure to high levels of air pollution during pregnancy has also been linked to low birth weight in Canada and other countries [26].

Health effects of short term exposure to air pollution are typically assessed using time-series studies, where associations between daily variations in air pollution levels and daily counts of deaths in a given area are estimated by Poisson regression models. One of the central issues in statistical modeling of time-series studies is adequate control for potential confounding. Confounders typically controlled for are those that change over relatively short periods such as seasonality, day of the week, and weather variables that are associated with both pollution levels and health outcomes. Control for confounding is usually achieved using smooth functions for time and temperature variables in the Poisson regression models [27]. For calendar time, penalized splines (PS) or natural cubic splines (NS) are commonly used as the smoothing functions. The degree of smoothing is determined by the degrees of freedom (df) allowed in the smoothing functions [28]. This must be selected carefully to ensure that there is neither over-fitting (too many df), which would fit the "noise", nor under-fitting (too few df), which would not remove the confounding effects of potential confounders in order to avoid bias in the effect estimate. Typically, 3 - 12 df per year have been used. Another important consideration in time-series models is the lag structure, which refers to the period between exposure to the pollutant and the event (health outcome). Lag periods used can be described using single-day lag models, which allow for a lag period of a number of days, or combined lag models. In combined models, the pollution exposure levels are averages of multiple single-day lags. Distributed lag models look at the effect of cumulative exposure to pollution over the course of several days, thus allowing each day to have an effect on health outcomes.

There have been many advances in time-series models since they were first used to study short-term effects of exposure to air pollution in the 1980s [28]. Early statistical approaches included standard regression models, which have now been replaced by semi-parametric models. The two main statistical models currently used are based on Generalized Linear Models (GLM) with parametric splines or Generalized Additive Models (GAM) with non-parametric splines. GAMs offer increased flexibility in estimating the smooth component of the model relative to GLM, and had been preferred over GLMs until 2002. It was then discovered that these methods underestimated the standard errors of linear terms in the model (the air pollution regression coefficients) and over-

estimated the effect of air pollution on health outcomes [28-30]. The discovery of these methodological and computational issues came at the time when the United States Environmental Protection Agency (EPA) was finalizing its most recent review of the epidemiologic evidence on particulate matter air pollution. As a result, all of the findings from time-series studies that had been based on GAMs were re-evaluated using alternative methods. Approximately 40 original studies from the US, Canada and Europe were reanalyzed and then peer reviewed by a panel appointed by the Health Effects Institute (HEI) [31]. The HEI re-analysis report stated that no optimal analytic method could be recommended to estimate the air pollution health effects. Studies that have compared different approaches have found that, although there may be some sensitivities in the results, the effects remain statistically significant with the common approaches used [11].

The purpose of the present paper is to quantitatively assess the impact of fine particulate matter (PM$_{2.5}$) and ozone on mortality (total, cardiovascular and respiratory) in Canada, and explore the sensitivity of the air pollution effects estimates to different model specifications. A secondary objective is to explore potential effect modification of socioeconomic and demographic variables on the effect of air pollution and health.

2. Materials and Methods

2.1. Data

2.1.1. Location, Exposure and Outcome

Air pollution and mortality data were analyzed for the following 12 Canadian cities: Calgary, Edmonton, Halifax, Hamilton, Montreal, Ottawa, Toronto, Quebec City, St John, Vancouver, Windsor and Winnipeg. Data sets were previously assembled and provided by Health Canada. The air pollution data were obtained through the National Air Pollution Surveillance (NAPS) program administered by Environment Canada, which is subject to an extensive quality assurance program. A single daily measurement for each pollutant was available and represented the average of the measurements of all monitors in that city. On days when one or more monitors were not functioning, daily measurements were derived from the remaining monitors. Daily ozone concentrations collected include the 1-hour (1-hr) and the 8-hour (8-hr) maximum concentrations. The 1-hr maximum, available on a daily basis from 1980-2001, was used in the analyses to facilitate comparison of results with previous findings. Particulate matter measurements represent the 24-hour average cumulative mass measurements from all the monitors in one city. PM$_{2.5}$ was measured once every six days. However, the data had occasional random periods with missing data in many of the cities. In general, the time period with data available for each city varied between 6 - 16 years since 1984. Records of the daily mean temperature for the time-series were also available.

Health outcome data for this study were obtained from the Canadian Institute for Health Information (CIHI). To ensure the quality of data collected, CIHI regularly performs quality checks of its databases. Deaths were classified using the ICD-9 (International Statistical Classification of Diseases) codes. Records that had been classified using the ICD-10-CA scheme were converted to the ICD-9 classification scheme by CIHI and were then subject to a quality assurance program. Mortality data for the 12 Canadian cities were available for the 20-year period 1981-2001. The databases included information on residence (city), age, date of death, and single underlying cause of death.

2.1.2. Potential Effect Modifiers

Data on 29 ecologic covariates representing city-level demographic, socioeconomic, health care, and lifestyle determinants were used to explore effect modification. The data were initially compiled for use in the international study, Air Pollution and Health: A combined European and North American Approach (APHENA) [32], but was of limited use due to the lack of uniform data of variables in the US and Europe. To explore effect modification, city-specific risk coefficients, βs, were regressed on the city-level covariates, by weighted linear regression with weights inversely proportional to the variance of each city's β. Twenty-nine variables that might modify the exposure-mortality association were considered by including them in the time series models individually.

2.1.3. Analyses

1) First stage (city-specific estimates)

Poisson regression models allowing for over dispersion were used to estimate the associations of ozone and PM$_{2.5}$ with mortality. The city-specific model is presented in Equation (1):

$$\log E\left(\mu_t\right) = s_1\left(time, df\right) + s_2\left(temp_t, 3df\right) \\ + \beta P_t + \left(DOW\right) + \left(holiday\right) \quad (1)$$

where $E(\mu_t)$ is the expected value of the Poisson distributed variable μ_t, which represents the daily counts of events (deaths) on day t. The term $s_1(time, df)$ controls for seasonality, where s_1 is a smooth function with natural cubic splines as basis functions for the time variable and df is the degrees of freedom that allows s_1 to take various functional forms. The function s_1 models the non-linear association between time-varying covariates, calendar time, and daily mortality. To control for weather, the term $s_2(temp, 3)$ was included, where s_2 is a smoothing function of temperature on day t with 3 df. P is the

pollutant concentration (ozone in ppb or PM$_{2.5}$ in μg/m^3) on day t; DOW and holiday are dummy variables included in the model representing the day of week and holidays. The regression coefficient β represents the log relative increase (if β is positive) in the number of events in the target population per unit increase in pollutant concentration. Time-series studies generally report results as percent change in mortality per 10 units change in pollutant concentration (This value is simply obtained by multiplying the regression coefficient β by a factor of 1000).

Three mortality outcomes for each age group were considered based on the ICD-9 codes: <800 for total mortality corresponding to all non-accidental causes of death, 390 - 459 for cardiovascular causes of death, and 460 - 519 for respiratory causes of death.

Sensitivity analyses exploring the effect of degrees of freedom allowed for seasonality control in the smooth function of calendar time and the effect of varying the lag period were conducted. These analyses were limited to ozone, which had daily data. Effects estimates for ozone on all outcomes across all ages were determined with models allowing for 1 - 20 df for the time variable per year of data available. For temperature, three df were allowed in all analyses, based on previous findings that indicate that results are robust to varying degrees df used for temperature. Models included the same lagged term for temperature variable as was used for the pollutant under consideration.

To compare the effect of natural splines and penalized splines, a sensitivity analyses on the risk estimates was carried out on the Toronto data set. After the results of the df and lag period analyses were examined, three values of the df (4, 8 and 12 per year) and three lag periods (for ozone: lag1, lag02 and dist02) were selected for analyses of the data for the remaining cities. Combined lag structures could not be applied to PM$_{2.5}$ since data were available for every sixth day; thus, three single-day lag periods were selected for this pollutant (lag0, lag1, lag2).

2) Second stage (pooled estimates)

City-specific estimates were combined to arrive at pooled estimates by applying fixed effects (FEM) and random effects (REM) regression models. In the fixed effects approach, effects estimates ($\beta's$) were assumed to be normally distributed around an overall estimate and were pooled using inverse variance weighting, with weights proportional to the inverse variance of each city's β. In the random effects regression approach, the city-specific βs were assumed to form a sample of independent observations from a normal distribution with the same mean and with variance equal to the sum of the between-city variance and the variance of β. The between-city variance is added to the city-specific variance and is estimated using the maximum likelihood estimation (MLE) method [33].

2.2. Effect Modification

Heterogeneity between city specific estimates was assessed by the I^2 index, which is a measure of the total variability among effect sizes that can be attributed to true heterogeneity (between-city variability) [34]. In general, an I2 less than 25% suggests low heterogeneity between cities. To explore potential effect modification, weighted linear regression of the city-specific estimates was performed onto each ecologic covariate. Weights were inversely proportional to the variance of each city specific risk estimate (β). Models where the data showed a significant linear association at the 95% confidence level between the potential effect modifier and the risk estimates were assumed to potentially modify the pollutant-health outcome relationship.

All analyses were completed using R statistical software for Windows version 2.6.1 [35].

3. Results

Descriptive statistics for the mortality data are provided **Tables 1** and **2**. There were a total of approximately 1.6 million deaths among all age groups between 1981 and 2000 across the 12 Canadian cities considered. The total exposed population was approximately 9.1 million. Individual city population ranged from 100,000 in St John to 2.3 million in Toronto, based on the 1991 Census. Mean daily death counts varied between 0 (respiratory mortality) and 48 deaths (total mortality), depending on the size of the city.

Descriptive statistics of the exposure database are presented in **Table 3**. Mean annual temperatures for the 12 cities ranged from 2.4°C (Edmonton) to 10.6°C (Van-

couver). Ozone measurements were available on a daily basis during the period 1981-2000, with few missing data (except for Halifax). The mean measurements of the 1-hr maximum ozone levels were in the range 27 - 37 ppb. For $PM_{2.5}$, the mean 24-hour levels varied between 9 $\mu g/m^3$ and 16 $\mu g/m^3$. The time periods during which $PM_{2.5}$ data was available were not uniform across cites. $PM_{2.5}$ was generally measured every sixth day for most cities, with occasional intermittent missing data across longer periods. Ozone and $PM_{2.5}$ levels were not strongly correlated, with the highest correlation coefficients being 0.46 and 0.41 for Windsor and St John, respectively.

The city level socio-demographic, health services, and lifestyle ecological variables were assessed for potential effect modification of the association between air pollution and mortality. Several of the variables listed were highly correlated.

Table 1. Total number of death counts in the 12 Canadian cities in the 1981-2000 period.

Outcome/Age group	Total counts
All-cause mortality	
all ages	1,564,583
75 and over	748,498
under 75	815,978
Cardiovascular mortality	
all ages	641,072
75 and over	369,177
under 75	271,855
Respiratory mortality	
all ages	134,663
75 and over	85,971
under 75	48,683

Table 2. Summary of the population size and mean number of daily mortality counts by cause and age group in the 12 Canadian cities.

City	Population (×1000)	All-cause mortality			Cardiovascular mortality			Respiratory mortality		
		All ages	*75 and over*	*Under 75*	*All ages*	*75 and over*	*Under 75*	*All ages*	*75 and over*	*Under 75*
Calgary	711	10	5	5	4	2	2	1	1	0
Edmonton	617	11	5	6	5	3	2	1	1	0
Halifax	231	6	3	3	2	1	1	1	0	0
Hamilton	319	10	4	5	4	2	2	1	0	0
Montreal	1776	48	22	26	19	10	9	4	2	2
Ottawa	880	15	7	8	6	4	3	1	1	0
Quebec City	540	17	8	9	7	4	3	1	1	1
St John	103	3	1	1	1	1	1	0	0	0
Toronto	2276	47	22	24	18	11	8	4	3	1
Vancouver	1832	29	15	14	12	8	4	3	2	1
Windsor	191	6	3	3	3	2	1	0	0	0
Winnipeg	615	14	7	7	6	4	2	1	1	0

Table 3. Descriptive statistics for the study period, air pollutants and temperature data used in the analyses of mortality outcomes.

City	Calgary	Edmonton	Halifax	Hamilton.	Montreal	Ottawa	Quebec City	St John	Toronto	Vancouver	Windsor	Winnipeg
Temp (°C)												
Mean	4.5	3.0	6.5	8.0	6.6	6.4	4.4	5.2	8.1	10.5	9.8	3.1
25th centile	−1.9	−5	-0.8	0.1	−2.1	−2.7	−4.6	−1.7	0.2	6.3	1.5	−7.4
Median	5.6	4.7	6.9	8.2	7.6	7.45	5.4	6.1	8.3	10.3	10.2	4.7
75th centile	13	13	14.6	17	16.7	17	14.8	13.7	17.2	15.3	19.1	15.4
Ozone (ppb)												
Time period (month/year)	01/81-12/00	01/81-12/00	01/81-12/00	01/81-12/00	01/81-12/00	01/81-12/00	01/81-12/00	01/81-12/00	01/81-12/00	01/81-12/00	01/81-12/00	01/81-12/00
No. obs.[1]	7305	7302	5945	7290	7304	7303	7208	7144	7305	7292	7225	7159
No. missing[2]	0	3	1360	15	1	2	97	161	2	13	80	146
Mean (1 hr max)	33.1	31.2	29.0	34.8	28.7	28.8	28.8	34.5	34.2	27.0	36.9	30.0
Maximum (1 hr max)	94	110	100	129	115.7	105	135	160	144.1	104.6	159	99
25th centile	25	22	21	22.3	18.8	20	20	26	22.3	19.1	21	21.5
Median	33	30.5	28	31	26.2	27	28.5	32.3	30.5	26.6	31.5	29
75th centile	41	40	35	44	35.9	35.7	35.0	40.0	42.3	34	49	37.5
PM$_{2.5}$ (µg/m^3)												
Time period (month/year)	12/84-12/00	12/84-12/00	12/84-12/96	01/95-12/00	12/84-12/00	12/84-12/00	12/85-12/00	09/92-09/99	12/84-12/00	12/84-12/00	12/87-12/00	09/84-09/00
No. obs.	891	791	657	418	1180	807	524	1125	1537	1082	1031	816
No. missing	6414	6514	6648	6887	6125	6498	6781	6180	5770	6223	6274	6489
Mean	10.2	10.1	11.0	15.3	14.7	10.7	11.3	7.7	14.7	11.8	16.3	9.0
Maximum	66.1	64.0	45.5	74.1	72.0	53.8	50.4	53.2	71.0	67.0	85.6	71.3
25th centile	5.7	5.3	6.1	7.7	7.8	5.1	6.0	3.8	7.3	6.7	8.7	5.2
Median	8.3	8	9.15	12.5	12	8.32	9	6.3	12.34	9.8	13.7	7.3
75th centile	12.1	12	13.5	20.3	18.8	13.8	14.0	9.9	19.5	14.1	20.7	11.0

[1]Total number of observations; [2]Number of missing observations.

Figure 1 presents the pooled percent increase (random effects) in mortality outcomes across all ages associated with an increase of 10 ppb in the previous day's ozone concentrations (lag1). The number degrees of freedom allowed per year of data available was varied between 1 and 20 in the smooth function of time (natural splines) in the city-specific models. Estimates stabilized after allowing 4 - 5 df per year, displaying slight decreases as the degrees of freedom increased. In the absence of substantial heterogeneity among city specific estimates, fixed effects and random effects models gave comparable results across cities. Estimates were positive and statistically significant across all outcomes.

Results showing the effect of varying the lag period between exposure to ozone and day of death are presented in **Figure 2**. The figure represents the pooled results (random effects) for all ages at 4, 8 and 12 df per year of data. Higher estimates were observed for combined lag models relative to single day lags. Wider confidence intervals for respiratory mortality compared to other mortality outcomes were observed, a result of the low daily counts for this outcome. For total and cardiovascular mortality, effects estimates were significant at all lag structures and df examined.

The mortality effects estimates for ozone and PM$_{2.5}$ across all age groups with eight degrees of freedom for seasonality control and three lag structures are summarized in **Tables 4** and **5**, respectively. For ozone, results were statistically significant across all age groups, with higher estimates observed with the combined lags rela-

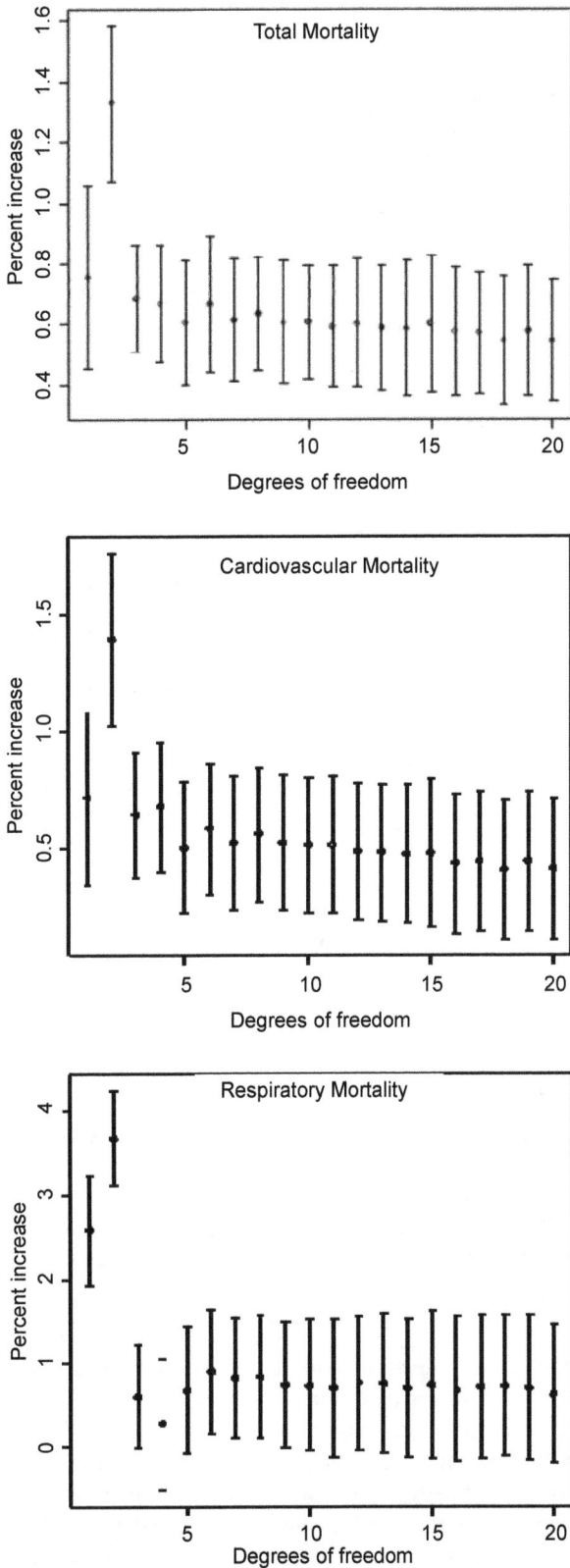

Figure 1. Pooled percent increase in mortality and 95% CI associated with a 10 ppb increase in the 1-hr maximum ozone levels in 12 Canadian cities with a 1-day lag with varying degrees of freedom per year for seasonality control.

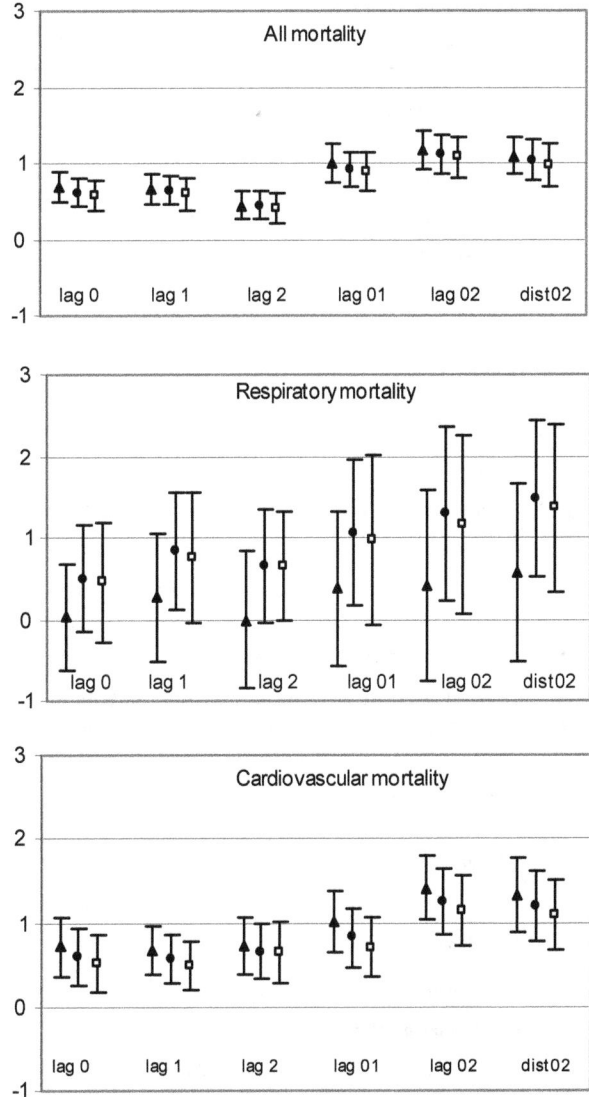

Figure 2. Pooled percent change in mortality (all ages) associated with an increase of 10 ppb in the 1-hr maximum ozone concentrations in 12 Canadian cities. Results are from random effects models with various df allowed for seasonality control. (Δ 4 df; ● 8 df; □ 12 df).

tive to the single day lag period. Effects estimates were generally stronger among the elderly. However, respiratory mortality estimates were stronger for the <75 age group compared to the ≥75 group, contrary to what might be expected.

For PM$_{2.5}$, three single-day lag structures were evaluated with exposure on the same day (lag0), the previous day (lag1) and two days prior (lag2). Effects represent the percent increase in the outcome associated with a 10 μg/m^3 increase in the 24-hour average PM$_{2.5}$ levels. Generally, fewer outcomes were statistically significant in relation to PM$_{2.5}$, compared to ozone. Estimates at 1- or 2-day lag periods were higher relative to effects at same day exposure, indicating that PM$_{2.5}$ may have a delayed

Table 4. Pooled percent increase in mortality outcomes associated with a 10 ppb increases in the 1-hour maximum ozone levels for three lag structures. Results are from fixed effects models with 8 df allowed for seasonality control.

Outcome/Age group	Percent increase (95% CI)		
	lag1	lag02	dist02
All mortality			
All ages	0.64 (0.45 - 0.82)	1.12 (0.87 - 1.38)	1.03 (0.77 - 1.30)
Under 75	0.59 (0.33 - 0.84)	1.01 (0.66 - 1.36)	0.95 (0.59 - 1.31)
75 and over	0.69 (0.42 - 0.96)	1.25 (0.87 - 1.62)	1.05 (0.66 - 1.45)
Cardiovascular mortality			
All ages	0.56 (0.27 - 0.84)	1.25 (0.85 - 1.65)	1.20 (0.79 - 1.61)
Under 75	0.07 (−0.36 - 0.51)	0.76 (0.16 - 1.37)	1.63 (0.01 - 1.25)
75 and over	0.93 (0.55 - 1.31)	1.62 (1.09 - 2.15)	1.40 (0.84 - 1.96)
Respiratory mortality			
All ages	0.86 (0.21 - 1.51)	1.34 (0.44 - 2.25)	1.49 (0.55 - 2.43)
Under 75	0.98 (−0.08 - 2.04)	2.14 (0.67 - 3.62)	2.47 (0.95 - 3.98)
75 and over	0.79 (−0.04 - 1.62)	0.86 (−0.29 - 2.01)	1.58 (0.33 - 2.84)

Table 5. Pooled percent increase in mortality outcomes associated with a 10 μg/m³ increase in the 24-hr average $PM_{2.5}$ concentrations. Results are from fixed effects models with 8 df allowed for seasonality control.

Outcome/Age group	Percent increase (95% CI)		
	lag0	lag1	lag2
All mortality			
All ages	0.35 (−0.23 - 5.94)	1.43 (0.84 - 2.31)	0.98 (0.39 - 1.57)
Under 75	0.57 (−0.26 - 1.39)	0.91 (0.08 - 1.73)	0.14 (−0.68 - 4.97)
75 and over	0.12 (−0.72 - 5.96)	1.98 (1.14 - 2.81)	1.85 (1.01 - 2.68)
Cardiovascular mortality			
All ages	−0.23 (−1.18 - 0.72)	1.03 (0.09 - 1.97)	1.77 (0.83 - 2.71)
Under 75	0.11 (−1.37 - 1.68)	−1.44 (−4.84 - 1.96)	0.88 (−0.61 - 2.36)
75 and over	−0.45 (−1.68 - 0.78)	2.14 (0.92 - 3.35)	2.39 (1.17 - 3.61)
Respiratory mortality			
All ages	−1.12 (−3.19 - 1.95)	0.30 (−1.72 - 2.33)	1.31 (−0.75 - 3.36)
Under 75	0.04 (−3.80 - 3.99)	−0.27 (−3.87 - 3.33)	−1.77 (−5.53 - 2.88)
75 and over	−0.95 (−3.54 - 1.64)	0.73 (−1.74 - 3.21)	3.17 (0.61 - 5.72)

effect on health outcomes. Effects for mortality on the same day of exposure (lag0) were not statistically significant. Positive and significant effects were seen for total and cardiovascular mortality in 1- and 2-day lag models. Respiratory mortality effects were only significant at 2-day lag and among the elderly. Effects were consistently higher for older age groups.

There was no substantial heterogeneity between city-specific estimates in the majority of models applied. Two outcomes that displayed the highest heterogeneity based on the I^2 index were selected for effect modification analyses: cardiovascular mortality with ozone (assessed at lag02, 8 df for time, for <75 years, I^2 index 22%) and cardiovascular mortality with $PM_{2.5}$ (assessed at lag0, 8 df for time, for <75 years, I^2 index 16%). **Table 6** presents statistically significant results (at the 95% confidence level) of the effect modification analysis of $PM_{2.5}$ and ozone. Results represent the percent increases in daily number of deaths associated with an increase of 10 units in $PM_{2.5}$ or ozone at two different values for the effect modifier, corresponding to the 25th and the 75th percentiles of the city-specific distribution that variable.

Table 6. Percent change in mortality associated with a 10 µg/m^3 increase in PM$_{2.5}$ concentrations at the 25th and 75th percentile of the city-specific distributions of covariates that displayed significant effect modification.

Outcome/ Effect modifier	Percent change (95% CI)		p-value
	25th centile	75th centile	
Cardiovascular mortality and PM$_{2.5}$			
Area	−0.72 (−3.45, 2.02)	−0.17 (−2.79, 2.46)	0.05
Unemployment, males	1.72 (−1.32, 4.77)	0.16 (−2.44, 2.76)	0.05
Manufacturing	2.75 (−0.88, 6.38)	−0.24 (−2.85, 2.37)	0.04
Stress	1.6 (−0.89, 4.09)	0.47 (−1.8, 2.75)	0.01

Results can be seen as showing pollutant effects in a city characterized by a level of the effect modifier corresponding to the 25th or 75th percentile.

4. Discussion

This study presents results of the short-term effects of ozone and PM$_{2.5}$ exposure on mortality across 12 Canadian cities. Statistically significant associations were observed across the three mortality outcomes, with estimates being generally higher among the elderly. Risk estimates were robust to seasonality control when more than five degrees of freedom per year of data available were allowed. Sensitivity of risk estimates was observed to varying lag structures with higher estimates when using combined lag structures for ozone, and with 1 or 2 day lags for PM$_{2.5}$. Analyses of socio-demographic, health services and lifestyle covariates did not identify any potential effect modifiers that warrant further investigation.

4.1. Degrees of Freedom for Seasonality

Across the three mortality outcomes, effects estimates were found to stabilize beyond five degrees of freedom per year in the smoothing function of calendar time. Results obtained are in agreement with other studies that have explored the sensitivity of degrees of freedom for seasonality control. Peng et al. [20] conducted a simulation study that compared various methods commonly used to adjust for seasonal and long-term trends. By examining the variability of the regression coefficient, β, using 1 - 20 df per year, results indicated that the bias in β was only serious for df between 1 and 4 with natural splines (and between 1 and 6 df with penalized splines) and was stable afterwards. Another study in California found that effects estimates decreased with increasing df when evaluated at 4, 8 and 12 df per year with a greater reduction observed going from 4 to 8 df [13].

Although there is no preferred method to choose the optimal degrees of freedom, Touloumi et al. [36] suggest that the approach followed in NMMAPS (7 df per year)

yields conservative air pollution effects estimates, since this value of df is large enough to ensure adequate control for seasonal and long-term trends. Many previous studies have selected a fixed value for df (generally between 4 - 12 df per year) to be used in analyses, based on sensitivity analyses or previous results [13,14,37,38]. The analyses in this study evaluated all the outcomes for different age groups at 4, 8 and 12 df per year to compare the effects across this range.

Natural cubic splines were used in all analyses presented. Penalized splines have also been used in time-series studies and both methods are believed to yield comparable results. Mortality risk estimates associated with ozone obtained from both approaches were compared in this study using the data set for one of the 12 cities (Toronto). Risk estimates varied between −4% and 6% when comparing both approaches at values of 6 - 14 df per year for time, confirming the comparability of risk estimates based on natural splines and penalized splines.

The effect of varying the df for the temperature variable was not explored in this study. It is generally accepted that the effects estimates are not as sensitive to the method used to control for temperature as they are to controlling for calendar time [39]. The approach followed in APHENA for controlling for temperature was adopted in this study, where three degrees of freedom were allowed in the smooth function of temperature in all models.

4.2. Effects Estimates

The use of the 1-hr maximum daily average for ozone facilitated the comparison of results with previous findings, as many of previous studies used this measure. The World Health Organization suggests that the 8-hour average may be a better indicator for respiratory function and lung inflammation [40]. However, correlation coefficients between the 1-hr and 8-hr maximum ozone levels were in the 0.94 - 0.97 range across the 12 Canadian cities. Thus, similar results are to be expected using either measurement. Results of both measures were compared in the project Air Pollution and Health—A European Approach 2 (APHEA 2), and have been reported to give comparable results [11].

Pooled effects estimates across cities for ozone were positive and significant for total and cause specific mortality. Significant results observed were in the range of 0.56% to 2.47% increase in mortality for a 10 ppb increase in the 1-hr maximum ozone levels. In general, higher effects were obtained for the ≥75 age group. These results are comparable with previous studies [9,10,15,16,32]. A slightly stronger effect was detected for respiratory mortality, which is also consistent with other studies. In APHEA 2, a stronger association between ozone and respiratory mortality was found (2.26%) relative to other mortality outcomes (0.9% cardiovascular mortality and 0.66% for total mortality) per 10 ppb increase in the 1-hr maximum ozone levels [11]. In a study within NMMAPS that looked at 95 US urban communities, a positive and significant association (0.64% increase) per 10 ppb increase in the previous week's ozone levels was estimated for respiratory and cardiovascular mortality, slightly higher than the estimate for total mortality (0.52%). Further, in an Australian study of four cities, a significant association was obtained for respiratory mortality (2.2% increases per 10 ppb increase), but not for other outcomes (Simpson *et al.* 2005). The greatest effects estimates were observed with combined lag models in this study. The higher effects estimated with the 3-day average (lag02) and the distributed lag models (dist02) suggest that the effect of ozone may not only depend on same day exposure, but also on the exposure over the previous 2 days. This indicates that single day lag models may underestimate the cumulative effect of ozone on mortality due to repeated exposure to high levels of ozone. Hence, the combined and distributed lag models may be more appropriate for estimating ozone health effects. This is in agreement with previous findings that suggest multi-day exposure lags are higher than single day lags [9,10,32,41]. Studies that investigated lag models taking into account the previous week's ozone levels in 95 US cities and found that effects were consistently higher than those of single day lag models [9,10]. Meta-analyses that have looked at the health effects of ozone have found positive effects for both total and cardiovascular mortality [8] or only total mortality [16].

Estimates from this study are lower than the Canadian estimates from APHENA for all-cause mortality and cardiovascular mortality outcomes for ozone as the exposure pollutant. The data sets used in APHENA represent a subset of the data used in this study, covering a shorter time period (1987-1996). It is hypothesized that the use of a shorter time-series and the inclusion of additional covariates in the models (two terms for temperature control) in APHENA, may have led to different results. A sensitivity analysis exploring the effect of varying the number of temperature terms included in the time series was carried out. Results show that the use of one term (same day temperature (temp0)) or two terms (same day and previous day temperature (temp01)) for the temperature variable produced comparable risk estimates. Further investigation is needed to more fully explain the difference in results between this study and the Canadian APHENA results.

Fine particulate matter showed statistically significant effects estimates combined across cities. For total and cardiovascular mortality, significant estimates were in the range of 0.91% - 2.39% increase per 10 µg/m^3 increase in $PM_{2.5}$. Results of this study are in agreement with previous study findings where estimates reported have generally been in the range of 0.8% - 2.4% increase [42]. A number of previous studies have reported comparable results [12-14], although others have found no significant effect on mortality [8,15]. The only significant estimate detected for respiratory mortality was a 3.17% increase for ≥65 at a 2-day lag. It is unusual to detect a relatively strong association for this age group when other groups considered did not show any significant effects. Compared to respiratory diseases, cardiovascular diseases are more prevalent which leads to increased power to detect weak associations [43]. It is possible that due to low number of respiratory related deaths, the models applied were not able to detect the weak association and that the 3.17% increase observed was obtained by chance.

Effects were consistently higher for older age groups, supporting the hypothesis that the elderly may be more susceptible to the effects of $PM_{2.5}$; this may be a result of exacerbation of pre-existing conditions that are more prevalent among individuals in this age group or due to reduced antioxidant defenses [44].

Effects of mortality on the same day of exposure (lag0) were not significant. Rather, across all mortality outcomes and two age groups (all ages and ≥75), the effect of $PM_{2.5}$ was strongest at 1-day lag (and sometimes at 2-day lag) compared to the effect of same day exposure, suggesting a delayed $PM_{2.5}$ effect. As with the case of ozone, findings for $PM_{2.5}$ reported in the literature have been inconsistent. For example, a study in Montreal found that cardiovascular mortality was more affected by exposure to $PM_{2.5}$ in previous days [18], whereas a study in 10 US cities found a stronger same-day exposure effect [41]. Results have also been inconsistent for respiratory deaths. Previous studies have reported stronger effects on day exposure levels [38] or exposure in the prior 1 or 2 days [45]. This inconsistency may be a result of the different chemical components of the $PM_{2.5}$ mixture with different chemicals responsible for immediate or delayed responses in individuals across the various study locations. This may also be explained by the different population structures where certain subpopulations are more vulnerable to air pollution [46].

4.3. Effect Modification

City-specific results in general did not display significant heterogeneity across outcomes based on the I^2 index, which was generally in the range of 0% - 25%. The lack of heterogeneity between estimates of Canadian cities is supported by findings of APHENA [32]. Two mortality outcomes that showed some level of heterogeneity among cites were examined for effect modification by 29 ecological variables. None of the variables were found to modify the ozone-mortality relationship ($p > 0.05$). With PM$_{2.5}$, the association was statistically significant for four variables: area of city, percent of unemployed males, percent manufacturing and percent of population stressed. The PM-mortality association is not likely to be affected by the geographic area of city *per se*, but by other factors associated with it. The remaining city-level variables identified were found to modify the effect in the opposite sense of what would be expected. For example, mortality was seen to decrease with higher percentage of stress levels.

Previous findings on effect modification have been inconsistent, with several studies concluding that the effect of air pollution is not modified by city-level variables or reporting only geographical variations [11,12,16, 38]. However, Ostro *et al.* [13], reported that effect of PM$_{2.5}$ was higher among females, whites, diabetics, or persons with less than high school education. In addition, Bell and Dominici [10] looked at effect modification patterns in 98 US communities and reported that higher estimates were associated with higher unemployment, fraction of African American population, public transportation use, lower temperature and lower prevalence of central air conditioning. In APHEA 2, life expectancy was identified as an effect modifier.

The use of only 12 cities in this study may have limited the effect modification analyses. Although the PM$_{2.5}$ effect was found to be modified by several city-specific characteristics, results cannot be considered as providing strong evidence of effect modification. As several previous studies have reported, it is possible that effect modification with the covariates considered does not occur in the case of short-term exposure to air pollution. Repeating this analysis with a greater number of cities would give greater power to detect heterogeneity—if present—and allow stronger conclusions to be made regarding effect modification.

4.4. Biological Mechanisms

The exact biological mechanisms by which air pollution

leads to morbidity and premature deaths remain under active investigation. However, much of the current evidence suggests that exposure to ozone and PM induces oxidative stress and inflammation in the lung tissue that lead to local and systemic events. The inflammatory response in the lungs has been demonstrated in animal and controlled human studies [47-49]. Inflammation in the lungs triggers the release of cytokines and chemokines that lead to sub-clinical systemic inflammation that may alter the vascular system [48,50,51].

Observed cardiovascular effects can be partially explained by activation of pulmonary neural reflexes that result from interactions between pollutants and lung receptors. Increases in fibrinogen levels and reductions in heart rate, two risk factors for cardiac diseases that lead to hospital admissions, have been associated with exposure to air pollution. Reductions in heart rate can lead to decreased parasympathetic input, which may in turn lead to arrhythmia and cardiovascular mortality [52,53]. Lung inflammation is also believed to exacerbate underlying lung diseases by weakening lung defense mechanisms. Animals with chronic obstructive pulmonary diseases (COPD) or chronic lung inflammation have been found to be more vulnerable to combustion particles compared to normal animals [52,53]. Influenza infections have also been shown to be exacerbated by air pollution in experiments [54,55]. Further, studies on mice and humans indicate that PM$_{2.5}$ may accelerate the development of atherosclerosis [48,56]. Other studies have detected PM in the heart muscle and brain cells indicating its ability to diffuse into the bloodstream which may lead to direct toxic effects [48,50].

4.5. Strengths and Limitations

This study examined the associations of two ambient air pollutants and health outcomes in 12 Canadian cities, with a total exposed population of 9 - 10 million Canadians. Statistical methods applied were uniform across all cities enabling the direct pooling of city-specific results.

The literature on the health effects of short-term exposure to PM$_{2.5}$ is somewhat limited [38], as its use in time series studies is relatively recent. Many previous studies have focused on larger particles rather than fine PM due to the availability of data, or have used conversion factors to convert between the two particle fractions. This study adds to the literature quantitative evidence of the significant effects of fine PM. This study was based on measurements of PM$_{2.5}$ as recorded by fixed monitors in each city; hence, errors inherent in conversion factors were not introduced into the measurements. Further, analyses in previous studies have sometimes been hindered by the different measurements methods that were used for each city [57]. However, air pollution and mortality data were collected under a common framework and subject to the same quality assurance programs across the 12 Canadian cities included in this study.

Concentrations of ambient air pollution obtained from fixed outdoor monitors throughout each city were used as a surrogate measure for the population average personal exposure. This method assumes that exposure among all individuals in a given area is identical and does not take into account the differences in activity patterns (such as time spent outdoors) [58]. The feasibility of obtaining such data, usually collected for regulatory purposes, at low cost and no burden to study subjects, has made it convenient to use in time-series studies. As a result, time-series studies are believed to be subject to exposure misclassification, especially among subpopulations that are at a higher risk since their activity patterns may differ from that of the general population [58,59]. However, the use of fixed monitor measurements is supported by previous studies that have shown a strong correlation between outdoor, indoor and personal exposure to particulate matter [60,61]. It has also been reported that the presence of error in measurement of the exposure would lead to non-differential misclassification of exposure and hence underestimation of health effects [62-64]. Jerret et al. [63], showed that the health effects were three times higher when analyses were based on individual's proximity to high traffic regions compared to using community average concentrations. Models that correct for measurement error have also been developed [64].

Another limitation in this study was the systematically missing exposure data for $PM_{2.5}$. The incomplete exposure data may have resulted in an underestimation of the true effect estimates, based on the findings of a recent study by Samoli et al. [65] that showed systematically missing daily PM_{10} and ozone data gave considerably lower effect estimates. Having $PM_{2.5}$ measurements for every sixth day may have also led to effect estimates with greater uncertainty than those calculated for ozone (which had daily data). In Canada, $PM_{2.5}$ data have been collected on a daily basis since the late 1990s. The unavailability of daily PM data may have also contributed to greater heterogeneity between city estimates, thereby increasing the possibility of observing spurious associations as effect modifiers.

Confounding of co-pollutants in the $PM_{2.5}$ effect has been looked at in other studies. $PM_{2.5}$ is highly correlated with other co-pollutants, and it is often difficult to disentangle which component of the air pollution mixture is the one responsible for the observed health effects [66,67]. This study did not look at potential confounding by co-pollutants beyond ozone. Some previous studies have looked at the effect of seasonal variation in the levels of ozone, where higher effects were detected in the summer when ozone levels are typically higher [11,66,68]. This

was not explored in this study.

Further, the power to detect heterogeneity between city estimates and consequently potential effect modifiers was limited by the low number of cities. It is recommended to repeat effect modification analyses with a larger number of cities to arrive at more conclusive results regarding potential effect modifiers.

Finally, the potential biases associated with the use of mortality data obtained from death certificates needs to be considered. A Canadian study by Stieb et al. [69], looked at the classification of cardio-respiratory diseases in emergency department visits. Findings found a fair degree of agreement in the diagnosis of seven independent assessments, with no evidence of diagnostic bias in relation to daily air pollution. The databases in this study have been subject to quality control by CIHI. Nevertheless, if errors were present in the management of data, this would result in non-differential misclassification bias, as such errors would not likely be related to variation in air pollution levels.

4.6. Public Health Implications

Results of this study indicate a substantial public health burden from ozone and $PM_{2.5}$ pollution. Further reductions in the levels of these two pollutants would bring considerable health and economic benefits to Canadians. For example, based on the calculated effects in this study, a 10 ppb increase in 1-hr maximum ozone levels would correspond to an additional 1,368 (95% CI, 985-759) premature deaths each year in the 12 cities considered in this study (based on lag02 model with 8 df and average annual mortality between 1980 and 2000). Similarly, a 10 $\mu g/m^3$ increase in daily average of $PM_{2.5}$ would correspond to 1148 (95% CI, 521-2319) premature deaths annually in these cities (based on lag2 model with 8 df and the average annual mortality between 1980 and 2000). These figures will be higher when considering the total Canadian population and the inclusion deaths associated with long-term exposure to these pollutants. Long-term effects related to ozone and $PM_{2.5}$ exposure have been reported to be much greater than short-term effects [3].

Previous studies have looked at the exposure-response relationship between air pollutants and mortality in an attempt to identify a threshold concentration, below which air pollution would not lead to increases in deaths [32,70,71]. However, recent epidemiologic findings have consistently detected associations at low ambient pollution levels, without clear evidence supporting the existence of a threshold concentration [32,66,71].

5. Conclusion

This study supports previous findings that have linked

short-term exposure to ozone and PM$_{2.5}$ with mortality. Effects estimates were robust to confounding adjustment of seasonality but sensitive to lag structures. Statistically significant central estimates of the increase in mortality associated with a 10 ppb increase the 1-hr maximum ozone ranged from 0.56% (95% CI 0.27-0.84) to 2.14% (95% CI 0.95-3.98). For PM$_{2.5}$, significant central estimates of the increases in mortality ranged from 0.91% (95% CI 0.08-1.73) to 3.17% (95% CI 0.61-5.72). Although estimated effects are relatively weak, they represent a substantial health burden given the size of the exposed population. Based on these results, it is reasonable to assume that reductions in air pollution would likely lead to health benefits by reducing premature mortality.

REFERENCES

[1]   World Health Organization, "Tackling the Global Clean Air Challenge," 2011.
      http://www.who.int/mediacentre/news/releases/2011/air_pollution_20110926/

[2]   J. Evans, A. van Donkelaar, R. Martin, R. T. Burnett, D. Rainham, N. Birkett and D. Krewski, "Estimates of Global Mortality Due to Particulate Air Pollution Using Satellite Imagery," *Environmental Research*, Vol. 120, 2013, pp. 33-42.

[3]   Canadian Medical Association, "No Breathing Room—National Cost of Illness Study," Summary Report, 2008.
      http://www.cma.ca/multimedia/CMA/Content_Images/Inside_cma/Office_Public_Health/ICAP/CMA_ICAP_sum_e.pdf

[4]   S. Shang, Z. Sun, J. Cao, X. Wang, L. Zhong, X. Bi, H. Li, W. Liu, T. Zhu and W. Huang, "Systematic Review of Chinese Studies of Short-Term Exposure to Air Pollution and Daily Mortality," *Environment International*, Vol. 54, 2013, pp. 100-111.

[5]   A. Zanobetti, M. Franklin, P. Koutrakis and J. Schwartz, "Fine Particulate Air Pollution and Its Components in Association with Cause-Specific Emergency Admissions," *Environmental Health*, Vol. 8, 2009.

[6]   E. Samoli, M. Stafoggia, S. Rodopoulou, B. Ostro, C. Declercq, E. Alessandrini, J. Díaz, A. Karanasiou, A. G. Kelessis, A. Le Tertre, G. Randi, C. Scarinzi, S. F. Zauli-Sajani and F. Forastiere, "Associations between Fine and Coarse Particles and Mortality in Mediterranean Cities: Results from the MED-PARTICLES Project," *Environmental Health Perspectives*, Vol. 121, 2013, pp. 932-938.

[7]   R. D. Peng, E. Samoli, L. Pham, F. Dominici, G. Touloumi, T. Ramsay, R. T. Burnett, D. Krewski, A. Le Tertre, A. Cohen, R. W. Atkinson, H. R. Anderson, K. Katsouyani and J. M. Samet, "Acute Effects of Ambient Ozone on Mortality in Europe and North America: Results from the APHENA Study," *Air Quality, Atmosphere & Health*, Vol. 6, No. 2, 2013, pp. 445-453.

[8]   H. R. Anderson, R. Atkinson, J. L. Peacock, L. Martson

and K. Konstantinou, "Meta-Analysis of Times Series Studies and Panel Studies of Particulate Matter and Ozone," Report of a WHO Task Group, World Health Organization, Copenhagen, 2004.

[9]   M. L. Bell, A. McDermott, S. L. Zeger, J. M. Samet and F. Dominici, "Ozone and Short-Term Mortality in 95 US Urban Communities, 1987-2000," *Jama-Journal of the American Medical Association*, Vol. 292, No. 19, 2004, pp. 2372-2378.

[10] M. L. Bell and F. Dominici, "Effect Modification by Community Characteristics on the Short-Term Effects of Ozone Exposure and Mortality in 98 US Communities," *American Journal of Epidemiology*, Vol. 167, 2008, pp. 986-997.

[11] A. Gryparis, B. Forsberg, K. Katsouyanni, A. Analitis, G. Touloumi, J. Schwartz, E. Samoli, S. Medina, H. R. Anderson, E. M. Niciu, H. E. Wichmann, B. Kriz, M. Kosnik, J. Skorkovsky, J. M. Vonk and Z. Dortbudak, "Acute Effects of Ozone on Mortality from the 'Air Pollution and Health: A European Approach' Project," *American Journal of Respiratory and Critical Care Medicine*, Vol. 170, 2004, pp. 1080-1087.

[12] R. Klemm and R. Mason, "Replication of Reanalysis of Harvard Six-City Mortality Study," In: *Revised Analyses of Time Series Studies of Air Pollution and Health*, Special Report, Health Effects Institute, Boston, 2003, pp. 165-172.

[13] B. Ostro, R. Broadwin, S. Green, W. Y. Feng and M. Lipsett, "Fine Particulate Air Pollution and Mortality in Nine California Counties: Results from CALFINE," *Environmental Health Perspectives*, Vol. 114, 2006, pp. 29-33.

[14] S. Schwartz, "Daily Deaths Associated with Air Pollution in Six US Cities and Short-Term Mortality Displacement in Boston," *Revised Analyses of Time Series Studies of Air Pollution and Health*, Special Report, Health Effects Institute, Boston, 2003, pp. 219-226.

[15] R. Simpson, G. Williams, A. Petroeschevsky, T. Best, G. Morgan, L. Denison, A. Hinwood, G. Neville and A. Neller, "The Short-Term Effects of Air Pollution on Daily Mortality in Four Australian Cities", *Australian and New Zealand Journal of Public Health*, Vol. 29, 2005, pp. 205-212.

[16] D. M. Stieb, S. Judek and R. T. Burnett, "Meta-Analysis of Time-Series Studies of Air Pollution and Mortality: Effects of Gases and Particles and the Influence of Cause of Death, Age, and Season," *Journal of the Air & Waste Management Association*, Vol. 52, No. 4, 2002, pp. 470-484.

[17] Y. Chen, Q. Y. Yang, D. Krewski, Y. Shi, R. T. Burnett and K. McGrail, "Influence of Relatively Low Level of Particulate Air Pollution on Hospitalization for COPD in Elderly People," *Inhalation Toxicology*, Vol. 16, 2004, pp. 21-25.

[18] M. S. Goldberg, R. T. Burnett, J. C. Bailar, J. Brook, Y. Bonvalot, R. Tamblyn, R. Singh, M. F. Valois and R. Vin-

cent, "The Association between Daily Mortality and Ambient Air Particle Pollution in Montreal, Quebec 2. Cause-Specific Mortality," *Environmental Research*, Vol. 86, No. 1, 2001, pp. 26-36.

[19] S. Parodi, M. Vercelli, E. Garrone, V. Fontana and A. Izzotti, "Ozone Air Pollution and Daily Mortality in Genoa, Italy between 1993 and 1996," *Public Health*, Vol. 119, No. 9, 2005, pp. 844-850.

[20] R. D. Peng, F. Dominici and T. A. Louis, "Model Choice in Time Series Studies of Air Pollution and Mortality," *Journal of the Royal Statistical Society Series A—Statistics in Society*, Vol. 169, No. 2, 2006, pp. 179-198.

[21] R. T. Burnett, M. Smith-Doiron, D. Stieb, M. E. Raizenne, J. R. Brook, R. E. Dales, J. A. Leech, S. Cakmak and D. Krewski, "Association between Ozone and Hospitalization for Acute Respiratory Diseases in Children Less than 2 Years of Age," *American Journal of Epidemiology*, Vol. 153, No. 5, 2001, pp. 444-452.

[22] B. B. Jalaludin, T. Chey, B. I. O'Toole, W. T. Smith, A. G. Capon and S. R. Leeder, "Acute Effects of Low Levels of Ambient Ozone on Peak Expiratory Flow Rate in a Cohort of Australian Children," *International Journal of Epidemiology*, Vol. 29, No. 3, 2000, pp. 549-557.

[23] C. A. Lin, M. A. Martins, S. C. L. Farhat, C. A. Pope, G. M. S. Conceicao, V. M. Anastacio, M. Hatanaka, W. C. Andrade, W. R. Hamaue, G. M. Bohm and P. H. N. Saldiva, "Air Pollution and Respiratory Illness of Children in Sao Paulo, Brazil," *Paediatric and Perinatal Epidemiology*, Vol. 13, No. 4, 1999, pp. 475-488.

[24] A. Peters, D. W. Dockery, J. Heinrich and H. E. Wichmann, "Short-Term Effects of Particulate Air Pollution on Respiratory Morbidity in Asthmatic Children," *European Respiratory Journal*, Vol. 10, 1997, pp. 872-879.

[25] Q. Y. Yang, Y. Chen, Y. Shi, R. T. Burnett, K. McGrail and D. Krewski, "Association between Ozone and Respiratory Admissions among Children and the Elderly in Vancouver, Canada," *Inhalation Toxicology*, Vol. 15, No. 13, 2003, pp. 1297-1308.

[26] Health Canada, "Canada-United States Border Air Quality Strategy: Great Lakes Basin Airshed Management Framework," 2006. http://www.hc-sc.gc.ca/ewh-semt/air/out-ext/great_lakes-grands_lacs-eng.php#exposure

[27] T. F. Bateson, B. A. Coull, B. Hubbell, K. Ito, M. Jerrett, T. Lumley, D. Thomas, S. Vedal and M. Ross, "Panel Discussion Review: Session Three—Issues Involved in Interpretation of Epidemiologic Analyses—Statistical Modeling," *Journal of Exposure Science and Environmental Epidemiology*, Vol. 17, 2007, pp. S90-S96.

[28] F. Dominici, A. McDermott and T. J. Hastie, "Improved Semiparametric Time Series Models of Air Pollution and Mortality," *Journal of the American Statistical Association*, Vol. 99, 2004, pp. 938-948.

[29] T. O. Ramsay, R. T. Burnett and D. Krewski, "The Effect of Concurvity in Generalized Additive Models Linking Mortality to Ambient Particulate Matter," *Epidemiology*, Vol. 14, 2003, Vol. 18-23.

[30] J. M. Samet, F. Dominici, A. McDermott and S. L. Zeger, "New Problems for an Old Design: Time Series Analyses of Air Pollution and Health," *Epidemiology*, Vol. 14, 2003, pp. 11-12.

[31] Health Effects Institute, "Revised Analyses of Time-Series Studies of Air Pollution and Health—Special Report," Health Effects Institute, Boston, 2003.

[32] K. Katsouyani, J. M. Samet, H. R. Anderson, R. W. Atkinson, A. Le Tertre, S. Medina, E. Samoli, G. Touloumi, R. T. Burnett, D. Krewski, T. Ramsay, F. Dominici, R. D. Peng, J. Schwartz and A. Zanobetti, "Air Pollution and Health: A European and North American Approach (APHENA)," Health Effects Institute, 2009.

[33] C. S. Berkey, D. C. Hoaglin, A. Antczak-Bouckoms, F. Mosteller and G. A. Colditz, "Meta-Analysis of Multiple Outcomes by Regression with Random Effects," *Statistics in Medicine*, Vol. 17, 1998, pp. 2537-2550.

[34] T. B. Huedo-Medina, J. Sanchez-Meca, F. Marin-Martinez and J. Botella, "Assessing Heterogeneity in Meta-Analysis: Q Statistic or I-2 Index?" *Psychological Methods*, Vol. 11, No. 2, 2006, pp. 193-206.

[35] R Development Core Team, "R: A Language and Environment for Statistical Computing," R Foundation for Statistical Computing, Vienna, 2007.

[36] G. Touloumi, R. Atkinson, A. Le Tertre, E. Samoli, J. Schwartz, C. Schindler, J. M. Vonk, G. Rossi, M. Saez, D. Rabszenko and K. Katsouyanni, "Analysis of Health Outcome Time Series Data in Epidemiological Studies," *Environmetrics*, Vol. 15, No. 2, 2004, pp. 101-117.

[37] F. Ballester, P. Rodriguez, C. Iniguez, M. Saez, A. Daponte, I. Galan, M. Taracido, F. Arribas, J. Bellido, F. B. Cirarda, A. Canada, J. J. Guillen, F. Guillen-Grima, E. Lopez, S. Perez-Hoyos, A. Lertxundi and S. Toro, "Air Pollution and Cardiovascular Admissions Association in Spain: Results within the EMECAS Project," *Journal of Epidemiology and Community Health*, Vol. 60, No. 4, 2006, pp. 328-336.

[38] F. Dominici, R. D. Peng, M. L. Bell, L. Pham, A. McDermott, S. L. Zeger and J. M. Samet, "Fine Particulate Air Pollution and Hospital Admission for Cardiovascular and Respiratory Diseases," *Journal of the American Medical Association*, Vol. 295, No. 10, 2006, pp. 1127-1134.

[39] L. J. Welty and S. L. Zeger, "Are the Acute Effects of Particulate Matter on Mortality in the National Morbidity, Mortality, and Air Pollution Study the Result of Inadequate Control for Weather and Season? A Sensitivity Analysis

Using Flexible Distributed Lag Models," *American Journal of Epidemiology*, Vol. 162, No. 1, 2005, pp. 80-88.

[40] World Health Organization, "Health Aspects of Air Pollution with Particulate Matter, Ozone and Nitrogen Dioxide," Report on a WHO Working Group. WHO Regional Office for Europe, Copenhagen, 13-15 January 2003, pp. 1-92. http://www.euro.who.int/__data/assets/pdf_file/0005/112199/E79097.pdf

[41] A. L. F. Braga, A. Zanobetti and J. Schwartz, "The Lag Structure between Particulate Air Pollution and Respiratory and Cardiovascular Deaths in 10 US Cities," *Journal of Occupational and Environmental Medicine*, Vol. 43, No. 11, 2001, pp. 927-933.

[42] EPA, "Fourth External Review for Air Quality Criteria for Particulate Matter (Draft)," 2003.

[43] Y. H. Zhang, W. Huang, S. J. London, G. X. Song, G. H. Chen, L. L. Jiang, N. Q. Zhao, B. H. Chen and H. D. Kan, "Ozone and Daily Mortality in Shanghai, China," *Environmental Health Perspectives*, Vol. 114, No. 8, 2006, pp. 1227-1232.

[44] F. J. Kelly, "Oxidative Stress: Its Role in Air Pollution and Adverse Health Effects," *Occupational and Environmental Medicine*, Vol. 60, No. 8, 2003, pp. 612-616.

[45] J. Schwartz, D. W. Dockery and L. M. Neas, "Is Daily Mortality Associated Specifically with Fine Particles?" *Journal of the Air & Waste Management Association*, Vol. 46, No. 10, 1996, pp. 927-939.

[46] H. Kim, Y. Kim and Y. C. Hong, "The Lag-Effect Pattern in the Relationship of Particulate Air Pollution to Daily Mortality in Seoul, Korea," *International Journal of Biometeorology*, Vol. 48, No. 1, 2003, pp. 25-30.

[47] S. Ahmad, A. Ahmad, G. McConville, B. K. Schneider, C. B. Allen, R. Manzer, R. J. Mason and C. W. White, "Lung Epithelial Cells Release ATP during Ozone Exposure: Signaling for Cell Survival," *Free Radical Biology and Medicine*, Vol. 39, No. 2, 2005, pp. 213-226.

[48] N. Kunzli, M. Jerrett, W. J. Mack, B. Beckerman, L. LaBree, F. Gilliland, D. Thomas, J. Peters and H. N. Hodis, "Ambient Air Pollution and Atherosclerosis in Los Angeles," *Environmental Health Perspectives*, Vol. 113, No. 2, 2005, pp. 201-206.

[49] I. S. Mudway, N. Stenfors, A. Blomberg, R. Helleday, C. Dunster, S. L. Marklund, A.J. Frew, T. Sandstrom and F. J. Kelly, "Differences in Basal Airway Antioxidant Concen-Trations Are Not Predictive of Individual Responsiveness to Ozone: A Comparison of Healthy and Mild Asthmatic Subjects," *Free Radical Biology and Medicine*, Vol. 31, No. 8, 2001, pp. 962-974.

[50] R. D. Brook, B. Franklin, W. Cascio, Y. L. Hong, G. Howard, M. Lipsett, R. Luepker, M. Mittleman, J. Samet, S. C. Smith and I. Tager, "Air Pollution and Cardiovascular Disease—A Statement for Healthcare Professionals from the Expert Panel on Population and Prevention Science of the American Heart Association," *Circulation*, Vol. 109, No. 21, 2004, pp. 2655-2671.

[51] W. P. Watkinson, M. J. Campen, J. P. Nolan and D. L. Costa, "Cardiovascular and Systemic Responses to Inhaled Pollutants in Rodents: Effects of Ozone and Particulate Matter," *Environmental Health Perspectives*, Vol. 109, No. S4, 2001, pp. 539-546.

[52] P. S. Gilmour, D. M. Brown, T. G. Lindsay, P. H. Beswick, W. MacNee and K. Donaldson, "Adverse Health Effects of PM(10) Particles: Involvement of Iron in Generation of Hydroxyl Radical," *Occupational and Environmental Medicine*, Vol. 53, No. 12, 1996, pp. 817-822.

[53] R. J. Pritchard, A. J. Ghio, J. R. Lehmann, D. W. Winsett, J. S. Tepper, P. Park, M. I. Gilmour, K. L. Dreher and D. L. Costa, "Oxidant Generation and Lung Injury after Particulate Air Pollutant Exposure Increase with the Concentrations of Associated Metals," *Inhalation Toxicology*, Vol. 8, No. 5, 1996, pp. 457-477.

[54] E. W. Spannhake, S. P. M. Reddy, D. B. Jacoby, X. Y. Yu, B. Saatian and J. Y. Tian, "Synergism between Rhinovirus Infection and Oxidant Pollutant Exposure Enhances Airway Epithelial Cell Cytokine Production," *Environmental Health Perspectives*, Vol. 110, No. 7, 2002, pp. 665-670.

[55] J. T. Zelikoff, M. W. Frampton, M. D. Cohen, P. E. Morrow, M. Sisco, Y. Tsai, M. J. Utell and R. B. Schlesinger, "Effects of Inhaled Sulfuric Acid Aerosols on Pulmonary Immunocompetence: A Comparative Study in Humans and Animals," *Inhalation Toxicology*, Vol. 9, No. 8, 1997, pp. 731-752.

[56] Q. H. Sun, A. X. Wang, X. M. Jin, A. Natanzon, D. Duquaine, R. D. Brook, J. G. S. Aguinaldo, Z. A. Fayad, V. Fuster, M. Lippmann, L. C. Chen and S. Rajagopalan, "Long-Term Air Pollution Exposure and Acceleration of Atherosclerosis and Vascular Inflammation in an Animal Model," *Journal of the American Medical Association*, Vol. 294, No. 23, 2005, pp. 3003-3010.

[57] A. Le Tertre, S. Medina, E. Samoli, B. Forsberg, P. Michelozzi, A. Boumghar, J. M. Vonk, A. Bellini, R. Atkinson, J. G. Ayres, J. Sunyer, J. Schwartz and K. Katsouyanni, "Short-Term Effects of Particulate Air Pollution on Cardiovascular Diseases in Eight European Cities," *Journal of Epidemiology and Community Health*, Vol. 56, No. 10, 2002, pp. 773-779.

[58] M. L. Bell, J. M. Samet and F. Dominici, "Time-Series Studies of Particulate Matter," *Annual Review of Public Health*, Vol. 25, 2004, pp. 247-280.

[59] M. S. Goldberg, R. T. Burnett and J. R. Brook, "Counterpoint: Time-Series Studies of Acute Health Events and Environmental Conditions Are Not Confounded by Personal Risk Factors," *Regulatory Toxicology and Pharmacology*, Vol. 51, No. 2, 2008, pp. 141-147.

[60] N. A. Janssen, J. J. de Hartog, G. Hoek, B. Brunekreef, T. Lanki, K. L. Timonen and J. Pekkanen, "Personal Exposure to Fine Particulate Matter in Elderly Subjects: Relation between Personal, Indoor, and Outdoor Concentrations," *Journal of the Air & Waste Management Association*, Vol. 50, No. 7, 2000, pp. 1133-1143.

[61] D. Kim, A. Sass-Kortsak, J. T. Purdham, R. E. Dales and J. R. Brook, "Associations between Personal Exposures and Fixed-Site Ambient Measurements of Fine Particulate Matter, Nitrogen Dioxide, and Carbon Monoxide in Toronto, Canada," *Journal of Exposure Science and Environmental Epidemiology*, Vol. 16, No. 2, 2006, pp. 172-183.

[62] G. Hoek, B. Brunekreef, S. Goldbohm, P. Fischer and P. A. van den Brandt, "Association between Mortality and Indicators of Traffic-Related Air Pollution in the Netherlands: A Cohort Study," *Lancet*, Vol. 360, No. 9341, 2002, pp. 1203-1209.

[63] M. Jerrett, R. T. Burnett, R. J. Ma, C. A. Pope, D. Krewski, K. B. Newbold, G. Thurston, Y. L. Shi, N. Finkelstein, E. E. Calle and M. J. Thun, "Spatial Analysis of Air Pollution and Mortality in Los Angeles," *Epidemiology*, Vol. 16, No. 6, 2005, pp. 727-736.

[64] S. L. Zeger, D. Thomas, F. Dominici, J. M. Samet, J. Schwartz, D. Dockery and A. Cohen, "Exposure Measurement Error in Time-Series Studies of Air Pollution: Concepts and Consequences," *Environmental Health Perspectives*, Vol. 108, No. 5, 2000, pp. 419-426.

[65] E. Samoli, R. D. Peng, T. Ramsay, M. Touloumi, F. Dominici, R. W Atkinson, A. Zanobetti, A. Le Tertre, H. R. Anderson, J. Schwartz, A. Cohen, D. Krewski, J. M. Sa-

met and K. Katsouyani, "What Is the Impact of Systematically Missing Exposure Data on Air Pollution Health Effect Estimates?"

[66] P. Bellini, M. Baccini, A. Biggeri and B. Terracini, "The Meta-Analysis of the Italian Studies on Short-Term Effects of Air Pollution (MISA): Old and New Issues on the Interpretation of the Statistical Evidences," *Environmetrics*, Vol. 18, No. 3, 2007, pp. 219-229.

[67] M. Goldberg, "On the Interpretation of Epidemiological Studies of Ambient Air Pollution," *Journal of Exposure Science and Environmental Epidemiology*, Vol. 17, No. S2, 2007, pp. S66-S70.

[68] R. T. Burnett, J. R. Brook, W. T. Yung, R. E. Dales and D. Krewski, "Association between Ozone and Hospitalization for Respiratory Diseases in 16 Canadian Cities," *Environmental Research*, Vol. 72, No. 1, 1997, pp. 24-31.

[69] D. M. Stieb, R. C. Beveridge, B. H. Rowe, S. D. Walter and S. Judek, "Assessing Diagnostic Classification in an Emergency Department: Implications for Daily Time Series Studies of Air Pollution," *American Journal of Epidemiology*, Vol. 148, No. 7, 1998, pp. 666-670.

[70] M. J. Daniels, F. Dominici, J. M. Samet and S. L. Zeger, "Estimating Particulate Matter-Mortality Dose-Response Curves and Threshold Levels: An Analysis of Daily Time-Series for the 20 Largest US Cities," *American Journal of Epidemiology*, Vol. 152, No. 5, 2000, pp. 397-406.

[71] J. Schwartz, "Assessing Confounding, Effect Modification, and Thresholds in the Association between Ambient Particles and Daily Deaths," *Environmental Health Perspectives*, Vol. 108, No. 6, 2000, pp. 563-568.

4

Mapping Highly Cost-Effective Carbon Capture and Storage Opportunities in India

Richard A. Beck[1], Yolanda M. Price[2], S. Julio Friedmann[3], Lynn Wilder[3], Lee Neher[3]

[1]Department of Geography, University of Cincinnati, Cincinnati, USA; [2]International Center for Water Resources Research, Central State University, Wilberforce, USA; [3]Lawrence Livermore National Laboratory, Livermore, USA.

ABSTRACT

Carbon dioxide (CO_2) is the primary anthropogenic greenhouse gas (GHG). India's CO_2 emissions are expected to increase 70% by 2025. Geologic carbon storage (GCS) offers a way to reduce CO_2 emissions. Here we present the results of a search for the most cost-effective GCS opportunities in India. Source-Sink matching for large and concentrated CO_2 sources near geological storage in India indicates one very high priority target, a fertilizer plant in the city of Narmadanagar in Bharuch District of Gujarat Province, India that is <20 km from old oil and gas fields in the Cambay Basin. Two pure CO_2 sources are <20 km from deep saline aquifers and one is <20 km from a coal field.

Keywords: Global Warming; Carbon Dioxide; CO_2; Carbon Capture and Storage; CCS; Geologic Carbon Sequestration; GCS; India; Source-Sink Matching

1. Introduction

1.1. Global Context of Geologic Carbon Sequestration

Fossil-fuels are crucial to our global economy and standard of living and will remain so for several more decades. Until recently, nearly all fossil-fuel consumption emitted carbon dioxide to the atmosphere. Carbon dioxide (CO_2) is the primary anthropogenic greenhouse gas (GHG) [1]. The rate of CO_2 emissions has steadily increased with global industrialization. Global CO_2 concentrations in the atmosphere have increased 25% since 1850 and 15% since 1956. By 2011, global CO_2 emissions totaled a record 34 billion metric tons [2]. These emissions were expected to grow 1.9% per year until at least 2025 [3] with the developing world's emissions exceeding those of the current industrial nations in approximately 2020. The concentration of carbon dioxide in the atmosphere is expected to double by mid-century if the current emissions trend continues.

The biosphere and hydrosphere absorb approximately 1/3 of CO_2 emissions [4] on time scales of hundreds to thousands of years. The remaining 2/3 are currently vented to the atmosphere. Nearly all scientists expect this increasing carbon dioxide concentration in the atmosphere to enhance mean global temperatures via the greenhouse effect. This steady, observed increase in the concentration of CO_2 in the atmosphere coupled with the well established greenhouse effect presents a risk of excessive warming [5].

Recent globally averaged temperature measurements suggest that this risk is significant. Nine of the ten warmest years since 1850 have occurred between 1995 and 2004 [6,7]. Globally, 2012 was the ninth warmest year since 1979 but 11 of the warmest years since 1979 have occurred since 2001. The warmest year on record since 1901 was 2005. The second warmest was 2010 [8,9]. The earth is now nearly 0.5 degrees C warmer than it was in the period between 1961 and 1990 [7]. As one would expect, the influence of the observed temperature increase is greatest and most easily observed in the Arctic and Antarctic [10-12]. There is increasing evidence that warming is beginning to influence climate at mid-latitudes as well [13,14]. Regardless of one's interpretation of observed increases in atmospheric CO_2 and temperature, it would seem prudent to seek ways to mitigate the risk of excessive warming of the Earth by diverting the 2/3 of the carbon dioxide emissions not absorbed by the biosphere back into the Earth's crust or the oceans [15].

Given scientific uncertainties, observed acidification and political questions associated with deep ocean storage of carbon dioxide, this and many other studies are focused on finding the best ways and places to store carbon dioxide underground [15]. Not surprisingly, most of the technology and fundamental data necessary to capture, transport and store carbon dioxide underground (Geological Carbon Storage or GCS) appears to be available in the hydrocarbon and coal industries. Indeed, GCS began in 1972 in the Permian Basin of West Texas. There are now more than 75 GCS projects that inject more than 40 million tons of CO_2 per year worldwide [16]. This study seeks to find the most economical places to sequester carbon from the largest and most concentrated sources of carbon dioxide in the subsurface of South Asia. Given India's overwhelming predominance in the population and economy [17] and total and per capita CO_2 emissions of South Asia [17,18], its political stability, rapid economic development, and high level of technical expertise, this study focuses on India.

1.2. Carbon Dioxide Capture and Storage as a Means of Mitigating the Risk of Excessive Warming

Carbon dioxide capture and storage (CCS) has emerged as a critical technology pathway towards the stabilization and reduction of greenhouse gas emissions [1,19,20]. A necessary step to deployment of CCS/GCS at large scale is the accurate assessment of regional CO_2 storage potential [21,22]. Towards that end, private and governmental assessment of CO_2 storage potential in Australia and the Alberta basin [21,22] have served as the basis for research, planning, and policy. To help enable CCS/GCS technology deployment, the International Energy Agency (IEA) Greenhouse Gas R&D Programme has sponsored assessment for North America [23] and Europe [24]. These have provided data of sufficient accuracy, richness, and detail to enable calculations of CO_2 storage cost curves to further help decision makers in planning CCS deployment.

Against this backdrop, many workers have noted that successful global reduction in greenhouse gas emissions will require CCS in developing countries, most notably China and India [25-30]. India presents a particularly interesting case for CCS/GCS deployment, due to the strength of existing legal and regulatory frameworks, the rapid, imminent expansion of coal power as part of India's electrification strategy, and the geographic distribution of potential sinks. While large portions of the subcontinent are not well suited to storage, there is a surprisingly good overall match. Finally, there may be opportunities for India to deploy CCS/GCS in the context of the Kyoto clean development mechanism (CDM), the Asian-Pacific Partnership (APP), through direct Industrial sponsorship, and through novel development of resources (e.g., underground coal gasification).

1.3. Carbon Dioxide Emissions of India

India has the world's second largest population of more than 1.2 billion people that is growing at a rate of 1.5% per year [30,31] and currently consumes energy at a per capita rate of 1/5 the global average. India's GNP is growing at 5% - 7% per year and commercial energy consumption is increasing at 5.5% per year and is expected to accelerate [32,33]. Indeed, India's per capita carbon emissions doubled between 1990 and 2011 from 0.8 to 1.6 metric tons per person. In 2011, India emitted approximately 1.97 million metric tons of carbon and is now the world's fourth largest fossil-fuel CO_2-emitting country [21,33-35].

Approximately 70% of India's fossil-fuel CO_2-emissions are from coal, 22% from oil and 4% from natural gas [33]. India has proven coal/lignite reserves of 118 billion tons, recoverable oil reserves are estimated at 640 million tons (10 year supply) and natural gas reserves are estimated at 850 billion cubic meters (30 year supply) with large potential for gas hydrate exploitation [31]. The energy mix for the year 2100 is predicted to be coal 50%, natural gas 25%, nuclear and renewables 25%. Therefore the critical technologies for India's energy future are expected to be clean coal, natural gas including hydrates, thorium-based nuclear and renewables [31]. It is clear that India's carbon dioxide emissions will increase substantially by 2100 in order to support an estimated 1.65 billion people [31]. India's CO_2 emissions are currently increasing at 3% - 6% per year [2] and are expected to increase 70% by 2025 [36]. Approximately 82% of CO_2 emissions in South Asia are from India [23,24] making India the focus of this study. India's CO_2 emissions have risen from 1113 million metric tons in 2004 [37] to 1586 million metric tons in 2009 [18]. Of these emissions approximately 51% were from large (>0.1 million metric tons/year) stationary sources [18,23,24]. Given sufficiently cost-effective capture and transport technologies, there is clearly enormous scope for CO_2 sequestration from stationary sources in India.

1.4. Second Order Geological Carbon Capture and Storage Assessment for India

First order assessments of the CCS/GCS potential of the Indian Subcontinent have been published [19,23,26-28, 38,39] that began the search for CCS sites in India. This paper builds on these studies and the IEA GHG database to begin a second order assessment of the region's CCS/GCS potential and attempts to identify the "very lowest hanging fruit" [38] in terms of maximum impact on CO_2 emissions and lowest cost in preparation for CCS/GCS pilot projects in India.

According to Damodaran [40] CCS and GCS in particular have yet to be implemented on a large scale due to risks of leakage, access to technology, additional costs and financing difficulties. He also noted that CCS "can be economically feasible under specific conditions. This may be the case, for example, if CO_2 is captured from low-cost sources, such as gas processing or ammonia plants, and used towards a productive end, such as enhancing oil recovery at a nearby oil field". In particular, he noted that "high CO_2 flue gas concentration enhances efficiency of capture and enables low unit costs". Furthermore, Damodaran noted that "The costs of transport and storage are less onerous in comparison, and can be minimized by achieving economies of scale and/or by siting emitting plants close to potential storage sites such as oil and gas reservoirs". Indeed, a 2012 CCS/GCS scoping study by the Global CCS Institute [41] revealed that safety, energy penalties and cost issues remain prohibitive in the minds of most CCS/GCS stakeholders in India. This study seeks to identify those large (>0.1 million tons CO_2/yr) and concentrated (>30%) CO_2 flue gas sources most amenable to cost-effective CO_2 capture and storage that are very near probable storage reservoirs to identify the most cost-effective GCS opportunities in India.

2. Method

We used Internet searches to find missing locations for about one third of the International Energy Agency's (IEA) major CO_2 sources for India. Our study then used a Geographic Information System (GIS) to answer two questions. 1) Where are the major CO_2 sources, sinks and risks in India? 2) Where should near-term geologic CO_2 sequestration efforts be focused in India? We then used the Internet to locate 126 new CO_2 sources and added them to the IEA GHG CO_2 source database for India. We then captured and in most cases digitized two dozen cultural and physical map layers including sources, sinks and risks for geologic CO_2 sequestration in India. These layers were then used to find which major (relatively) pure CO_2 sources (>0.1 million tons CO_2/yr) are within 20 km of high volume sequestration sites including old oil and gas fields, deep saline aquifers and coal fields to focus our storage capacity assessment.

3. Results

3.1. Large Stationary Carbon Dioxide Sources in India

Van Bergen *et al.* [38] and Damen *et al.* [26] restricted their search for early carbon sequestration opportunities to CO_2 enhanced oil recovery (CO2-EOR) and CO_2 enhanced coal bed methane (CO$_2$-ECBM) sites due to their additional revenue potential. We expanded our search to include saline aquifers and depleted gas and oil reservoirs given that economic incentives are more likely ten

years after their initial study. We began with the IEA GHG/Ecofys Carbon Source database. The IEA GHG database includes stationary sources greater than 0.1 million tons of CO_2 per year with considerable detail regarding industrial sector, ownership, CO_2 gas concentration etc. Given our focus on India we filtered both IEA and LLNL CO_2 source databases for those greater than 0.1 million tons of CO_2 per year within India. The IEA GHG database for South Asia lacked geographic locations for approximately one third of major CO_2 sources. A major part of this effort to refine source-sink matching for South Asia involved Internet searches by company name, sector and other industrial details to find the names of the nearest populated place in order to obtain an approximate geographic location (latitude and longitude).

We then used a detailed map of India to make sure that the geographic location was reasonable and matched the district, province listed. In some cases the maps were not detailed enough so we were limited to providing a District level location. The average size of a district in India is 4300 km. Clearly those candidates which appear to be good candidates for CCS will require site visits and GPS locations to verify their exact locations and suitability. We were able to find geographic locations for 126 major CO_2 sources and added them to the IEA GHG CO_2 source database for India. We then engaged in some quality control work based on past field work in India (1983-2009) that included correcting obvious location errors. The refined IEA GHG database for South Asia indicated that approximately 82% of the CO_2 emissions were from India. We continued to focus our efforts on India accordingly (**Figure 1**). CCS including GCS is most efficient when the CO_2 streams are relatively concentrated. In order to find the most cost-effective candidates for CCS/GCS, we then filtered the IEA and LLNL CO_2 sources for South Asia for by industrial sector in order to isolate those CO_2 sources that usually have high purity CO_2 flue gas (ammonia, ethylene oxide, hydrogen, liquefied natural gas (LNG) and cement) (**Figure 2**).

3.2. Geological Carbon Dioxide Sinks in India

We collected geographic locations for three types of geologic carbon sequestration opportunities in India, oil and gas reservoirs, coal fields and saline aquifers. We did not include mineral trapping of CO_2 in flood basalts such as the Deccan Traps in this study simply because this is a relatively new field of study [42].

Oil and Gas Fields—Oil and gas fields are attractive because they have demonstrably contained oil and gas at high pressures for millions of years. They have demonstrable porosity, permeability and seal characteristics for which there is often direct data in the form of well logs and seismic data and a pool of labor with technical ex-

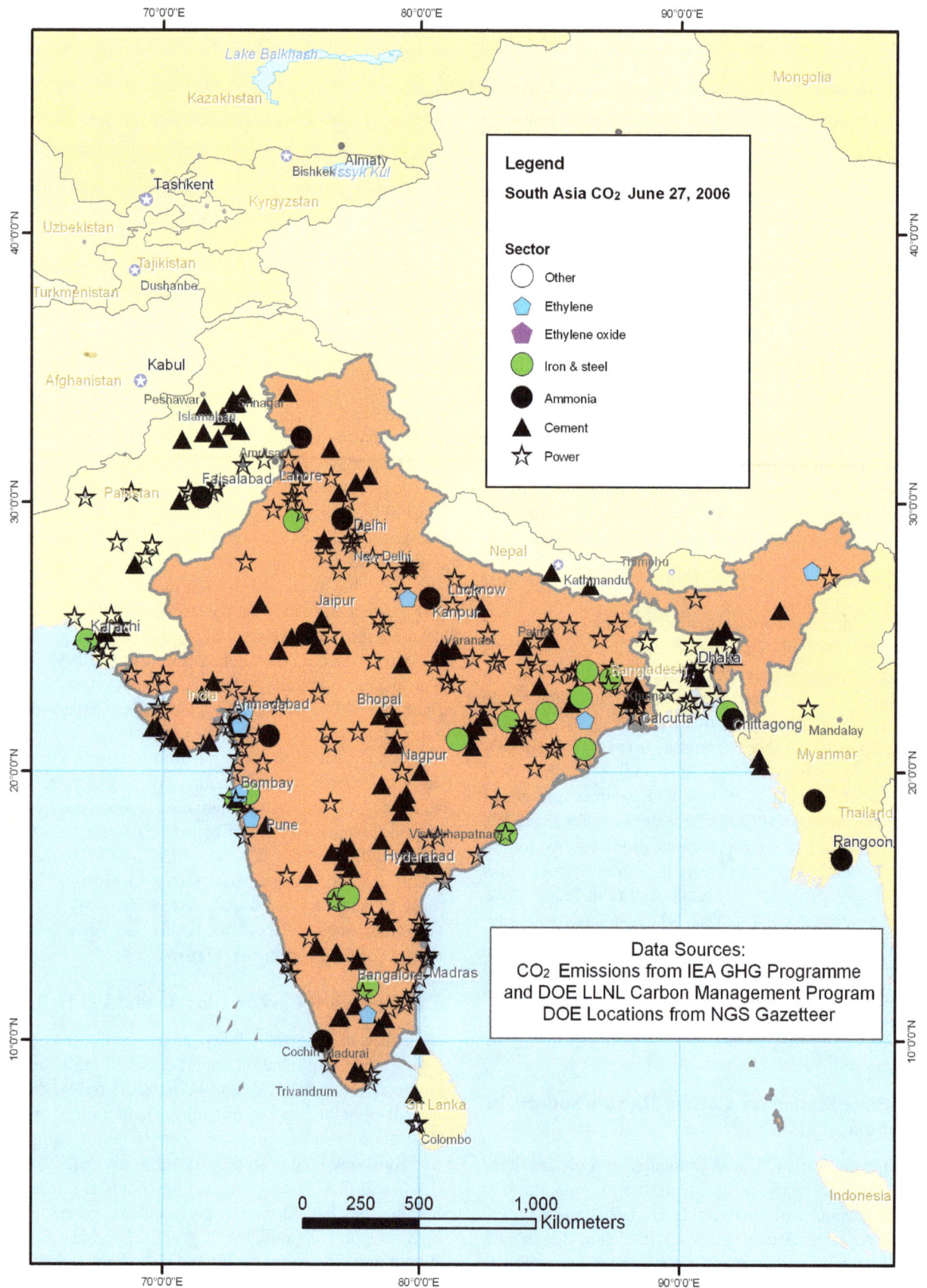

Figure 1. Map of South Asian IEA and LLNL GHG major stationary CO$_2$ sources with LLNL additions.

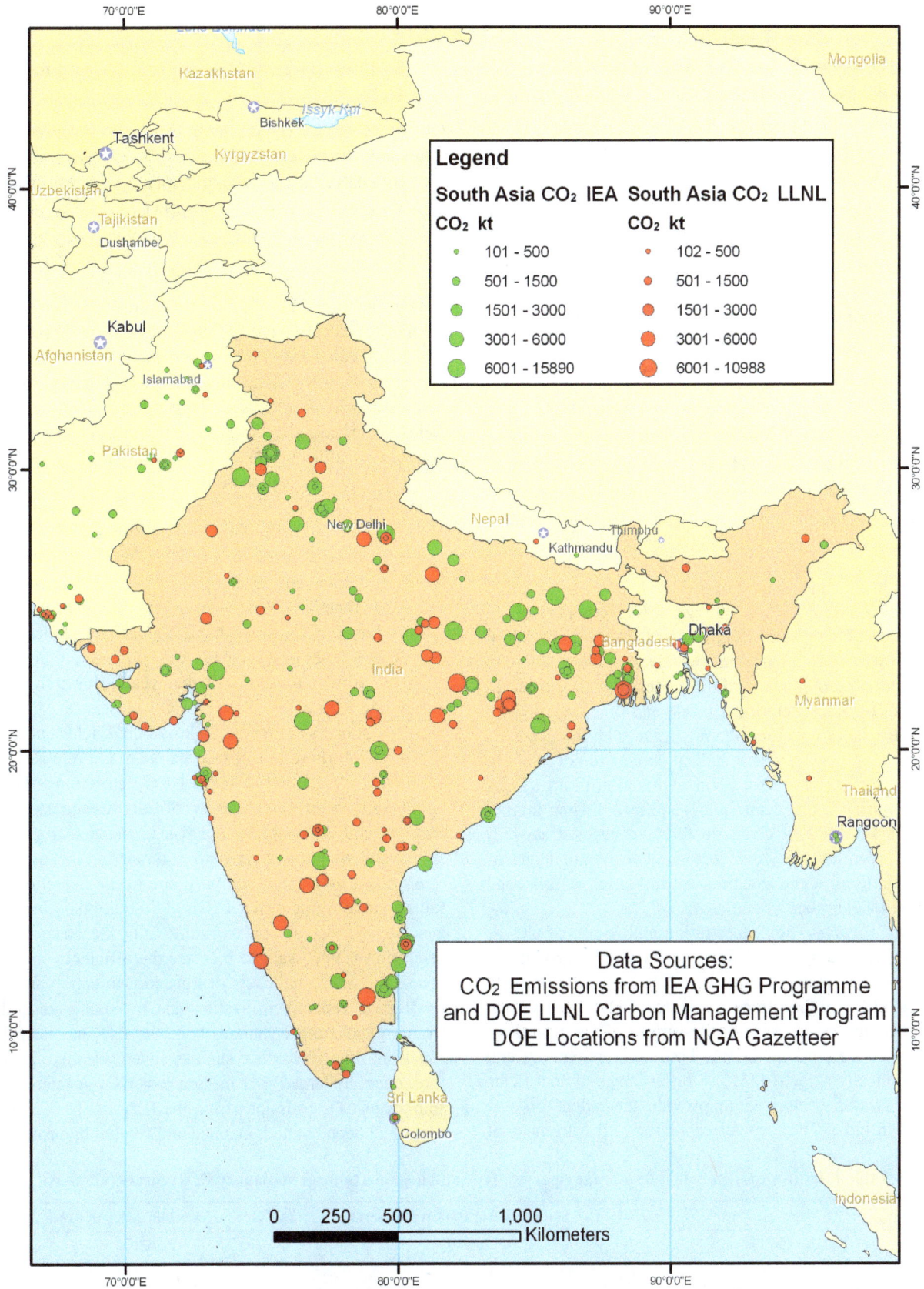

Figure 2. Map of large South Asia CO$_2$ source types (>0.1 million tons CO$_2$/yr) from IEA and LLNL stationary CO$_2$ sources by industrial sector and by implication, effluent CO$_2$ concentration in South Asia.

perience with these same fields. Moreover, the oil and gas industry has more than thirty years of experience with CO_2 injection into oil and gas reservoirs for CO_2-enhanced oil recovery (EOR) and much of the infrastructure in terms of pipelines, compressors and wells already exists or is readily available [43].

Van Bergen et al. [38] and Damen et al. [26] began their identification of early opportunities for CO_2 sequestration with the USGS World Oil and Gas Assessment database [44] in order to find sites for CO_2-EOR. As they noted, these assessment units and total petroleum systems are several hundred kilometers wide and not detailed enough for even first-order source sink matching. The USGS assessments do provide a guide as to where to begin looking for geologic carbon sequestration opportunities and contain a wealth of subsurface data for some sites.

We scanned, georeferenced, and vectorized maps of individual known oil and gas fields from maps within these assessment units [45-49]. We augmented this database with maps of oil and gas fields from the Directorate General of Hydrocarbons (India) and the Oil and Gas Regulatory Authority (Pakistan), BP-Amoco and Exxon-Mobil in a manner similar to that of Holloway et al. [28].

Deep Saline Aquifers—We are only beginning our assessment of deep saline aquifers in India. For our initial search we focused on large, very deep (>2 km) aquifers capable of providing confining pressures sufficient to hold CO_2 in a supercritical phase and that are on-shore for the sake of cost [50-52]. We began with the Gangetic Siwalik aquifer because it is deep, large and not likely to be of interest for hydrocarbon exploration (with risk of depressurizing the deep saline aquifer). Raster images from a synthesis of Gangetic foreland basin strata [53] were geo-referenced and vectorized to obtain isopachs (similar to structure contours in this case) of this sand rich alluvial system.

Coal Fields—The sequestration and re-use of CO_2 as part of enhancing coal bed methane (ECBM) production has been proposed by Stanton et al. [54] and van Berggen et al. [38]. In this process injected CO_2 displaces the methane that is adsorbed to the surface of the coal along fractures. This makes ECBM both environmentally and economically attractive [55]. A raster image of coal fields in India was vectorized to provide the extent of coal fields in India. The map corresponded well with maps of

individual coal fields [56].

3.3. Second-Order Source-Sink Matching for India

Geological carbon sequestration is most cost effective for large stationary sources [18,23,24]. Ideally such sources are near GCS sinks in order to reduce transport costs [40]. Therefore our next step was to filter the geo-referenced database for large CO_2 sources (>0.1 million tons CO_2/yr) that are less than 20 km from high volume geological carbon sequestration sites including old oil and gas fields, deep saline aquifers and coal fields to focus our search for highly cost-effective GCS opportunities in India (**Figure 3** and **Table 1**).

Approximately 5.6 million tons/yr of CO_2 are generated within 20 km of well understood old oil and gas fields. Approximately 40.6 million tons/yr of CO_2 are generated within 20 km of saline aquifers. Both old oil and gas fields and saline aquifers are likely to have significant long-term storage capacity (under evaluation by our team). In addition, about 94.8 million tons/yr of CO_2 are generated within 20 km of coal fields which may also have significant capacity for GCS.

Approximately 30 sources that generate more than 0.1 million tons of CO_2 are within 20 km of well understood oil and gas fields and major saline aquifers and emit almost 50 million tons of CO_2 per year. This represents nearly one fifth of India's total CO_2 emissions. An additional 47 sources are within 20 km of coal fields and emit nearly 100 million tons of CO_2 per year and represent an additional two fifths of India's CO_2 emissions. These results will focus our evaluation of CO_2 storage capacity, porosity and permeability and risk in India. Our general source sink matching results are summarized in **Table 1**.

As noted by Damodaran [40], one of the objections of Indian stake holders to GCS is the additional financial and energy cost of concentrating CO_2 flue gases. Concentration is necessary to increase the efficiency and decrease the additional costs of transport, injection and sequestration capacity utilization and to enable some desirable geochemical changes once the CO_2 has been injected into the subsurface such as supercriticality. Therefore, once the large (>0.1 million tons CO_2/yr) stationary sources of CO_2 emissions from the IEA and LLNL databases had been located, mapped and filtered by proximity

Table 1. India's CO_2 emissions by storage reservoir type and distance to large stationary CO_2 sources (kilotons).

CO_2 Reservoir Type	CO_2 Sources within 100 km of Reservoir	CO_2 Sources within 20 km of Reservoir
Oil and Gas Fields	76,660	5621
Deep Saline Aquifers	105,752	40,623
Coal Fields Total for All Types	304,747	94,869
Total for All Types	487,159	141,113

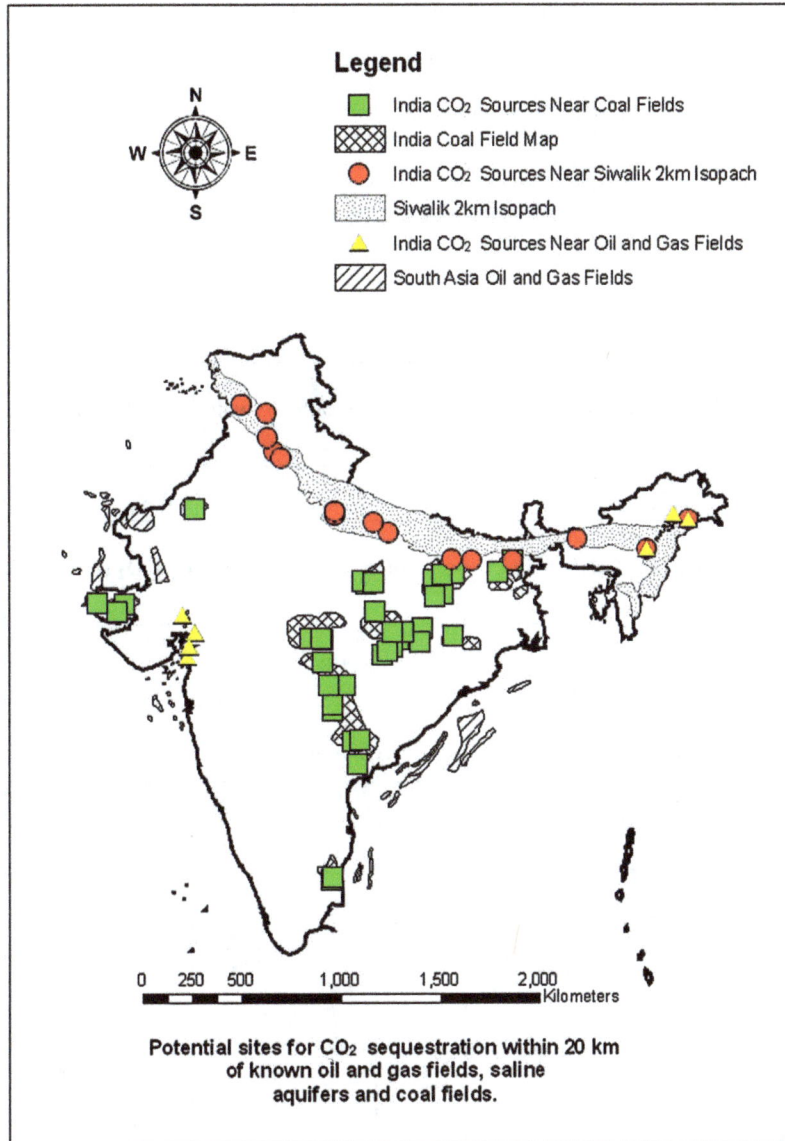

Figure 3. Map of major India (>0.1 million tons CO₂/yr) CO₂ sources <20 km from sinks classified by adjacent geological sink type in India.

(<20 km) to potential geological carbon storage sinks, we then filtered them by industrial sector type to isolate those with typically high (>30%) CO_2 flue gas concentrations.

We then filtered those large and concentrated stationary CO_2 sources by proximity to each type of GCS opportunity considered in this study (oil and gas fields, deep saline aquifers, and coal beds) (**Tables 2-5**).

We then mapped (**Figures 4-7**) a short list of what are likely to be some of the most cost-effective carbon sequestration opportunities in India by GCS opportunity type. The next step is for local Indian stake holders to evaluate each of these candidates in detail so that future proposals for GCS in India meet local requirements and budgets.

4. Conclusions

Our analysis indicates that approximately 5.6 million tons/yr of CO_2 are generated within 20 km of well understood old oil and gas fields in India. Approximately 40.6 million tons/yr of CO_2 are generated within 20 km of saline aquifers in India. Both old oil and gas fields and saline aquifers are likely to have significant long-term storage capacity. In addition, about 94.8 million tons/yr of CO_2 are generated within 20 km of coal fields which may also have significant capacity for geologic CCS.

Approximately 30 sources that generate more than 0.1 million tons of CO_2 are within 20 km of well understood oil and gas fields and major saline aquifers and emit almost 50 million tons of CO_2 per year. This represents

Table 2. India's large (>0.1 million tons CO_2/yr) and concentrated (>30% by volume) CO_2 sources by sector).

Sector	Name of Company	Lat	Long	CO_2 (KT)
ETH. OXIDE	RELIANCE	29.32	75.08	132
AMMONIA	DEEPAK FERTILIZERS AND PETROCHEM.	21.32	74.11	128
AMMONIA	GUJARAT NARMADA VALLEY FERTILIZERS	21.70	72.97	489
AMMONIA	DCM SHRIRAM FERTILIZERS & CHEMICALS	25.11	75.58	217
AMMONIA	STEEL AUTHORITY OF INDIA	22.20	84.88	163
AMMONIA	NATIONAL FERTILIZERS LTD.	29.38	76.97	328
AMMONIA	DUNCANS INDUSTRIES LIMITED	32.43	75.37	454
AMMONIA	FACT	29.97	76.23	327
AMMONIA	DUNCANS INDUSTRIES LIMITED	26.45	80.35	115
AMMONIA	INDIAN OIL CORPORATION LTD	26.09	91.58	149
TOTAL EMISSIONS				2502

Table 3. India's large (>0.1 million tons CO_2/yr) and concentrated (>30% by volume) CO_2 sources <20 km from oil and gas fields).

Sector	Name of Company	City	Lat	Long	CO_2 (KT)
AMMONIA	GUJARAT NARMADA	NARMANDANAGAR	21.70	72.98	489

Table 4. India's large (>0.1 million tons CO_2/yr) and concentrated (>30% by volume) CO_2 sources <20 km from deep saline aquifers in the Siwaliks.

Sector	Name of Company	City	Lat	Long	CO_2 (KT)
AMMONIA	DUNCANS INDUSTRIES	KANPUR	32.43	75.37	454
HYDROGEN	INDIAN OIL CORPORATION	GAWAHATI	26.09	91.58	149

Table 5. India's large (>0.1 million tons CO_2/yr) and concentrated (>30% by volume) CO_2 sources <20 km from coal fields.

Sector	Name of Company	City	Lat	Long	CO_2 (KT)
AMMONIA	STEEL AUTHORITY OF INDIA	ROURKELA	22.20	84.88	163

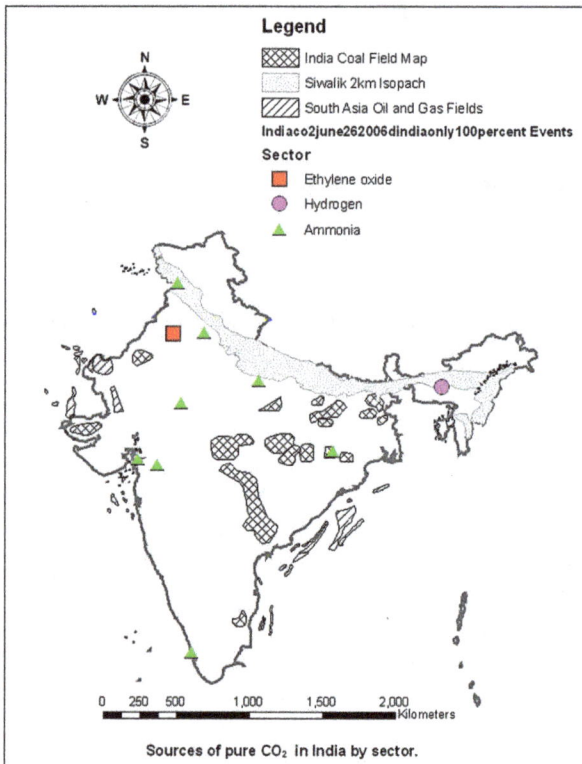

Sources of pure CO_2 in India by sector.

Figure 4. Map of large (>0.1 million tons CO_2/yr) and concentrated (>30% by volume) CO_2 sources by sector in India.

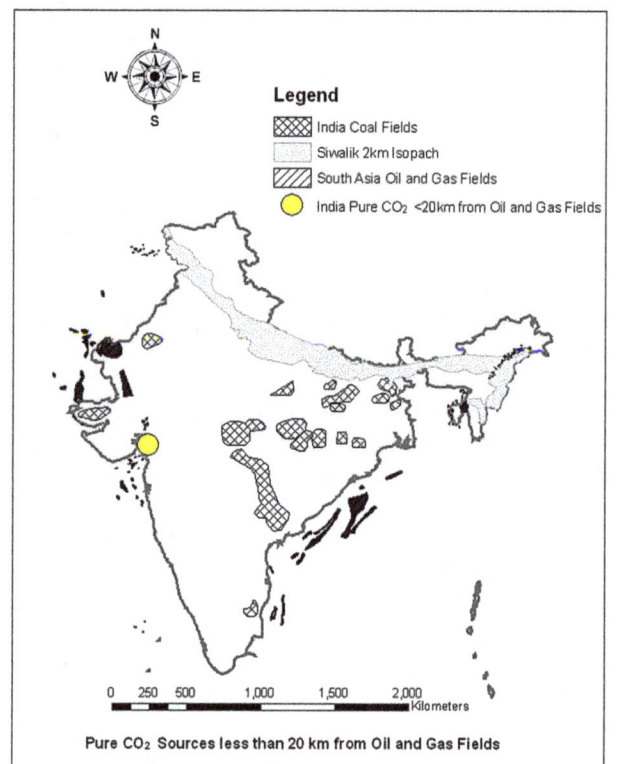

Pure CO_2 Sources less than 20 km from Oil and Gas Fields

Figure 5. Map of large (>0.1 million tons CO_2/yr) and concentrated (>30%) CO_2 sources <20 km from oil and gas fields in India.

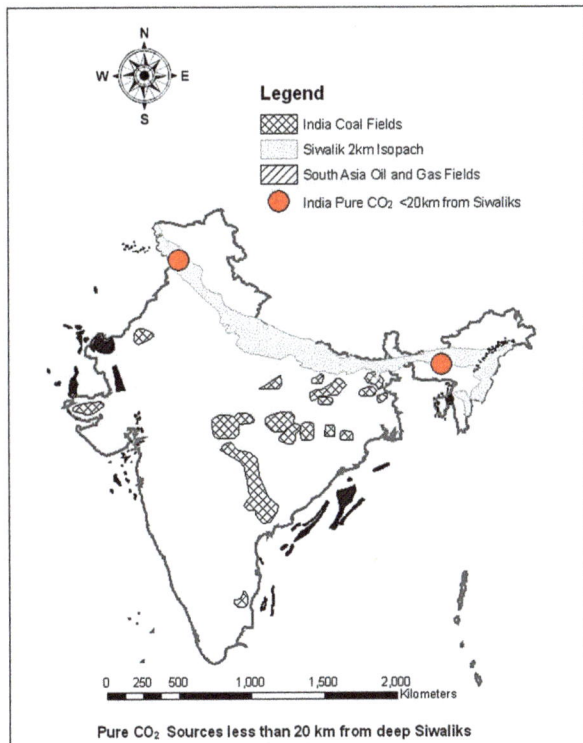

Figure 6. Map of large (>0.1 million tons CO$_2$/yr) and concentrated (>30% by volume) CO$_2$ sources <20 km from deep saline aquifers in the Siwalik Formations of the Gangetic Foreland Basin of India.

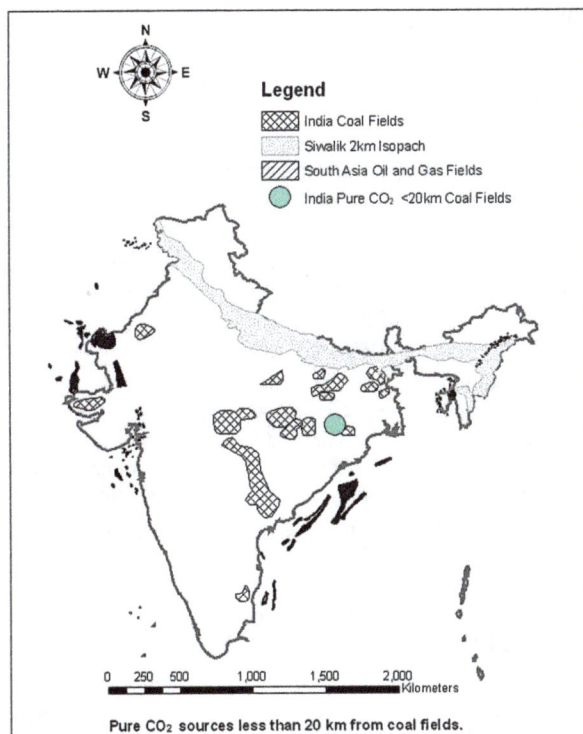

Figure 7. Map of large (>0.1 million tons CO$_2$/yr) and concentrated (>30% by volume) CO$_2$ sources <20 km from coal fields in India.

nearly one fifth of India's total CO$_2$ emissions. An additional 47 sources are within 20 km of coal fields and emit nearly 100 million tons of CO$_2$ per year and represent an additional two fifths of India's CO$_2$ emissions.

Source-Sink matching for large concentrated CO$_2$ sources in India indicates one very high priority target, a fertilizer plant in the city of Narmadanagar, Bharuch District, Gujarat Province, India, that is <20 km from old oil and gas fields in the Cambay Basin. Two pure CO$_2$ sources are <20 km from deep saline aquifers and one is <20 km from a coal field.

5. Acknowledgements

We thank Lawrence Livermore National Laboratory for financial support, office space and much of the data used in this analysis. Additional CO$_2$ source and sink data sets were contributed by the International Energy Agency and the United States Geological Survey.

REFERENCES

[1] Intergovernmental Panel on Climate Change, "IPCC Special Report on Carbon dioxide Capture and Storage," Interlachen, 2005. http://www.ipcc.ch/

[2] J. G. J. Olivier, G. Janssens-Maenhout and J. A. H. W. Peters, "Trends in Global CO$_2$ Emissions," PBL (Netherlands Environmental Assessment Agency), 2012, 40 p. http://edgar.jrc.ec.europa.eu/CO2REPORT2012.pdf

[3] Energy Information Administration, "Greenhouse Gases, Climate Change and Energy," 2006. http://www.eia.doe.gov/oiaf/1605/ggccebro/chapter1.html

[4] NASA, "The Biosphere," 1998. http://www.gsfc.nasa.gov/gsfc/service/gallery/fact_sheets/earthsci/eos/biosphere.pdf

[5] E. Mills, "The Coming Storm: Global Warming and Risk Management," *Risk Management Magazine*, Vol. 45, No. 5, 1998, pp. 20-27.

[6] P. D. Jones, M. New, D. E. Parker, S. Martin and I. G. Rigor, "Surface Air Temperature and Its Changes over the Past 150 Years," *Reviews of Geophysics*, Vol. 37, No. 2, 1999, pp. 173-199.

[7] P. D. Jones and A. Moberg, "Hemispheric and Large-Scale Surface Air Temperature Variations: An Extensive Revision and an Update to 2001," *Journal of Climate*, Vol. 16, No. 2, 2003, pp. 206-223.

[8] T. M. Smith, *et al.*, "Improvements to NOAA's Historical Merged Land-Ocean Surface Temperature Analysis (1880-2006)," *Journal of Climate*, Vol. 21, No. 10, 2008, pp. 2283-2293.

[9] NOAA, "NCDC: Global Surface Temperature Anomalies: Global Mean Temperature Estimates," 2013. http://www.ncdc.noaa.gov/cmb-faq/anomalies.php#mean

[10] M. C. Serreze, J. E. Walsh, F. S. Chapin III, T. Osterkamp, M. Dyurgerov, V. Romanovsky, W. C. Oechel, J. Morison, T. Zhang and R. G. Barry, "Observational Evidence of Recent Change in the Northern High Latitude Environment," *Climatic Change*, Vol. 46, No. 1-2, 2000, pp. 159-207.

[11] J. T. Houghton, Y. Ding, D. J. Griggs, M. Noguer, P. J. van der Linden, X. Dai, K. Maskell and C. A. Johnson, "Climate Change (2001)—The Scientific Basis," Geneva —Intergovernmental Panel on Climate Change (IPCC), 2001.

[12] C. D. Keeling and T. P. Whorf, "Atmospheric CO_2 Records from Sites in the SIO Air Sampling Network," In: Trends: A Compendium of Data on Global Change, Carbon Dioxide Information Analysis Center, Oak Ridge National Laboratory, US Department of Energy, Oak Ridge, 2004. http://cdiac.esd.ornl.gov/trends/co2/sio-mlo.htm

[13] Q. Fu, C. M. Johanson, J. M. Wallace and T. Reichler, "Enhanced Mid-Latitude Tropospheric Warming in Satellite Measurements," *Science*, Vol. 312, No. 5777, 2006, p. 1179.

[14] J. A. Francis and S. J. Vavrus, "Evidence Linking Arctic Amplification to Extreme Weather in Mid-Latitudes," *Geophysical Research Letters*, Vol. 39, 2012, Article ID: L06801.

[15] S. M. Benson, "Carbon Dioxide Capture and Storage in Underground Geologic Formations, The 10 - 50 Solution: Technologies and Policies for a Low-Carbon Future," The Pew Center on Global Climate Change and the National Commission on Energy Policy, 2004. http://www.pewclimate.org/docUploads/10%2D50%5FBenson%2Epdf

[16] GCCSI, "Projects," Global CCS Institute, 2013. http://www.globalccsinstitute.com/projects/browse

[17] World Bank, "Regional Data at a Glance," 1999. http://www.worldbank.org/html/extdr/spring99/regionaldata.pdf

[18] IEA—International Energy Agency, "CO_2 Capture and Storage: A Key Carbon Abatement Option," OECD/IEA, 2008.

[19] J. Gale, "Overview of CO_2 Emission Sources, Potential, Transport and Geographical Distribution of Storage Possibilities," 2002. http://arch.rivm.nl/env/int/ipcc/docs/css2002/ccs02-01.pdfs

[20] S. Pacala and R. Socolow, "Stabilization Wedges: Solving the Climate Problem for the Next 50 Years Using Current Technologies," *Science*, Vol. 305, No. 5686, 2004, pp. 986-972.

[21] J. Bradshaw, G. Allison, B. E. Bradshaw, V. Nguyen, A. J. Rigg, L. Spencer and P. Wilson, "Australia's CO_2 Geological Storage Potential and Matching of Emission Sources to Potential Sinks," In: Greenhouse Gas Control Technologies, *Proceedings of the 6th International Conference on Greenhouse Gas Control Technologies*, Kyoto, 1-4 October 2002, p. 2003.

[22] S. J. Friedmann, J. J. Dooley, H. Held and O. Edenhofer, "The Low Cost of Geological Assessment for Underground CO_2 Storage: Policy and Economic Implications," *Energy Conversion Management*, Vol. 47, No. 13-14, 2006, pp. 1894-1901.

[23] IEA—International Energy Agency Greenhouse R&D Programme, "Building the Cost Curves for CO_2 Storage: North America," Technical Report 2005/03, 2005.

[24] IEA—International Energy Agency Greenhouse R&D Programme, "Building the Cost Curves for CO_2 Storage: European Sector," Technical Report 2005/02, 2005.

[25] J. J. Dooley and S. J. Friedmann, "A Regionally Disaggregated Global Accounting of CO_2 Storage Capacity: Data and Assumptions," Pacific Northwest National Laboratory, Report PNWD-3431, 2004.

[26] K. Damen, A. Faaij, F. van Bergen, J. Gale and E. Lysen, "Indentification of Early Opportunities for CO2 Sequestration—Worldwide Screening for CO_2-EOR and CO_2-ECBM Projects," *Energy*, Vol. 30, No. 10, 2005, pp. 91931-1952.

[27] A. K. Singh, V. A. Mendhe and A. Garg, "CO_2 Storage Potential of Geologic Formations in India," *8th Greenhouse Gas Technology Conference*, Trondheim, 19-22 June 2006, pp. Session A1-A2.

[28] S. Holloway, A. Garg, M. Kapsche, A. S. Pracha, S. R. Khan, M. A. Mahood, T. N. Singh, K. L. Kirk, L. R. Applequist, A. Deshpande, D. J. Evans, Y. Garg, C. J. Vincent and J. D. O. Williams, "A Regional Assessment of the Potential for CO_2 Storage in the Indian Subcontinent," British Geological Survey, 2007, 201 p.

[29] R. V. Kapila and R. S. Haszeldine, "Opportunities in India for Carbon Capture and Storage as a Form of Climate Change Mitigation," Energy Prodedia, GHGT-9 Volume, 2008, 8 p.

[30] R. V. Kapila, H. Chalmers and M. Leach, "Investigating the Prospects for Carbon Capture and Storage Technology in India," Report for Christian Aid, 2009, 137 p.

[31] P. K. Kaw, "Short Report on the Energy Situation in India," Report on Research and Development of Energy Technologies, IUPAP Working Group on Energy, International Union of Pure and Applied Physics, 2004. http://www.iupap.org/wg/energy/annexb.pdf

[32] G. Marland, T. A. Boden and R. J. Andres, "Global, Regional, and National CO_2 Emissions," In: Trends: A Compendium of Data on Global Change. Carbon Dioxide Information Analysis Center, Oak Ridge National Laboratory, US Department of Energy, Oak Ridge, 2000.

[33] G. Marland, T. A. Boden and R. J. Andres, "Global, Regional, and National CO_2 Emissions," In: Trends: A Compendium of Data on Global Change. Carbon Dioxide Information Analysis Center, Oak Ridge National Laboratory, US Department of Energy, Oak Ridge, 2008.

[34] C. Le Quéré, *et al.*, "The Global Carbon Budget 1959-2011," *Earth System Science Data Discussions*, Vol. 5, No. 2, 2012, pp. 1107-1157.

[35] G. P. Peters, R. M. Andrew, T. Boden, J. G. Canadell, P. Ciais, C. Le Quéré, G. Marland, M. R. Raupach and C.

Wilson, "The Challenge to Keep Global Warming below 2°C," *Nature Climate Change*, Vol. 3, 2012, pp. 4-6.

[36] B. Walsh, "The Impact of Asia's Giants, How China and India Could Save the Planet—Or Destroy It," 2006. http://www.time.com/

[37] Energy Information Administration, "Annual Energy Review," 2004. http://www.eia.gov/totalenergy/data/annual/archive/038404.pdf

[38] F. van Bergen, J. Gale, K. J. Damen and A. F. B. Wildenborg, "Worldwide Selection of Early Opportunities for CO_2-Enhanced Oil Recovery and CO_2-Enhanced Coal bed Methane Production," *Energy*, Vol. 29, No. 9-10, 2004, pp. 1611-1621.

[39] P. Bumb and R. Vasant, "Carbon Capture and Storage (CCS) in Geological Formations as Clean Development Mechanism (CDM) Projects Activities (SBSTA)," 2009, 10 p. http://cdm.unfccc.int/about/ccs/docs/CCS_geo.pdf

[40] A. Damodaran, "Carbon Capture and Storage: India's Concerns, Carbon Sequestration," Leadership Forum, New Delhi, 2008. http://www.cslforum.org/publications/documents/IndiaPresentationFITFMeetingNewDelhi1208.pdf

[41] A. Gibson, "CCS Too Much for India Now?" Global Carbon Capture and Storage Institute, 2012. http://www.globalccsinstitute.com/insights/authors/alicegibson/2012/02/16/ccs-too-much-india-now

[42] S. Plasynski, A. McNemar and P. McGrail, "CO_2 Sequestration in Basalt Formations," US DOE, National Energy Technology Laboratory, 2008. http://www.netl.doe.gov/publications/factsheets/project/Proj277.pdf

[43] S. H. Stevens, V. A. Kuuskraa and J. Gale, "Sequestration of CO_2 in Depleted Oil & Gas Fields: Global Capacity, Costs and Barriers," *Proceedings of GHGT-5*, Cairns, 13-16 August 2000, pp. 278-283.

[44] USGS World Energy Assessment Team, "U.S. Geological Survey World Petroleum Assessment 2000—Description of Results," USGS Digital Data Series DDS-60 Multi Disc Set Version 1.0, 2000 (CD-ROMs).

[45] C. J. Wandrey and B. E. Law, "Map Showing Geology, Oil and Gas Fields, and Geologic Provinces of South Asia," Version 2.0, Open File Report 97-470C, 1999, Version 2 (CD-ROM).

[46] C. J. Wandrey, "Region 8 Assessment Summary—South Asia," In: C. J. Wandrey, R. Milici and B. E. Law, Eds., *U.S. Geological Survey Digital Data Series* 60, 2000, 39 p.

[47] C. J. Wandrey, "Patala-Nammal Composite Total Petroleum System, Kohat-Potwar Geologic Province, Pakistan," Petroleum Systems and Related Geologic Studies in Region 8, South Asia, U.S. Geological Survey Bulletin 2208-B, 2000 (CD-ROM).

[48] C. J. Wandrey, "Sembar Goru/Ghazij Composite Total Petroleum System, Indus and Sulaiman-Kirthar Geologic Provinces, Pakistan and India," Petroleum Systems and Related Geologic Studies in Region 8, South Asia, U.S. Geological Survey Bulletin 2208-C, 2000 (CD-ROM).

[49] C. J. Wandrey, "Sylhet-Kopili/Barail-Tipam Composite Total Petroleum System, Assam Geologic Province, India, Petroleum Systems and Related Geologic Studies in Region 8, South Asia," U.S. Geological Survey Bulletin 2208-D, 2000 (CD-ROM).

[50] S. T. McCoy, "The Economics of CO_2 Transport by Pipeline and Storage in Saline Aquifers and Oil Reservoirs," "Department of Engineering and Public Policy" Paper 1, 2009. http://repository.cmu.edu/epp/1

[51] J. K. Eccles, L. Pratson, N. G. Newell and R. B. Jackson, "Physical and Economic Potential of Geological CO_2 Sequestration in Saline Aquifers," *Environmental Science Technology*, Vol. 43, No. 6, 2009, pp. 1962-1969. http://pubs.acs.org/doi/abs/10.1021/es801572e

[52] S. Sharma, K. Michael, G. Allinson, M. Arnot, P. Cook, J. Ennis-King and V. Shulakova, "CO_2 Storage in Saline Aquifers: Review of Recent Scientific Progress and Remaining Issues," IFP, Rueil-Malmaison, 27-29 May 2009. www.ifp.com/content/download/67989/1473855/file/33_Michael.pdf

[53] D. W. Burbank, R. A. Beck and T. Mulder, "The Himalayan Foreland," In: Y. An and M. Harrison, Eds., *Asian Tectonics*, Cambridge University Press, Cambridge, 1996, pp. 149-188.

[54] R. Stanton, R. Flores, P. D. Warwick, H. Gluskoter and G. D. Stricker, "Coal Bed Sequestration of Carbon Dioxide," U.S. Department of Energy, 2001. http://www.netl.doe.gov/publications/proceedings/01/carbon_seq/3a3.pdf

[55] S. Wong, W. D. Gunter, D. H.-S. Law and M. J. Mavor, "Economics of Flue Gas Injection and CO_2 Sequestration in Coalbed Methane Reservoirs," In: D. Williams, B. Durie, P. McMullan, C. Paulson and A. Smith, Eds., *Proceedings of the 5th International Conference on GHG Control Technologies*, CSIRO Publishing, Cairns, 2001, pp. 543-548.

[56] M. S. Krishnan, "Geology of India and Burma," 6th Edition, Satish Kumar Jain for CBS Publishers and Distributors, Delhi, 1982, 536 p.

5

Drinking Water Quality Clinics and Outreach in Delaware Focusing on Educating Master Well Owners

Gulnihal Ozbay[1*], Amy Cannon[1], Amanda Treher[1], Stephanie Clemens[2], Albert Essel[1], Dyremple Marsh[1], John Austin[3]

[1]College of Agriculture & Related Sciences, Cooperative Extension, Delaware State University, Dover, Delaware, USA; [2]Master Well Owner Network, University Park, Pennsylvania, USA; [3]Office of Sponsored Research, Delaware State University, Dover, Delaware, USA.

ABSTRACT

The United States Environmental Protection Agency (EPA) has the authority to regulate the public water systems. The EPA does not have the jurisdiction to regulate private drinking water wells. This leaves approximately fifteen percent of the nation's population without any regulation being held in place to protect their source of drinking water. With that fifteen percent of the US population having private wells for drinking water, it makes the number of people whose drinking water is unprotected by regulation at a little over 15 million US households. This concern is even more acute in areas with groundwater that is close to the surface. Delaware residents live in a region with low elevation which is very close to the coast with low elevation and the shallow groundwater makes us concern about contaminated well water even more intense. As one of the Water Resources Program partners, we have offered free Drinking Water Quality Clinics to local well owners over the past 4 years in Delaware State University. Since 2009, over 400 Delaware residents have benefited from these clinics. At each clinic, an information session was offered in the evening, with an opportunity to hear from and speak with a drinking water well expert. Participants were given sample bottles and water testing performed the following day included pH, nitrite, nitrate, sulfate, alkalinity, fluoride, hardness, iron, lead, cadmium, arsenic, Total Coliform, and *E. coli*. Over half of the samples returned out of range values for pH, while 72 returned results positive for Total Coliform and *Escherichia coli* bacterium. Data are examined for correlations, and improved understanding of local well owners. These tests shared with local well owners insights into what may be wrong with their water. In addition, any tests that came back outside of the normal range were reported to homeowners in writing. Mailed with the written reports were also information specific to what test results were outside of the limits, and actions to take to correct the exact problem the well owners encountered. The data reported here are examined to discuss the correlations of information, and ways that the Drinking Water Quality Clinics have improved our understanding of local wells and ownerships. In conclusion, regular testing on a yearly basis is the most effective way to ensure that public health is maintained.

Keywords: Water Quality; Well Water; Drinking Water; Private Well Water; Water Sample Testing

1. Introduction

According to the World Health Organization (WHO), the quality of drinking-water is a strong environmental determining factor of health [1]. Drinking-water safety lays the groundwork for the prevention and control of water-borne diseases [1]. In 2008, the WHO created an International Network of Drinking-Water Regulators (RegNet) due to requests from the Member States [2]. Member States consist of any member of the United Nations that accepts the WHO Constitution [3]. This network was created to address regulatory issues in relation to drinking water in a more organized way [2]. RegNet is an international forum that allows for sharing and discussing the strategies to address all facets of guardianship and the care of public health as it relates to drinking water [2]. RegNet's mission is "to protect public health, as it relates to drinking water, through the promotion of excellence and the continual improvement of regulatory frameworks and systems [2].

In the Mid-Atlantic region, which includes Pennsylvania, Virginia, Maryland, West Virginia, and Delaware, over 5.5 million people have private drinking water supplies. These systems include well, cisterns, and springs [4]. When a homeowner has a private water supply, they are responsible for multiple aspects of their own water. These people must test their water on a regular basis to ensure that it is safe for drinking and cooking [5]. They are in charge of the care and maintenance of their system, and they also are accountable for solving their own water problems [6]. Many times, these citizens lack the knowledge and resources to make well-informed decisions concerning their water supply. They often wait until a problem arises to take action [6].

In looking at the different kinds of private water systems, there are wells, which are drilled or bored into the ground until the groundwater table is reached [6]. They are between 6 - 300 meters in depth and should be located at least 30 meters away from sources of contamination [7]. There is usually a casing, grouting, or a sanitary well cap to protect the water from contamination [6]. Cisterns are containers used for the catchment of rainwater to be used for private water sources [8]. Springs are formed when the side of a hill, a valley, or excavation meets groundwater. Spring water is highly susceptible to contamination [9].

Drinking Water Quality Clinics held here at Delaware State University have been an extension of the Master Well Owner Network (MWON) that began in 2003 [10]. In the MWON Program, volunteers were taught many subjects regarding private water supplies [10]. Topics included groundwater, well construction and maintenance, land use impacts, water testing, conservation, solution and treatment of water problems [10]. As a continuation of the MWON, we here at Delaware State University, as a part of our Cooperative Extension Program, have offered the public multiple opportunities to attend information sessions regarding their private water supplies. In this way, the free educational program can be shared from community about private drinking water supplies. In addition to the teachings, we offered the public an opportunity to have their water tested on the day following the Drinking Water Quality Clinic, on campus in our Aquatic Sciences Laboratory.

The clinics were held after funding was over for the MWON, and have been supported as part of the 1890 Water Resources Center Program partners, lead by Virginia State University. Since 2012, this program has been funded by the 1890 Water Resources Center Grant. **Map 1** illustrates the local Delmarva Peninsula region that includes Delaware to show the uses of land in the area.

The results are available in this article, regarding not only what was found in the testing of the water, but also the effectiveness of the teaching and learning of the seminar format in which homeowners learn about their wells and water quality. The water samples have been tested for pH first, which can be an indicator of other water quality problems [12]. Measurement of pH is important because low pH can allow toxic metallic elements and compounds to become mobile and "available" for uptake by people drinking the water [13]. Chlorine is also measured in the samples, because it can cause eye and nasal irritation, as well as stomach discomfort and potential nervous system effects in infants and children when found in drinking water [14]. We tested for Sulfates, which are important to measure due to a bitter taste that can be brought out in the water from Sulfates, as well as having laxative effects on humans and animals consuming them in water [15]. Water samples were tested for Nitrite due to the fact that infants below the age of six months who drink water containing nitrite in excess of the Maximum Contaminant Level could become seriously ill and, if untreated, may die [16]. Symptoms may show as shortness of breath and blue-baby syndrome [16]. Nitrate levels were also measured considering that Nitrites convert to Nitrates in a baby's stomach from a bottle made with drinking water, which can also be the cause of blue baby syndrome [17]. Alkalinity is a measure of the ability of water to neutralize acids [18]. This is an important factor in drinking water because a certain level of alkalinity will help raise pH, which will keep toxic metallic elements and compounds from becoming mobile in the drinking water to be consumed [13]. Hardness was also checked in the water samples to show levels of $CaCO_3$ (calcium carbonate) in the water [12]. Although hardness levels do not denote a true health hazard, at certain levels they do help prevent corrosion of metals in pipes and sink fixtures in the home [12]. Iron concentrations were determined to show possible reasons for metallic taste of drinking water, discolored water, and orange or brown stains on laundry [12]. Fluoride levels were tested in the samples because it can cause tooth discoloration and skeletal decay of bone structure [12]. Water samples were also evaluated for *E. coli* and Total Coliform Bacteria. The importance of measuring the levels of these last two can

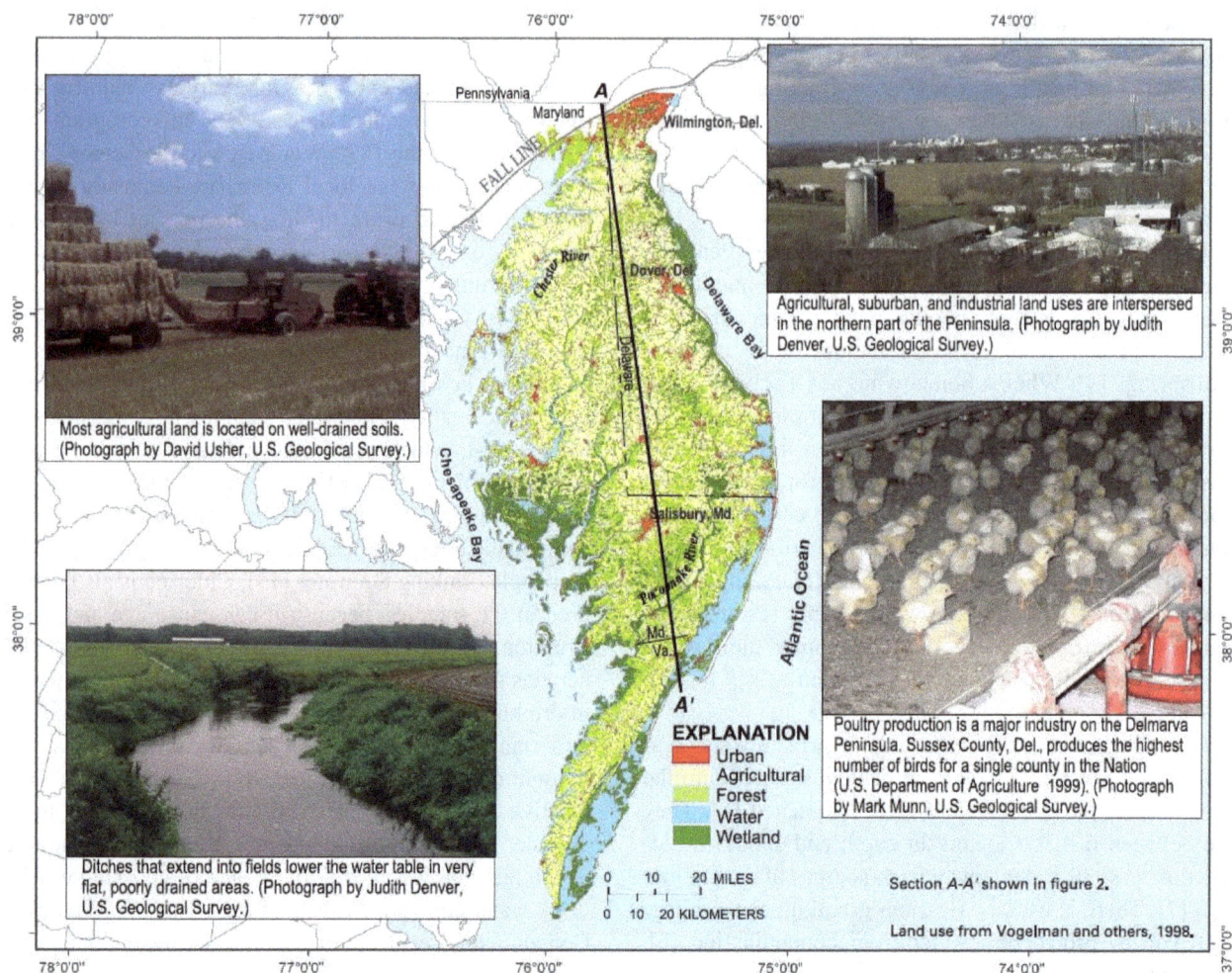

Map 1. The Delmarva Peninsula is primarily rural, with small towns and residences interspersed among agricultural and forested land. Major areas of urban growth are to the north and along the Atlantic coastline [11].

not be underestimated because coliform bacteria are often called "indicator organisms" due to the fact that they indicate the potential for disease-causing bacteria to be present in water [19]. *E. coli* presence in water samples shows a strong indication that human sewage or animal waste has contaminated the water supply [19]. A few strains of *E. coli* can produce a powerful toxin and can induce severe illness and death [19]. More common signs of infection are stomach upsets and general flu symptoms such as fever, abdominal cramps and diarrhea [19]. Lead, Arsenic and Cadmium are metals that were also tested for in selected drinking water samples. All of these tests will show homeowners if they have cause for concern in the water samples they bring from their homes, and then be able to address the concerns if they are present.

In regards to the effectiveness of the training workshop seminar, participants were given pre-tests and post-tests to garner how much information they learned from the Drinking Water Quality Clinics. The homeowners

were also asked to fill out a form sharing how close their wells are to potential sources of contamination. From these details, it can be determined what may be causing contamination by what sources are located close to the homeowners' wells. Workshop evaluation is provided to the Well Water Quality Clinic Participants to help us improve the way we reach out and effectiveness of information delivery during the workshops (please see the copy of the questionnaire in the appendices).

2. Purpose and Research Objectives

The purpose of the information reported here is to communicate the details and results of Drinking Water Quality Clinics held at Delaware State University for local well owners.

1) To define the information that was shared with participants in the clinics and why this information is relevant to drinking water.

2) To illustrate the results from the water sample tests, as well as the results from participant tests for informa-

tion and program evaluations.

3. Materials and Methods

Each of the Drinking Water Quality Clinics has had a keynote speaker who is an expert on drinking water and wells. This offers the public an insight into how important it is to have their water tested, and to learn more about their well systems. This also presents an opportunity for homeowners to participate in a question and answer session to find out more about their wells after the keynote speaker presents their material. At the end of the question and answer period, residents are offered two water bottles to take home and take water samples. The following morning, we provide a location next to our laboratory where we collect samples until noon. At that point, we begin the testing of each sample. Due to the number of samples, we are unable to provide testing in triplicate, but the tests are formatted, able to be replicated and accurate.

During the information sessions of the clinics, many facts were shared with well owners regarding their drinking water supply. Participants were advised how they are responsible for maintenance, testing, treatment and protection of their wells. The participants were offered information on where the water comes from - aquifers. The speakers also shared what kinds of aquifers are available in Delaware, and how these aquifers may become affected by surrounding above-ground activities. The importance of proper well construction was discussed, as well as proper well location to prevent drinking water contamination. In the well construction section, the significance of a sanitary well cap, grout and preventing creatures from living under the well cap was shared.

Members of the local community who joined in these sessions were informed of wellhead protection. The location where the well meets the air is important, especially in unconfined aquifers, like those found in most of Delaware, and all of Southern Delaware. In addition to wellhead protection, protecting the health of the families of the well owners was a crucial driving factor to make sure people know how important it is to have their water tested every year, at least. This will improve and protect the health of the people who drink this water. Because

many pollutants do not have a smell or taste, it is crucial to keep testing current. Primary and secondary pollutants were discussed, and details about each contaminant were given to participants at the information sessions, prior to having their water tested. **Picture 1** illustrates the public information sessions at the clinics, as well as some sample testing.

3.1. pH

The first test performed was for pH. This is performed by using the YSI EcoSense pH10 pH and Temperature Pen (YSI, Yellow Springs, Ohio). This instrument is calibrated first by using YSI Buffer 3821 for pH 4.00. The second buffer is YSI Buffer 3822 for pH 7.00, and the third is YSI Buffer 3823 for pH 10.00. Once calibrated, the implement is placed into each sample individually to measure the pH. Once the data is recorded, the equipment is cleaned with de-ionized water to prepare it for the next sample. The instrument is then wiped clean with a Kimwipe, and the next sample pH is recorded until all samples have been tested for pH.

3.2. Nitrites

Nitrites are the next thing that was measured in the drinking water from wells. These were found by using HACH Method 8507, the Diazotization Method for Low Range concentrations from 0.002 to 3.000 mg/L of Nitrite (NO_2^-) in solution of drinking water. Powder pillows of reagent are used in this method, being added to a 10 mL cylindrical cuvette. The results are read by the HACH DR 2500 spectrophotometer (HACH Co. Loveland, Colorado).

3.3. Nitrates

Next, Nitrates were measured from the well drinking water samples. They were measured using HACH Method 8171, the Cadmium reduction method. This test is for mid-range levels of nitrate, from 0.1 to 10.0 mg/L concentrations in solution. Once powder pillow reagents are added, and the allotted time is taken for the test to complete its reaction, the HACH DR 2500 spectrophotometer was used to measure the levels of Nitrate (NO_3^-) in solution.

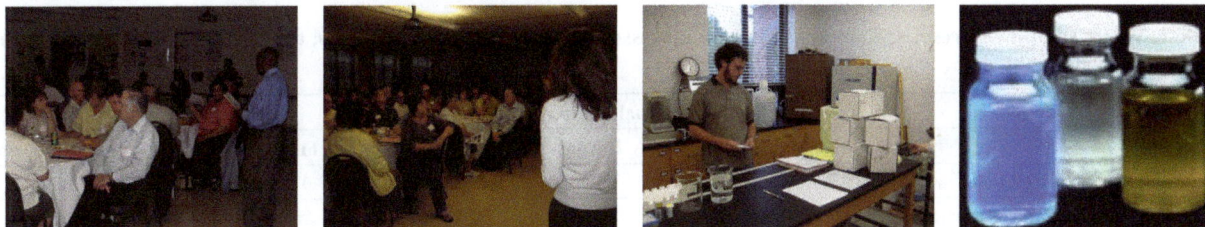

Picture 1. (Left-right) workshop attendees being addressed by the Associate Dean of Cooperative Extension, Dr. Essel and Master Well Owner Network employee, Mrs. Clemens, samples being collected for testing and test bottles for total coliform and *E. coli*.

3.4. Hardness

The following test on the drinking water samples is for Hardness. This test is performed with the Palintest PM 254 Method, the photometer method with automatic wavelength selection. The measurement of hardness is tested over a range of 0 - 500 mg/L $CaCO_3$. Water hardness is caused by the presence of calcium and magnesium salts [20], so this test measures these cations. The results are focused upon Ca^{2+} part of the $CaCO_3$, which indicates the actual hardness and potential corrosiveness of the water.

3.5. Sulfate

Testing for Sulfate was performed in the samples received. This is measured using HACH Method 8051, USEPA SulfaVer 4 Method. This method is equivalent to USEPA method 375.4. The levels measured are from 0 to 700 mg/L of Sulfate in the samples. Results are found by the HACH DR 2500 Spectrophotometer set to a wavelength of 450 nm. This reads the turbidity that is proportional to the amount of Sulfate (SO_4^{2-}) in sample solution.

3.6. Iron

Iron is also measured in the drinking water for each sample collected. It is calculated using the Palintest PM 155 Method, the photometer method with automatic wavelength selection. The levels are measured in the low range, from 0 - 1.0 mg/L of Iron in solution. These results are found by using the YSI 9500 Photometer (YSI, Yellow Springs, Ohio).

3.7. Fluoride

Another component that was assessed was Fluoride levels in the well water. This was analyzed using HACH Method 8029. This is the SPADNS Method that measures 0.02 - 2.00 mg/L Fluoride (F^-). The reagent solution was used. There is also an option to use the AccuVacAmpuls, but this was the second option that was not chosen to be used. The reagent solution was chosen to be utilized instead. The results from this test were read by the HACH DR 2500 to measure Fluoride in well drinking water solution.

3.8. Total Coliform and *Escherichia coli*

Total Coliform tests were performed on a presence or absence basis in the well water samples. They were analyzed using IDEXX's patented Defined Substrate Technology* (DST). Essentially, a bottle with reagent is used to collect the sample the morning of the test. The homeowner brings the sample bottle, with reagent already inside to the lab for the actual test. It is stored for 24 hours at 35°C in the lab, and then examined for results. A yellow color denotes presence of Total Coliform bacteria. If the yellow color fluoresces under a black light, then this denotes the presence of *E. coli* bacteria. This method has been approved by the United States Environmental Protection Agency (USEPA), as well as other nations who are currently using this method for detection. This test does not provide quantitative outcome, but rather denotes the presence or absence of bacterium.

3.9. Lead, Arsenic and Cadmium

A Perkin Elmer AAnalyst 600 Graphite Furnace Atomic Absorption Spectrometer was utilized to measure levels of the above three metals. The protocols used were from EPA Method 200.9: Determination of Trace Elements by Stabilized Temperature Graphite Furnace Atomic Absorption [21]. Not all samples were tested for metals. It was determined that samples with especially low pH to be tested for metals. The reason for this determination is that acidic solutions tend to make metals more soluble, so a low pH can be an indicator of potential heavy metals in water [11] and samples provided by residents living close to the industrial sites were given priority.

4. Results

Over a period of three years, four clinics were held, free of charge to the public. The data in **Table 1** shows the number of tests that showed irregular results, outside of the listed safety limits of the measured items. In addition, the tests in 2012 were enhanced to include metals testing. This was only done for the samples that had pH out of range. There was only one result that yielded unsafe results for Lead contamination. This information is analyzed further in the discussion section.

Table 1. Water sample test results. Common water quality issues include pH, iron, hardness, total coliform, and *E. coli* in the private well waters tested.

	pH	Nitrite	Nitrate	Hardness	Sulfate	Iron	Fluoride	*E. Coli*	Total Coliform
Drinking Water Quality Clinic Total Results									
Allowable Safe Limits	6.5 - 8.5	1 mg/L	10 mg/L	100 mg/L	250 mg/L	0.3 mg/L	4 mg/L	0 colonies per 100 mL	
Out of about 200 residents' water, # of water samples not within safe limits listed here	106 (53%)	0 (0%)	8 (4%)	12 (6%)	0 (0%)	32 (16%)	0 (0%)	28 (14%)	37 (19%)

4.1. pH

The importance of pH is noted as a potential indicator of threats to human health. The above data shows pH as having the highest occurrence among the tests for being out of a healthy range. Causes for this kind of are that natural and human processes determine the pH of water [22]. "High pH causes a bitter taste, water pipes and water-using appliances become encrusted with deposits, and it depresses the effectiveness of the disinfection of chlorine, thereby causing the need for additional chlorine when pH is high. Low-pH water will corrode or dissolve metals and other substances [22]. The main concern here is human health. Effective treatment options are available to help well owners with pH issues.

4.2. Nitrites

Nitrite testing is important because nitrites are used as food preservatives and in medicine to relieve cardiac pain [23]. They can cause infants below six months who drink water containing nitrite in excess of the maximum contaminant level (MCL) to become seriously ill and, if untreated, to even die [16]. Symptoms are indicated by shortness of breath and blue baby syndrome [16]. Fortunately, none of the samples tested in any clinic returned results of Nitrite over the recommended level, or MCL.

4.3. Nitrates

Elevated concentrations of nitrate in water from aquifers near the surface and used for domestic supply are of particular concern because many homeowners are not aware of possible risks [11]. Unlike public supply wells, domestic wells are not regularly monitored [11]. With this being the case, homeowners' well water from the clinics was tested for nitrates. Nitrates can come from many sources: runoff from fertilizer use, leaking from septic tanks, sewage, and erosion of natural deposits [24]. Only 8 out of 200 samples yielded results that were out of range for nitrates. This is a very encouraging number for our sample size.

4.4. Hardness

Hardness was measured as a test of the amount of calcium carbonate in the water samples [25]. Hardness or softness of water also is a personal preference, as it also relates to how people feel the water affecting their hair or skin when washing with their water [25]. This test is, therefore, not a measure of something that affects human health; as a matter of fact, calcium and magnesium salts that used to measure hardness are actually essential nutrients [25]. Out of 200 samples tested, only 12 showed results out of range for Hardness. The remaining samples were all considered to be soft, therefore having no poten-tial threat to build up any white scale on pipes or dishes [25].

4.5. Sulfate

Sulfate (SO_4^{2-}) is a substance that occurs naturally in drinking water [26]. Sulfates are made up from sulfur and oxygen and are a part of naturally occurring minerals in some rock formations that contain groundwater [15]. Health concerns regarding sulfate in drinking water have been raised because of reports that diarrhea may be associated with ingesting of water containing high levels of sulfate [26]. Of particular concern are sub-groups within the general population that may be at greater risk from the laxative effects of sulfate when they experience an abrupt change from drinking water with low sulfate concentrations to drinking water with high sulfate concentrations [26]. The results here were found to be all within the range that is indicated by the EPA to be safe limits, although no actual regulations are in place for sulfates. They are part of the Secondary Maximum Contaminants Level (SMCL) determinations set by the EPA [27]. SMCLs are set for nuisance chemicals that only have aesthetic concerns rather than health concerns [27]. High levels of sulfates usually result in a salty taste to the water [27].

4.6. Iron

The testing for Iron is a nuisance test [27]. The presence or absence of Iron does not indicate a health concern if found to be over the recommended levels [27]. This test shows the level of Iron in drinking water for aesthetic purposes only [27]. These purposes are for color, odor and taste [27]. Problems associated with high levels of Iron include rusty sediment, a bitter, metallic taste, brown-orange stains, iron bacteria, and discolored beverages [28]. Higher concentrations of dissolved Iron are considered typical in poorly oxygenated groundwater on the Delmarva Peninsula [11]. Out of 200 samples tested 32 tests returned with levels of Iron in them that are over the Secondary Maximum Contaminant Level (SMCL) designated by the EPA. This SMCL is not enforceable by any regulation, because the presence of Iron is not considered a risk to human health [27].

4.7. Fluoride

Fluoride is another nuisance chemical, as determined by the EPA, with only aesthetic concerns, rather than human health concerns when found in drinking water [27]. Fluoride helps with dental health so many water systems add small amounts to drinking water [6]. At the same time, excessive consumption of naturally occurring fluoride can damage bone tissue [6]. Levels of fluoride over the SMCL set by the EPA can cause tooth discoloration [27].

Out of the 200 samples that were tested since 2009, none of the samples were found to be over the SMCL recommended limit.

4.8. Total Coliform and *E. coli*

Total Coliform test indicates not a health threat itself, but rather is used as an indicator of whether or not potentially harmful bacteria may be present [29]. *E. coli* is a bacterium that indicates that the water may be contaminated with human or animal wastes [29]. Inside the gastrointestinal tract of humans and warm-blooded animals, *E. coli* poses no threat [30]. Disease-causing microorganisms (pathogens) in the wastes of humans or other animals can cause diarrhea, cramps, nausea, headaches or other symptoms upon ingestion [29]. These pathogens may pose a special health risk for infants, young children, and people with severely compromised immune systems [29]. Out of the 200 samples tested, 37 tested positive for the presence of Total Coliform bacterium, while 28 samples tested positive for the presence of *E. coli*.

4.9. Lead, Arsenic and Cadmium

Lead is a toxic metal that has been used for many years in products found in and around homes [31]. Lead is sometimes used in household plumbing components or in water service lines used to bring water from the main line into the home [31]. A prohibition on lead in plumbing lines has been in effect since 1986 [31]. Infants and young children who drink water containing lead in excess of the action level (greater than 15 parts per billion in more than 10% of water samples tested) could experience delays in their physical or mental development [31]. Children could also show slight deficits in attention span and learning abilities [31]. Adults who drink this water could develop kidney problems or high blood pressure [31].

Arsenic is a semi-metallic element in the periodic table [32]. Arsenic is odorless and it is tasteless [32]. It enters drinking water supplies from natural deposits in the earth or from agricultural and industrial practices [32]. People who drink water containing arsenic in excess of the Maximum Contaminant Level (MCL) for many years could experience skin damage or problems with their circulatory system, and may have an increased risk of getting cancer [32].

Cadmium is a metal found in natural deposits such as ores containing other elements [33]. Some people who drink water containing cadmium well in excess of the maximum contaminant level (MCL greater than 5 ppb) for many years could experience kidney damage [33]. The multiple health effects discussed are not intended to catalog all possible health effects for these metals [31]. Rather, it is intended to inform people of the most significant and probable health effects, associated with lead,

arsenic and cadmium in drinking water [31].

4.10. Surveys

At the start and the end of each information session, participants were asked to take a Pre-test and a Post-test, in order to find out what kind of information was relevant in sharing new knowledge with participants. Less than fifty percent of the well owners who attended the clinics actually returned these tests. Out of that less than fifty percent, only some of the questions were answered correctly. Below you will find the questions that gave the community members trouble in answering correctly before the clinic (**Table 2**). During the clinic, all of this information was available to them, and they were able to answer the questions correctly following the information sessions. The correct answers are highlighted in bold.

Throughout the clinics, participants were also asked to fill out evaluation forms. These forms helped us to understand how the participants viewed the information that was shared, as well as shaped the learning that was gained from participating in the clinics. One hundred percent of the evaluations returned gave positive feedback on the personal opinions of the participants about the clinics. Ninety-three percent of participants had their water tested after attending the information session part of the clinic. In addition to enjoying the clinics, and having their water tested, there were also questions about the interactive learning that played out in the clinics. Sixty-five percent of participants decided that they will be more careful about activities that happen within fifteen meters (fifty feet) of their wells. This shows not only learning, but a heightened awareness of the importance of keeping drinking water wells safe from potential contamination that would be caused by above-ground activities.

5. Discussion

For all of the testing performed since 2009, most of results that vary outside of the recommended limits are pH, Total Coliform and *E. coli*. We will take a closer look at the Mid-Atlantic region that Delaware is included within on the Delmarva Peninsula. Within this region we will examine the geological formations that run underground, where drinking water wells find their water sources. We will also examine the U.S. Geological Survey information on the geography, elevations and water quality assessments. In addition, we will place our attention on responses to survey questions asked of the Drinking Water Quality Clinic participants to lead us to some determination of the reasons for the test results as well as the success of the outcome of the testing. Below is **Figure 1** that contains an illustration of how the aquifers appear underground in the local region where water tests were performed, in addition to how wells are dug in these dif-

Table 2. Questions that well owners learned in the clinics. The table highlights few important questions regarding the well location, potential pollutants, and water treatment.

2. At a minimum, how far should sources of pollution (*i.e.* septic systems) be kept from a private water supply?

    A. At least 10 feet   **B. At least 100 feet**   C. At least 1000 feet

4. Type of drinking water standards that are associated with pollutants that cause aesthetic problems like tastes, odors, or stains.

    A. Primary   **B. Secondary**   C. Tertiary

8. Once a well is contaminated, water treatment is the only option for a homeowner.

    A. True   B. False

14. Which of the following is the #1 use of water in the home?

    A. Shower   **B. Flushing Toilet**   C. Laundry

15. Simple in home test kits are good enough for homeowners to test their drinking water supply for Coliform bacteria.

    A. True   **B. False**

16. Which of the following pollutants causes health problems if you consume too much?

    A. Hardness   B. Iron   **C. Lead**

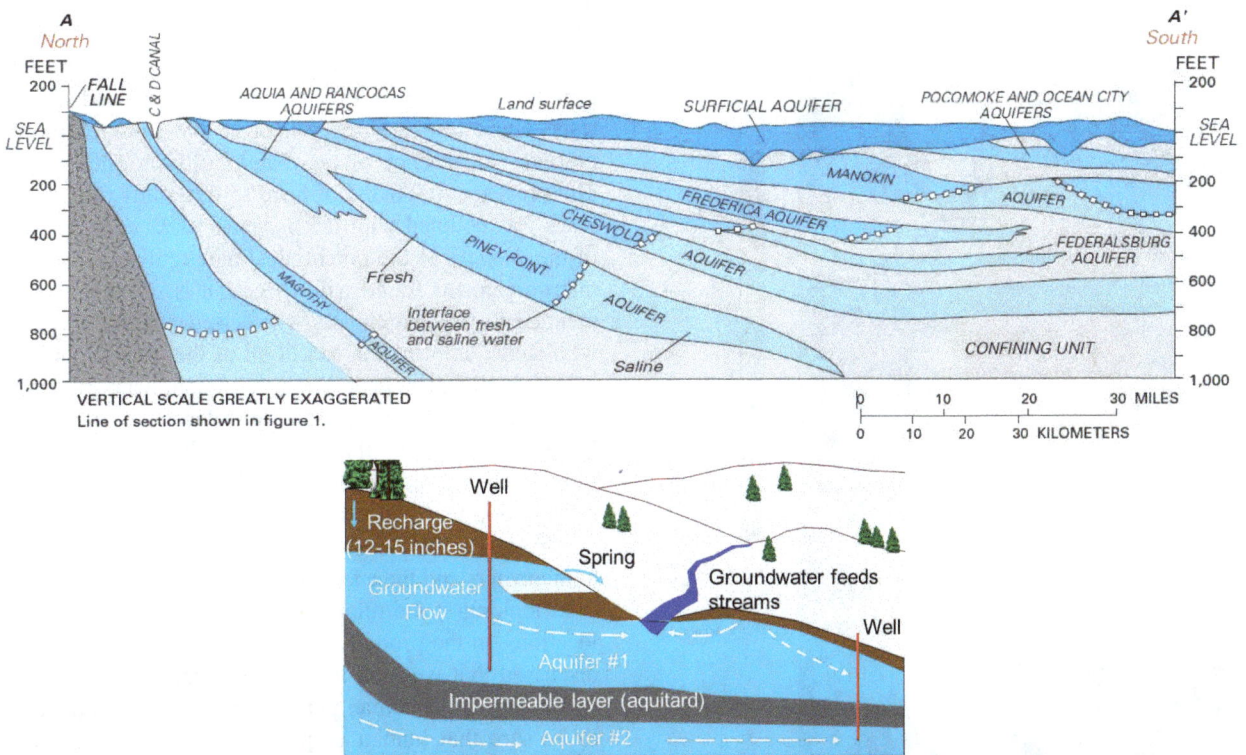

Figure 1. Sediments underlying the Delmarva Peninsula form an alternating series of confined aquifers and associated confining units, overlain by an extensive surficial aquifer that is under water-table conditions [11]. Well placement within aquifers to show how drinking water may be affected by having wells located in different kinds of aquifers.

ferent kinds of aquifers.

It is good to keep the above picture in mind in thinking of how the water may be obtained from well owners and delivered to their faucets. In addition, **Map 2** demonstrates the local elevations in the area, so that we can properly view all aspects of the geography and geology. The previous figure is a picture of the land and water below the surface. The following figure is a picture of the land and water on top of the surface.

In addition, the information obtained here must be considered in the light of where these contaminants may be originating, in order to appear in the well water of the

homeowners tested. Since most of the homeowners are in southern Delaware, **Table 3** shows sources of nutrients in the Inland Bays in southern Delaware. This would lead to an idea of where the nitrates and nitrites may be coming from. It is also not beyond reason to think that the *E. coli* and Total Coliform bacteria may be originating from agricultural sources if the homeowner does not have a septic tank nearby or on site. Of course, septic tanks may also be causing the bacterial problems.

In the view of the region from above the surface, it is easy to see that the elevations are low. This makes the people who pull their drinking water from wells more vulnerable, because the water they drink is a younger age and more susceptible to contamination [36]. Considering these factors, we must take into account the depths of the wells found in this study. Approximately half of the wells in this program were over 61 meters (200 feet). Most wells above that depth, at the shallower depths, are the wells where the drinking water may be more vulnerable. This makes an even split in the well depth information that has been obtained for this study.

Moving forward, we will examine pH, and why so many samples in the clinics with many participants showed almost half of the samples with out of range pH. When the pH samples were out of range, mostly all of the samples were acidic, below the minimum recommended 6.5 pH, rather than above 8.5 pH. A look at **Map 3** will show

how this is consistent with the pH range for the entire eastern seaboard of the US.

It can be seen how low the pH of the precipitation in the eastern United States is generally much more acidic than the west coast, and much lower the pH is than the low part of the values within range of pH 6.5 - 8.5. All of the samples that were counted as being out of range are actually directly in line with the natural precipitation, and therefore, the natural surface water aquifers of this area of the nation. There are certain times when the pH is within the range given. These samples may be affected by one of two things: 1) either the aquifer that the well is contained within is deep and old, and is not affected by surface water pH, or 2) the homeowners have a water treatment system that directly influences pH of the drinking water.

The other guideline that was out of range of the safe drinking water limits is the Total Coliform and *E. coli* tests. These tests are very sensitive and show either the presence or absence of the Total Coliform and *E. coli*. The presence of Total Coliform in this study, almost always was coupled with the presence of *E. coli*. Because of the dangers of the potential of having disease causing microorganisms found with *E. coli*, it is strongly recommended to the participants who tested positive to begin with shock chlorination to kill all of the microorganisms in their well water. In addition, we recommend following up with another test at the State of Delaware Public Health Laboratory to ensure that the bacteria has been removed. In addition, we recommend considering an ongoing treatment option that will keep them bacteria-free. The ongoing treatment will ensure the health and safety of the homeowners and their families.

In looking at the relationship between wells and their locations on the property of the homeowners, it was also discussed that we look towards the septic systems. If *E. coli* comes from human pets waste, then it is imperative to be sure that human or pets waste is not located too close to the drinking water well. There was no statistical correlation between the locations of the wells in relation to the locations of the septic systems. Only two of the wells found to have the presence of Total Coliform and *E. coli* were located closer than the recommended 15 meters

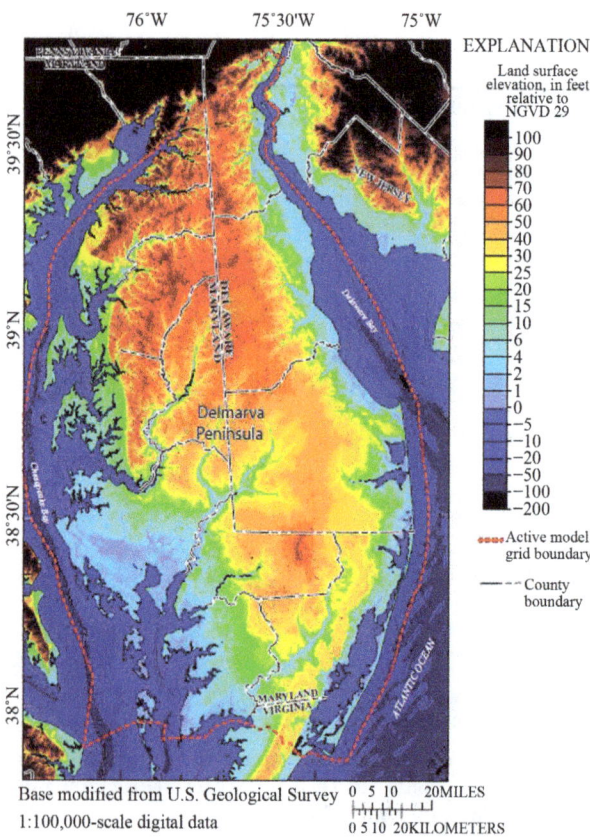

Map 2. Local elevations of Delmarva Peninsula [35].

Table 3. Nutrient sources as indicated in waters of Delaware Inland Bays [34].

Nutrient Sources	Indian River Bay		Rehoboth Bay		Little Assawoman Bay	
	Nitrogen	Phosphorus	Nitrogen	Phosphorus	Nitrogen	Phosphorus
Agriculture	44.6%	39.4%	33.0%	17.0%	54.7%	52.6%
Boating	<0.1%	<0.1%	<0.1%	<0.1%	<0.1%	<0.1%
Forest	11.0%	19.2%	7.4%	9.4%	6.7%	19.5%
Point Sources	12.5%	15.0%	27.3%	56.9%	0.0%	0.0%
Rainfall	6.2%	8.6%	8.8%	6.9%	12.8%	11.5%
Septic Tanks	16.0%	9.3%	11.2%	3.8%	14.6%	5.6%
Urban	9.8%	8.6%	11.7%	5.9%	11.2%	10.8%

Hydrogen ion concentration as pH of precipitation, 2002

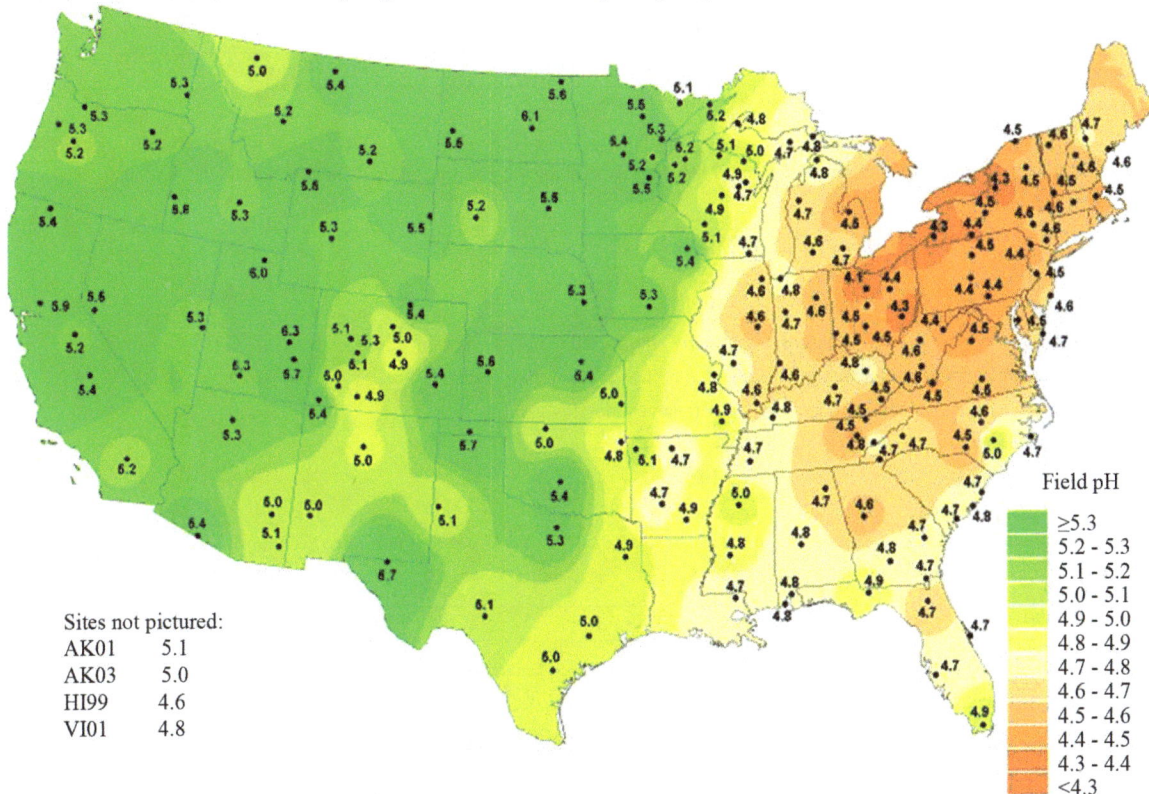

Sites not pictured:
AK01   5.1
AK03   5.0
HI99   4.6
VI01   4.8

Field pH
≥5.3
5.2 - 5.3
5.1 - 5.2
5.0 - 5.1
4.9 - 5.0
4.8 - 4.9
4.7 - 4.8
4.6 - 4.7
4.5 - 4.6
4.4 - 4.5
4.3 - 4.4
<4.3

National Atmospheric Deposition Program/National Trends Network
http://nadp.sws.uiuc.edu

Map 3. pH of precipitation in United States (National Atmospheric Deposition Program, http://nadp.sws.uiuc.edu).

(50 feet) away from the septic system. However, in the State of Delaware, it is recommended that the well be located at least 30 meters (100 feet) away from the septic system in an unconfined aquifer. The shallow surface aquifers in this local region are probably mostly unconfined, unless the wells are drilled or dig to a depth of more than 61 meters (200 feet). Approximately half of the wells tested with the presence of Total Coliform and *E. coli*. Although there is no statistical evidence that these two (well location and septic system location) are related, it is always best to be safe about the quality of the drinking water that is being consumed. It will be an area for future study and analysis to interpret all homeowners' location data and correlate this information with verifiable quantitative data regarding specific point and nonpoint sources of drinking water contaminants.

The Drinking Water Quality Clinics have been wellreceived at Delaware State University. Faculty, staff and students alike have attended the clinics to participate in the meetings and testing. In addition, the most impressive part of these clinics has been the contact and interaction with local well owners. Over 95% of the evaluations completed by participants had positive feedback and shared appreciation for the information (**Table 4**). In addition,

many participants were interested in returning to have their water tested in the future. That is the best way to protect the safety of the people who drink well water. At least once a year, it is important to have the water tested to ensure there are no health concerns floating around in the drinking well water.

6. Acknowledgements

We would like to thank the following individuals for their assistance during the workshops and/or water testing clinics: Frank Marenghi, Brian Reckenbeil, Benjamin Reining, Kenneth Hannum, BalajiBabu, RajuKhatiwada, and Dr. Lathadevi Karuna Chintapenta of the Aquatic Sciences Program in Delaware State University. We would like to extend our sincere thanks to Ms. Corrie Cotton and Mrs. Daphne Pee of the Mid-Atlantic Water Program EPA Region 3 for their support throughout the program. We also would like to thank Mrs. Erin Ling and Dr. Brian Benham of the Master Well Owner Networking Program at Virginia Tech for their support and providing information and resources for us to share with the citizens. Finally, we would like to thank Mrs. Troy Darden and Mr. Pablo Mojica for their assistance with the program advertisements, flyers, and posters. Last, we are grateful to

Table 4. Clinic evaluation form. This is the exact form that participants filled out to share their opinions of the clinics, as well as what was learned or taken away from the clinics.

EVALUATION

---

1. Circle the word below that best describes your opinion of today's program.

| Very | Poor | Poor | Fair | Good | Very | Good | Excellent |

2. After attending today's program, check any of the things listed below that you are <u>planning</u> on doing to your well in the next six months.

_____I plan on getting my water tested

_____I plan on shock chlorinating my well or spring

_____I plan on having a new well drilled

_____I plan on installing a sanitary well cap on my well

_____I plan on inspecting my well to ensure it is properly constructed to prevent contamination

_____I plan on being more careful with activities within 50 feet of my well or spring

_____I plan on seeking more information about water treatment devices

_____I plan on purchasing a water treatment devices

_____I plan on taking some other action (please specify) _____

_____I don't plan on taking any actions on my water supply

3. Please use the space below to give general comments about the Safe Drinking Water Program or to recommend topics for future Cooperative Extension Programs.

_____

_____

_____

_____

THANK YOU FOR YOUR COMMENTS!

Dr. Asmare Atalay of Virginia State University for providing the funding for us to continue our program activities. This program is funded by USDA-MAWP for Master Well Owner Network (MWON) Program, USDA-NIFA and USDA-NIFA CBG for 1890 Water Resource Center Program.

REFERENCES

[1] Water Sanitation Health, 2013. http://www.who.int/water_sanitation_health/dwq/en/

[2] International Network of Drinking-Water Regulators, 2012. http://www.who.int/water_sanitation_health/dwq/RegNet/en/index.html

[3] Countries, 2013. http://www.who.int/countries/en/

[4] E. Ling, S. Clemens, G. Ozbay, C. Cotton, B. Benham, and D. Pee, "The Mid-Atlantic Master Well Owner Network: Educating Private Water Supply Users and Protecting Ground Water,"2012. http://www.usawaterquality.org/conferences/2011/sessions/presentations/A/Ling.pdf

[5] Environmental Protection Agency Website, "Private Drinking Water Wells," 2013. http://water.epa.gov/drink/info/well/

[6] Environmental Protection Agency, "Pamphlet. Drinking water from Household Wells," EPA 816-K-02-003, 2002.

[7] Water: Private Wells. Basic Information. 2013. http://water.epa.gov/drink/info/well/basicinformation.cfm

[8] C. Renshaw and J. T. Trippe, "Fresh-Water Cistern," US

Patent and Trademark Office, Washington DC, US Patent No. 3, 1970, pp. 517-513.

[9] National Water Program Outcome Report. Projects of Excellence. A Partnership of USDA CSREES & Land Grant Colleges and Universities. 65 pages (In page 8), 2010.

[10] E. Ling, B. Benham, S. Clemens, C. Cotton and G. Ozbay, "Mid-Atlantic Regional Master Well Owner Network. National Water Program. A Partnership of USDA NIFA & Land Grant Colleges and Universities," Land Grant and Sea Grant National Water Conference Abstracts and Proceedings, Washington DC, Received 2010 Projects of Excellence.

[11] J.M. Denver, S. W. Ator, L. M. Debrewer, M. J. Ferrari, J. R. Barbaro, T. C. Hancock, M. J. Brayton and M. Nardi, "Water Quality in the Delmarva Peninsula, Delaware, Maryland, and Virginia, 1999-2001: Reston, Va.," US Geological Survey Circular 1228," 2004, 36 p.

[12] S. Clemens, "How to Interpret Water Test Reports: A Guide for Private Well Owners," Pennsylvania State University, University Park, 2009.

[13] EPA, "Water: Monitoring and Assessment. 5.4 Ph. What Is pH and Why Is It Important?" 2013. http://water.epa.gov/type/rsl/monitoring/vms54.cfm

[14] EPA, "Water. Basic Information about Regulated Drinking Water Contaminants. Basic Information about Disinfectants in Drinking Water: Chloramine, Chlorine and Chlorine Dioxide," 2013. http://water.epa.gov/drink/contaminants/basicinformation/disinfectants.cfm

[15] D. Varner, S. Skipton, P. Jasa and B. Dvorak, "Drinking

Water: Sulfates and Hydrogen Sulfide," University of Nebraska-Lincoln Extension, Institute of Agriculture and Natural Resources. Publication Number: G1275, 2004.

[16] Environmental Protection Agency Website, "Basic Information about Nitrite (Measured as Nitrogen) in Drinking Water," 2013. http://water.epa.gov/drink/contaminants/basicinformation/nitrite.cfm#three

[17] P. D. Robillard, W. E. Sharpe and B. R. Swistock, "Nitrates in Drinking Water. Penn State. College of Agricultural Sciences, Cooperative Extension, Agriculture and Biological Engineering, University Park," 2001.

[18] Environmental Protection Agency Website, "Total Alkalinity. What Is Alkalinity and Why Is It Important?" 2013. http://water.epa.gov/type/rsl/monitoring/vms510.cfm

[19] B.R. Swistock, W.E. Sharpe and P.D. Robillard, "Treating Coliform Bacteria in Drinking Water. Penn State. College of Agriculture Sciences. Cooperative Extension. Agricultural and Biological Engineering, University Park," 2001.

[20] Palintest Test Instructions for Hardness, PM 254, AP 254 AUTO.PHOT.15.AUTO, 2005.

[21] J. T. Creed, T. D. Martin and J. W. Odell, "Method 200.9. Determination of Trace Elements by Stabilized Temperature Graphite Furnace Atomic Absorption," Revision 2.2. Environmental Monitoring Systems Laboratory, Office of Research and Development, US Environmental Protection Agency, Cincinnati, 1994.

[22] H. Perlman, "Water Properties: pH. US Department of the Interior, US Geological Survey. The USGS Water Science School," 2013. http://ga.water.usgs.gov/edu/ph.html

[23] Encyclopædia Britannica, "Encyclopædia Britannica Online. Encyclopædia Britannica Inc.," 2013. http://www.britannica.com/EBchecked/topic/416124/nitrite

[24] Environmental Protection Agency Website, "Basic Information about Nitrate in Drinking Water," 2013. http://water.epa.gov/drink/contaminants/basicinformation/nitrate.cfm#six

[25] P. D. Robillard, W. E. Sharpe and B. R. Swistock, "Water Softening. Penn State. College of Agricultural Sciences. Cooperative Extension. Agriculture and Biological Engineering, University Park," 2001.

[26] Environmental Protection Agency Website, "Sulfate in Drinking Water," 2013. http://water.epa.gov/drink/contaminants/unregulated/sulfate.cfm

[27] Environmental Protection Agency Website, "Secondary Drinking Water Regulations," 2013. http://water.epa.gov/drink/contaminants/secondarystandards.cfm

[28] K. Mancl, P. E. Sailus and L. Wagenet, "Private Drinking Water Supplies: Quality, Testing, and Options for Problem Waters," Cooperative Extension Publication NRAES-47. Northeast Regional Agricultural Engineering Service. Cornell University, Ithaca, 1991.

[29] Environmental Protection Agency, Office of Water, "Water on Tap: What You Need to Know," Publication Number: EPA 816-K-03-007, 2003, 32 Pages.

[30] P. Feng, S. D. Weagant, M. A. Grant and W. Burkhardt, "Enumeration of Escherichia coli and the Coliform Bacteria," Bacteriological Analytical Manual, Chapter 4, 2002.

[31] Environmental Protection Agency Website, "Basic Information about Lead in Drinking Water," 2013. http://water.epa.gov/drink/contaminants/basicinformation/lead.cfm#content

[32] Environmental Protection Agency Website, "Basic Information about Arsenic in Drinking Water," 2013. http://water.epa.gov/drink/contaminants/basicinformation/arsenic.cfm#content

[33] Environmental Protection Agency Website, "Basic Information about Cadmium in Drinking Water," 2013. http://water.epa.gov/drink/contaminants/basicinformation/cadmium.cfm#content

[34] Delaware Inland Bays Estuary Program (DIBEP), Delaware Inland Bays Estuary Program Characterization Report. Science and Technical Advisory Committee, Dover, Delaware, 1993, pp. 1-3.

[35] W. E. Sanford, J. P. Pope, D. L. Selnick and R. F. Stumvoll, "Simulation of Groundwater Flow in the Shallow Aquifer System of the Delmarva Peninsula, Maryland and Delaware," US Geological Survey Open-File Report 2012-1140, 2012, 58 p.

[36] S. Eberts and D. Hebert, "Studies Reveal Why Drinking Water Wells Are Vulnerable to Contamination," 2013. http://www.usgs.gov/newsroom/article.asp?ID=2403

Note List of Main Abbreviations

EPA or USEPA—United States Environmental Protection Agency

pH—Power of Hydrogen. Used as a measure of molar concentration of hydrogen ions in a solution.

E. coli—Escherichia coli

WHO—World Health Organization

MWON—Master Well Owner Network

Mg/L—Milligrams per Liter. A measure of concentration of analyte in a solution.

MCL—Maximum Contaminant Level

SMCL—Secondary Maximum Contaminant Level

6

Urbanization and Its Impacts to Food Systems and Environmental Sustainability in Urban Space: Evidence from Urban Agriculture Livelihoods in Dar es Salaam, Tanzania

Wakuru Magigi

Moshi University College of Cooperative and Business Studies (A Constituent College of Sokoine University of Agriculture), Moshi, Tanzania.

ABSTRACT

Urbanisation is the key factor underpinning and catalysing changes in food systems, environmental quality, climate change and agriculture livelihoods in the overall urban ecosystem setting and its sustainability. The paper explores Dar es Salaam, a rapidly expanding city in Sub-Saharan Africa, and shows that urban agriculture provides urban ecosystem services and contributes to environmental sustainability. The interconnections of environmental justice, urban ecosystem services and climate change and variability found eminent feature that influence land governance, productivity and aesthetic value of the city. The study reaffirms the pivotal role urban agriculture which plays to enhance community health services and access to resources, with important implications on urban environmental sustainability and redistributive spatial land use planning policies and practices. The process of urbanisation, forms of urban agriculture and government strategies for enhanced urban food systems in the city economy have been highlighted. Equally, the process triggers the transformation of settlements from rural in character to modernity with an augmented land use conflicts. The results suggests that with increasing population, a clear spatial land use planning and management strategy is required to over- come the challenges and enhanced food systems and urban environmental sustainability in rapidly urbanizing cities like Dar es Salaam in Sub-Saharan Africa.

Keywords: Urbanization; Food Systems; Environmental Sustainability; Urban Space; Urban Agriculture; Livelihoods

1. Introduction

Urbanisation is the key factor underpinning and catalysing changes in land use, land transactions, increased rural-urban immigration and the overall urban agriculture land use in Dar es Salaam city. Unregulated urban agriculture implementation in urban and periurban land was found predominantly expanding in an unorganised manner in the city. Other factors include a reluctance of local and central government planning institutions to include the sector in land use plans for improved urban ecosystems interconnections with environmental conservation needs. The emerging urban agricultural land use pattern, by and large, indicates a mismatch with the widely cherished planning norms and standards and land market[1].

[1]The land market was observed to influence land allocations and resident choices and is therefore an important factor in urban agriculture integration in and use planning practise for improved system.

Urban land use planning instruments such as zoning and density distribution and principles including equitable provision of basic services in urban land development are important ingredients for integrating urban agriculture in land use planning processes for improved urban ecosystems and food security.

The urbanisation process is accompanied with expansion of the city boundary, which engulfs periurban settlements. The process of expansion of the city boundaries is resulting in periurban settlements coming within the city's "zone of influence". This is increasing interaction of residents including smallholder farmers' access to the city economy, in terms of capital outsourcing from relatives and friends. Similarly, it helps the provision of labour and exchange of goods and services. This is one fact of the villages, from rural in character to periurban char-

Urbanization and Its Impacts to Food Systems and Environmental Sustainability in Urban Space: Evidence
from Urban Agriculture Livelihoods in Dar es Salaam, Tanzania

63

acter where conflicts in resource use becomes common. Likely, influence capital earnings which encourage home ownership through housing construction, where friends and relatives can inhibit when they arrive in the city for social obligations. Changing demography and particularly the changing age structure of the population, a high rate of urbanization, and a faster rate of population growth in relation to economic growth are major drivers of environmental change in Africa, with significant impacts on the natural resource base. This indicates the rationale of understanding the rural-urban symbiotic relations existing between the city and its hinterlands for sustaining livelihoods of the people and urban environment.

The symbiotic relations existing in nature observed to be economically and ecologically viable in terms of investment attractions including housing development, greening the city and opening up of social services. Principally, urban expansion lowers the amount of land potential for periurban agriculture in the city growth processes. The reduction of land for this sector appears to intensify poverty of smallholder farmers who depend on it as a livelihood strategy and increased environmental degradation. This suggests the need for urban planning institutions, after land declaration to establish a clear development conditions to safeguard urban agriculture forms which play a great part to ensure city greening, ecological conservations and food systems [1]. Food system in the context refers to process of households to have a substantial amount of food which include low cost in its production sphere throughout the year. Each year, as the population increases in the city with climate change impacts on crop production, food prices and the amount of natural resources, which is going to sustain this population, to improve the quality of lives and to eliminate poverty remain finite and therefore increasing the challenge of sustainable development achievements.

Urban Planning and Land Declaration Need

Population growth presents a major challenge because of the patterns of production and consumption that shape the world, as well as the problems of pervasive poverty. In adapting to this challenge urban land use planning need to be accommodative. This among others means the process of land declaration and zoning in urban land use planning system needs to be transparent and involve different actors with interest in land property development. The involvement of actors in land declaration and subsequently, decision making, preparation, implementation and monitoring ensures urban environmental sustainability. Otherwise, the non-involvement of these actors in planning processes lead to difficulties in institutionalising, coordinating and controlling urban agriculture prac-

tice, effective resource utilization and misuse as well as difficulties in ensuring monitoring and evaluation of land use plans once implemented.

Conflict arising from landholders, fragmented types of cadastral survey prohibitive costs and increased housing densification are some features of ineffective land use planning. In addition, presence of land use conflicts and housing density results in decreased food production and they affect urban food systems. Increase price of various food types is an apparent feature in the city which affects the city poor and shifting to the periurban is the only alternative remain for the poor for survival in African cities. This is affecting residents' nutrition levels and subsequently labours productivity within the settlement and the city within environmental justice measure. Other factor enhancing the shift includes changes of the land market dynamics and prohibitive raise of life costs.

It is imperative to note that, the [2] shows that the 21st Century is the century of the city, although a significant portion of humanity's impact on the global environment originates in urban areas. It continues showing that there are also opportunities to mitigate and adapt to global environmental change through urban planning processes. Half of the world's population already lives in urban areas and by the Middle of this century, most regions of the developing World will be predominantly urban. In supporting this, [3] estimates that today 33% of Africa's population lives in cities and by 2050 this will be 60%. The report adds that Globally 800 million people are engaging in urban agriculture producing 15 - 20 percent of all food. More people noticed are employed in processing, marketing and transporting. These arguments unravel the complexities of these relationship is a demand towards enhancing food systems and urban environmental sustainability in cities of Africa [4].

2. Methods and Materials

The study analyses the urban agriculture practise Goba, Chang'ombe "A" and Ubungo Darajani in Dar es Salaam City. Various forms of urban agriculture including vegetable production are analysed with special emphasis on understanding the questions of food security, urban poverty, environmental conservation and urbanization consequences in urban development processes. A total of 450 stakeholders were interviewed during this study.

The selection of the case studies was guided by various factors including a long and outstanding settlement development history, rapidly urbanizing settlements in terms of housing, social demographic changes and presence of agricultural activities. Both preliminary field data collection, desk search and on spot observation were methods deployed in data collection. Key informant interviews were carried out in the selected cases so as to establish the historical evolution of agriculture activities, food supply, markets, resource use, prices, land use planning practices, challenges, planning institutions roles and linkages to climate change, environmental conservation and urban ecosystems Data collected were analyzed using Statistical Package for Social Science (SPSS) and Map info Software. Throughout the analysis, data were differentiated regarding member's perceptions, activities, institutions arrangement and their roles and linkages, nature of activities, potentiality and negative impacts of agricultural activities for improved livelihoods of the residents in the study settlements and urban ecosystems environments.

3. Results and Discussion

Debate on sustainable human settlements spins on understanding and facilitation process of involving various planning institutions in reducing health burden, chemical and physical hazards, achieve high quality urban environment, increase city productivity and Foreign exchange earnings in the urban land management system and governance. It should also strive to achieve sustainable consumption, ensure food security and minimizes transfer of environmental burdens in urban development and ecological systems. Urbanization is central to global environmental change among other factors that is necessary to be an integral part in understanding urban sustainability, urban ecosystems and towards transition towards sustainability.

3.1. Forms and Locations of Urban Agriculture in the Study Settlements

Different forms and locations of urban agriculture exist in Dar es Salaam city. The urban agriculture provides income, nutrition and often a safety-net function to the poorest sectors of society and therefore enhances urban food systems, access to resources and social services within environmental justice perspectives. 65% of the farmers in both settlements studies reported using their farm product for subsistence while the rest used it for income earning and for giving to their relatives and friends. The selling takes place in the producer's home, and city markets including Kariakoo, Tandale, Temeke sterio, Makumbusho. The distance to market areas limits access of smallholder farmers to selling their perishable vegetable products to the retailers. These retailers buy from smallholder farming premises and sell both in the streets and in the big markets within the city. The long distance traveled by these retailers fetching vegetables from the farming premises to the market in the city is one of the voices of smallholder farmers. This reduces their profits, as they have to balance the cost of buying and selling. This indicates the need for improving the transport facilities in periurban areas to ensure the accessibility to consumers including to market places and enhance food system flows to residents. This also appears meeting the [5] explicitly recommends that; *A paradigm shift in design and urban planning is needed that aims at reducing the Distance for transportation food by encouraging local food production, where feasible, within city boundaries and especially in immediate surroundings.*

In addition, opening new markets in periurban settlements helps in reducing traveling distance and also in reducing congestion in the city during daytime when vehicle traffic jams to the city center increases. As such, urban agriculture is an important vehicle for poverty reduction in Dar es Salaam and other cities in Sub-Saharan Africa. However, the forms of urban agriculture practiced in the city are documented in **Table 1**.

Exposing urban agriculture to the public the knowledge, risks and way for sustaining the activity in relation to resource management is important. Respondents involving in floriculture and tree seedling contribute much to city environmental conservation and recreation. For example, floriculture and other canopy trees provide areas for resting and aesthetic value of the settlements studies. The problem among others includes lack of designated areas for ornamental tree, floriculture and seedlings in the city, which calls for more intervention in planning

Table 1. Farming systems common to urban areas in Dar es Salaam.

Farming System	Product	Location and Markets
Aquaculture	Fish and Seafood, vegetables, fodder	Fish ponds, river streams, sewerage and wetland: Within the city
Horticultural	Vegetables, fruits, compost	Home lots, parks, right-of-ways, wetlands, railway lines and undeveloped and unfinished plots and houses
Ornamental	Flowers, Tree	right-of-ways, Home lots
Animal Husbandry	Milk and Eggs, Meats, Manure, Skin and hides	Zero grazing, right-of-ways, steep and gentle slopes, open spaces, urban fringe areas
Poultry Keeping	Egs, Meats, Manure, income	Zero grazing in home lots
Agri-forestry	Fuel, fruits and nuts, compost, building materials	Home sites, steep and gentle slopes, wetlands, forests parks, urban forest zones

Urbanization and Its Impacts to Food Systems and Environmental Sustainability in Urban Space: Evidence
from Urban Agriculture Livelihoods in Dar es Salaam, Tanzania

65

and design for enhancing environmental conservation, social services such as recreation and income levels of actors involving in the sector. The use of road strip for cultivation is common and is expanding smallholder farmers' vulnerability to risks of eviction by the city council. In addition, the further the smallholder farmer's proximity in the city centre the more vulnerable they become compared to the periurban farmers. The former, accompanied with city harassment and fear of eviction, which proved to destruct the urban environment and create a nuisance.

3.2. Urban Agriculture Requirement in the City and Interconnection with Environmental Change and Urban Ecosystem

Urban agriculture as a land use function requires land, water, labour, inputs (seeds, fertilizers etc). The abundant and adequacy of these resources are limited in urban areas and therefore pressure on it and to climate change rehabilitation and adoptability becomes apparent feature. Farming systems adopted by city residents has a adverse impact of urban environmental and health of the residents. For example, crop rotation and mixed farming adopted by smallholder farmers in the three study settlements help increase their productivity and at the same time conserve land. In case of Goba settlements, it has been noted that local leaders already established socially regulated norms of not clearing bushes along the river sources to conserve the soil. The leaders argue that vegetable and other annual crop farming exposes the soil after bush clearing. Thus, the enhanced deforestation resulted in soil and climate changes. One landholder in Goba settlement remarked:

"···in the 1990's we experienced cold and pleasing air in our settlement throughout the year, today weather has changed and it is too hot almost throughout the year. I think this is caused by bush clearing for residential and urban farming land use in our settlement."

The statement shows bush clearing for farming has resulted in changes of the micro-climate. This reinforces the need for environmental conservation when urban agriculture strategies are to be accommodated in the city setting. Equally, it is the author's view that such a statement may only show a subjective perception or respondent. In addition, in both the study settlements, crop rotation of amaranth farming is popular. The practise involves land preparation, planting, weeding and harvesting processes. The majority of respondents reported to produce vegetables crops, which suffer from fewer diseases (Goba 90%; Mabibo-72% and Chang'ombe "A"-68%).

In both settlements environmental risks were associated with poor management of inorganic wastes, the use of wastewater and the use of pesticides. The use of wastewater by some smallholder farmers and retailers before selling causes consumers to fear the health aspects associated with urban farming. How hygienically safe can this be? These sellers of green vegetables (mchicha) are washing their wares in the heavily polluted waters of the Mabibo River-Kibangu in Dar es Salaam. The water from this river is used in different sections of vegetable growing, especially during germination and in three weeks before harvesting. 87% of smallholder farmers show that even though are using this water, they used to leave using the water to one week before harvesting. One can wonder if the water in these rivers is polluted and to what extent? And if polluted, the time the smallholder farmers left without using the water as a preparation for harvest, does it have any health impacts? This shows training for smallholder farmers involved in farming product especially vegetables crops, as well more intervention on the raised question to clear fears voiced by consumers in the city for improved urban environmental justice. This means when people are trained will have equal awareness and see their potential in their involvement of all people regardless of race, color, sex, national origin, or income with respect to the development, implementation and enforcement of environmental laws, regulations, and policies in regard to urban agriculture implementation and legislative framework applications

3.3. Problems Affecting Urban Agriculture in the City

Urban agriculture faces a wide range of constraints in its production sphere. These include pest attacks, adverse weather conditions and timely access to inputs such as seeds and pesticides. Others include use of the suggested polluted river water for irrigation and land subdivision for residential premises. These are causing decline in the quantity, quality and safety of vegetables expected by consumers. Apart from that limited extension services to smallholder farmers to educate them on sustainable urban farming are hardly provided and therefore increases disease vulnerability. Respondents from Goba-64%, Ubungo Darajani-70% and Chang'ombe "A"-67% confess inadequate of extension services in their areas. Extension services provision to smallholder farmers, ensure information flows, marketing and dissemination of new innovation are factors appeared that could enhance their productivity in the city. Urban Agriculture livelihood found that the large proportion of producers rely on buyers for market information

Besides, fear of eviction of the respondents by the government is common and in one way affects crop cultivation in periurban areas such as Goba, which was declared as ripe for urban development. Problems of insecure to access land as an input plagues urban smallholder farmers almost everywhere in the city where different forms of urban agriculture practice discussed are conducted. Specific land problems arise because by-laws and regulations prohibit food production of certain types especially livestock rearing, but sometimes crops. Smallholder farmers are often pushed out of their land and it is taken for residential and commercial purposes. This affects the smallholder farmers who want to engage in improving their farm lots for social and economic benefits but lacks funds for improving their plots in urban areas as the study settlements shows.

3.4. Environmental and Climate Change Impacts Associated with Urban Agriculture

It was difficult for smallholder farmer respondents in the three cases to give a proper answer on how much sucks or harvests they get in quantity per month. This was due to a lack of record keeping that could enable them to understand how much he or she harvested per day, month or per year. The knowledge observed from smallholder farmers are calculations done through number of ridges they have and the cost of seeds purchased during germination period. 52% of smallholder farmers even forget how many lots or terraces they have harvested per season. The rest seem to remember number of lots in their heads with no record keeping. Similarly the question of labour and time spent in farming are not counted as they do not regard the activity to be a permanent employment and livelihood strategy. This shows that more training is needed in record keeping and costing inputs used in farming for the betterment of farmers.

The use of wastewater for vegetable farming in the study settlements is a common practise. Both domestic and industrial wastes are flowing in river channels, which are in turn used for urban vegetable production. According to [6], in countries where poor sanitation and hygiene

conditions exist and untreated waste water and excreta are widely used in agriculture, intestinal worms pose the most frequently encountered health risk (**Table 2**).

Decision makers including planners have negative altitude toward urban agriculture practise. Most of planners are questioning about the health aspects resulting from the use of waste water which seems to be heavily polluted from rivers for farming and washing products. Likewise, urban agriculture has been blamed for increasing malaria cases in cities, which according to recent studies is questionable[2]. Wastewater treatment is important[3], but may be difficult since even domestic wastes water and sanitation systems in the settlements are directed to these rivers, which increases more health risks of consumers. Thus, upgrading of these settlements and training of smallholder farmers can be important for improving their livelihood including urban food system in the urban setting increases the fairness of decisions to achieve environmental justice objectives.

Climate changes, water availability and perishable nature of amaranth vegetable products cause it to be seasonal activity in the city. This affects the sells and income level and adds pressure on water use during the dry season. The organisation of the sector in a proper land area becomes important where infrastructure facilities can be provided, and monitored and profits to the majority ensured. It is through land use planning that this can be achieved and land can be properly allocated to add value to agricultural land.

3.5. Environmental Justice Effects on Policy within the Communities

According to land development policy, the fragile and hazardous land needs to be zoned for conservation. This includes being conserved by planting trees to prevent erosion and land degradation. The reality observed is that these areas have been encroached upon by smallholder farmers and people developing their settlements. Inadequate policy and legal enforcement to guide these haz-

Table 2. Summary of health risks associated with the use of wastewater for irrigation.

Group exposed	Health threats		
	Helminths	Bacteria/viruses	Protozoa
Consumers	Significant risks of helminth infection for both adults and children with untreated waste water	Cholera, typhoid and shigellosis outbreaks reported from use of untreated wastewater; seropositive responses for Helicobacter pylori (untreated); increase in non-specific diarrhoea when water quality exceed 10 exponent 4 thermotolerant coliforms per 100ml	Evidence of parastic protozoa found on wastewater-irrigated vegetable surfaces, but no direct evidence of disease transmission

Source: [6].

---

[2]Study conducted by [7] in the city of Dar es Salaam shows that high number of Malaria parasites are not associated with urban agriculture, instead poor drainage and sanitation system are the cause.

[3]See also [8].

ardous lands as well as controlling housing development in these areas seems to be a major problem for smallholder farmers. Particularly, problems are resulting from continuous land subdivision for construction purposes, which limits urban agriculture practise. Smallholder farmer respondents (*i.e.* 72% in Goba, 78% in Ubungo Darajani and 84% in Chang'ombe "A") show that they were not aware of the use of such land and restrictions behind it. Likewise, no land use plans showing the use of that land are put in place at local level. Apart from that neither local leaders nor local authority staff informed them on such use if any. This indicates inadequate communication and information flows in coordinating and controlling land development in the city.

Farmers are handling the manure and composite on their farms without awareness of the health risks associated with faecal contaminations. Using manure and composite wastes can affect human health through contaminations, and therefore increases residents' vulnerability to diseases. This is caused by the handling of manure on their farm lots as the study shows.

Although the quantity of food produced by city farming does not match up to that outside the city, its impact is quite considerable when looking at its social aspects, especially in terms of its employment, nutritional values and food security for the families. It is the authors' view that although the national and municipal policy makers and urban managers do not acknowledge this important role, we need to consider it in modern urban planning and design decisions. This can help to minimises several challenges posed to urban farming including understanding if urban land is either not available or not accessible; and when available, it is suitable or not suitable for farming in designated sites. The market managers show that one aspect among others that the urban planners and other stakeholders doubt is the existence of vegetable wastes in market areas and in residential premises. They confirm that the agriculture wastes are beneficial. The wastes are easily decomposed and are used in existing farms and therefore employ people and at the same add nutrients to soils used for farming within the city.

According to [9], urban agriculture is classified in Use class P which according to this Act requires special treatment in its sitting. Using the same Ordinance, section 78 empowers the Minister responsible for planning and for putting in place regulation to protect land development. This shows that the [9] and use class P have been redefined to enable their integration in policy implementation. However, though this shows recognition of urban agriculture in policies that implement legislative framework, it does not provide prominence that would allow useful implementation. For example, it has been difficult to know which land type is suitable and how it can be inte-

grated in other land uses including residential areas in both urban and periurban settlements. Due to the importance and contribution of urban agriculture as seen in this context, location, type and scale determinants are preferred for proper coordination and control. Similarly, use of local leaders including deployment of Subward and Ten cell leaders in coordinating and controlling such development are vital to ensure sustainable use of land and enhancing urban ecological systems.

3.6. Challenges Influencing Urban Agriculture Setting and Production in Connections with Climate Change, Environment and Urban Ecosystem Services

3.6.1. Contractual Relations Are Very Important to Build Trust

In the three settlements explored, people access land for urban agriculture and other land use function through buying, inheritance and granted right of occupancy. The study showed that 78% of the respondents interviewed are not aware of official procedure requirements for getting land for urban agriculture. This is because they have little contact with government agencies and the English language is used for these policy and legal documents. For example, planning regulations, standards and administrative procedures are published in English while the majority of landholders and tenants cannot read or speak English (92%) in all settlements explored. Only 8% of smallholder farmer respondents could speak and write English. In this regard use of the Swahili language to communicate these procedures would be preferable.

The granted right of occupancy has built confidence in the smallholder farmers involving in plot farming in the study settlements. These farmers understand that they are secure in terms of not being evicted by the local authority, or must receive high compensation value in case of eviction or compulsory land acquisition for public interests. These farmers are aware of the by-laws guiding urban agriculture particularly on restricting high raised crop farming such as maize in their plots in Ubungo Darajani and Chang'ombe "A". This has increased vegetable, mushroom and floriculture farming for improved income and social well-being.

In the three settlements examined, smallholder farmers were not well organised. The smallholder farmers understand the potential of their farming activity for improving their social well-being, but they have not united to form an organization. This is also augmented by different tenure status, reasons for farming and plot size. For example, some farmers depend on only farming for their household subsistence while others for petty business. Some farmers observed are secured in terms of land ownership while others are not. Similarly, other smallholder farmers use inplot farming while others are offplot including a fragile land. Forming an organization can enhance their rights to use the land in whatever tenure to be granted and to negotiate on land user rights for increasing their productivity. Correspondingly, this will help to present their feelings and perceptions in improving the sector to politicians including Ward Councilors who are the policy makers and representative in their area. The link of smallholder farmers with this political leadership is minimal, which enhances marginalisation of smallholder farmers. The organization of smallholder farmers would help fight for better prices for their by-products and other matters in the production sphere, and would improve urban land governance for city development.

3.6.2. Inadequate and Inaccessibility to Information and Transparent

Access to information, communication and credit facilities are resources that are important for improving smallholder farmer welfare. The study indicates that smallholder farmers are experiencing inadequate access to land, information, delivery systems and rights to get involved in land use planning. This limits the improvement and understanding of their rights available opportunities, and interaction with planning institutions. This results resulting in difficulties in decisions undertaken at different levels, which affect urban agriculture livelihoods, needs for food storage system and maintaining a range of urban food supply sources, which are important elements is sustaining lives of poor resourced persons and achieving environmental justice.

The availability of credit for urban farming including amaranth farming activities is one factor that can pull the respondents out of poverty and improve urban land governance. 93.8% of respondents in the study settlements had no access to credit. Only 2.2% of respondents in Goba reported having received credit of up to 400 USD. In Chang'ombe "A" and Ubungo Darajani settlements the figure is only 4%. This indicted the need for putting in place and strengthening credit scheme access that can help to improve the lives of smallholder farmers. Likely, insecurity of land tenure as collateral enhances this. This character seems common in other African cities including Nairobi and Uganda as reported by [10].

The system of access to land includes both statutory and customary tenure occur for a number of reasons. In some cases, the original settlements were developed under traditional or customary tenure systems on land which was at the time outside urban administrative boundaries, but was later absorbed within them as the urban areas expanded. At such times, various statutory tenure categories, such as formal granted right of occupancy have been introduced, to which later construction had to conform.

The cases of Chang'ombe "A" and Goba show this. Alternatively, early developments may have changed their tenure status as legislation changed, or regularisation programmes enabled previously informal settlements to become legal as in the case of Ubungo Darajani.

In addition, 6% of smallholder farmer respondents owning land in the three study settlements were observed to have *short-term titles* of 2 - 21 years offered by Local authority. This limits smallholder farmers' access to financial institutions for credit. For instance one landholder said:

"···I have got a title deed, but I have been constrained by the short term nature of title given (i.e. 2 years). Credit institutions refuse to provide a loan arguing that the title has low betterment value."

This quote shows that even though the residents are aspiring to get titles as a tool for asset betterment, the lease is too short and unattractive to credit institutions. Therefore, there is a need to consider granting a long-term title *i.e.* 33, 66 and 99 years leasehold to attract financial betterment to smallholder farmers who want to legalise their land towards poverty for community development. Providing a range of tenure options could be the most effective means of enabling the urban poor to improve their living conditions and livelihood opportunities and therefore improve urban land governance. In addition, existence of both customary and granted right of occupancy by having both urban and village administration working together need more intervention including deregistration of the later.

Increasing housing density

Demand for land for housing and other urban investment including industrial construction is increasing in the city and the settlement in particular. Informal land transaction and parcelling, land subdivision in mixed residential areas including the study settlements has caused the decrease of agriculture land and encroachment upon what would be public leisure spaces. All this results from increasing investment in the city and homeownership needs of urbanities, where each needs to have a decent house in the city. This calls for urgent action of the local authorities and other planning authorities to collaborate and control the increasing density resulting from informal land transactions taking place in the study settlements and the city as a whole. If this is unchecked, the future urban environment can be degraded and therefore endanger the fate of the incoming generation.

Public open space utilisation

In the planned parts of Goba, Ubungo Darajani and Chang'ombe "A" where urban land use plans exist, urban agriculture is dominant and it is practiced in plots and offplots. In parts of these settlements, informal settling exists which seems to encroach upon unplanned open

Urbanization and Its Impacts to Food Systems and Environmental Sustainability in Urban Space: Evidence
from Urban Agriculture Livelihoods in Dar es Salaam, Tanzania

69

spaces, which are zoned as conservation lands or hazardous lands. Development control of the open spaces is hardly done and therefore allows soil and environmental degradation, and increasing land use conflicts. Checking these impacts to avoid future settlement degradation is rational for sustainable use of urban land.

3.7. Planning Institutions Linkages

A notable problem is that the municipal council's operations in the city are inconsistent with the principles stated in the policy and legal documents which require integration of urban agriculture in urban planning system. Equally, the [11] seem not to be mentioning who will be responsible for coordinating this sector. The policy and legal statement is therefore observed to be too abstract to anchor the institutional body that will be accountable. For example, interviews with different institutions show the following responsibilities and the way they perceive urban agriculture (**Table 3**).

Table 3 indicates that urban agriculture has been integrated in policy and legal settings as one of the urban land uses. But one wonders whether it belongs to Ministry of Local Government or Ministry responsible for Lands. These Ministries as planning institutions are responsible for land use preparation and policy, regulation and order making and approve the same. Principally, they are the one who can help coordinate and control farming by connecting with other line Ministries. One wonders why they are hesitant. This is the policy enforcement gap, which is important to address in order to promote urban agricul-

ture. This suggests the need for the government to form a separate body within the Ministry and for the Municipality to deal with the issue of urban agriculture and link with the Ministry responsible for Agriculture and Livestock, Health and Water. The officials in the Ministries mentioned are hesitant on dealing with the issue, as it will add more administration cost.

In supporting the foregoing, the author argues that if the government wishes to adopt such arrangement, conditions for developing such land are important rather than exposing the free land market to take place. These conditions can entail ways of building construction and orientation, in plot activities including allowed forms of urban agriculture and others planning principles discussed. Agriculture land is part of settlement upgrading and livelihood strategy for the poor, which in practises are excluded by not having its designated land.

Planning standards

The formal land use planning standards and regulations, for example those being used in new formal settlement development, would generally not be appropriate for upgrading many informal settlements. These will impose severe payment burdens on residents, or the building of a particular type of housing which is difficult to afford by smallholder farmers. These farmers would find it very difficult to remain in a settlement upgraded according to conventional regulations. Besides, they fear that they cannot keep up with payments of land property tax and do not have the means to begin building an approved permanent house because of their unsustainable

Table 3. Planning institutions responses in coordinating urban agriculture.

Central government	Respondents		Responses		Remarks
	Male	Female	Yes	No	
Ministry responsible for lands and human settlement development	3	2	V		It is a form of land use in urban areas but hardly incorporated in planning schemes due to legal restriction. But argues the Ministry responsible for agriculture should be responsible
Ministry responsible for health	1	-		X	Have no legal mandate to coordinate urban agriculture, but affects the urbanities and argued is the cause of cholera and cause of malaria diseases
Ministry responsible for livestock and agriculture	1	1		X	The official says they don't deal with urban farming. Only deal with rural urban agriculture improvement as the urban context remains with Ministry responsible for land
Ministry responsible for water and energy	1	-		X	Have no legal mandate to coordinate urban agriculture
Ministry responsible for tourism	1	-		X	They confess its great potential for greening the city but it is the role of the Ministry responsible for lands, local government and agriculture and not theirs
Ministry responsible for local government	1	1		X	Argues urban agriculture gives meager profit to urban development and therefore even the reforms may change its to belong to cooperative department and not as such the way it looks
President office of environment	1			X	Argue to deal with urban farming when it brings environmental nuisance otherwise they are less concern
Ministry for planning and privatization	-	1		X	Looks urban area as area for investment of other profitable land use rather than urban agriculture

income levels. This is affecting their living standards and livelihood opportunities in the urban areas and shifting to periurban land is common.

Planning standards and regulations interact closely with those for the design and construction of buildings, the provision of infrastructure and access to formal credit and plot size needs. Official standards on plot size are often based on arbitrary assumptions regarding individual family needs, rather than the costs or impact on density levels and urban land market intricate. Given the commercialisation of urban land markets that have accelerated during the last two decades, low densities plots seem to be resulting from large plots subdivisions. Allocating of smaller plots for individual household occupation particularly after plan approval was a strategy adopted.

3.8. Infrastructure

Dar es Salaam city water demand is influenced by pressure on its use and increasing population. The major water sources include Ruvu chini and Juu plants, Mtoni plants and the availability of deep and shallow water grounds in residential neighbourhoods. Water consumption per capita per day is 187 litres in the city. The breakdown of the water system, power interruptions and the poor state of the water system pipes are among critical problems facing the sector. Increased pressure on water use for industry and domestic needs results from population increase in the city affects the settlement environment. This results from both the limited sources of water and from climate change, which reduces water levels especially during the summer season. The expanding pressure on water use demands due to increasing population, limits the utility agency to formalise water use for urban agriculture. This restriction has not been adhered to by residents and illegal water connection is common in the case studies, which reduces the amount of water reaching to other parts of the city.

Lack of sanitation facilities including drainage, sewerage system and waste collection points in the three settlements have led to increased dumping of wastes including domestic, industrial, constructional and agricultural in rivers and in open spaces. This is caused by poor land use planning, inadequate policy and legal enforcement and poor awareness of urbanities on managing urban land as the study settlements shows. This has degraded the quality of the settlements in terms of increasing air, water and land pollution. Solid waste heaps and domestic water drains towards river lines are common.

3.9. The Impact of Urbanisation

Urbanisation is the key factor underpinning and catalysing changes in land use, land transactions, increased ru-

ral-urban immigration and the overall urban agriculture land use in Dar es Salaam. It was observed that conflict arising from landholders, fragmented types of cadastral survey prohibitive costs and increased housing densification in both Chang'ombe "A" and Goba are the key features of ineffective land use planning within urbanisation processes. In addition, presence of land use conflicts and housing density may result in decreased food production. This is affecting residents' nutrition levels and subsequently labours productivity within the settlement and the city as a whole.

The health risks observed in the study settlements are those related to the use of wastewater resulting in excreta-related infectious diseases. In countries where poor sanitation and hygiene conditions exist and untreated waste water and excreta are widely used in agriculture, intestinal worms pose the most frequently encountered health risk [6]. Other excreta-related pathogens may also pose health risks, as indicated by high rates of diarrhoea, other infectious diseases, such as typhoid and cholera, and incidence rates of infection with parasitic protozoa and viruses. Thus, precautions are needed once agriculture practise is institutionalised in urban land development setting. For balancing smallholder farmers livelihoods while maintaining the city environment, health and economic returns of urban agriculture sector, training, partnership and involvement of actors in decision making is crucial. Thus, technical assistance is needed.

3.10. Urban Agriculture: A Livelihood Strategy for Food Security and Poverty Reduction in Environmental Justice Context

Different forms of urban agriculture persist in the City of Dar es Salaam. These forms of urban agriculture are providing nutritional food, employment, recreation, leisure and are greening the city. Urban agriculture contributes to poverty reduction and sustainable use of land.

Important factors influencing these forms of urban agriculture and therefore the urban environment sustainability include scarcity of land for farming, unstable crop prices, inadequate market for agriculture produce, poor packaging and poor record keeping, inadequate capital and long travel distance to the market. Other factors include absence of technology in wastewater treatment for irrigation and no clear designated areas for farming. These influencing factors create fear to smallholder farmers of involving in intensive production. Similarly, harassment from the local authorities is a common occurrence. This affects fresh food production and consumption and lowers income levels.

A desire to increase income and to supplement family consumption, nutritional value, income generation, employment alternatives and sign of prestige/hobby were in both settlements mentioned as motivation for actors' involvement in urban agriculture activities. This finding indicates the same as provided by [12]. 78% of smallholder farmers involving in this activity are poor (*i.e.* live under 1 USD per day). Likewise, people with high economic integrity involve in livestock keeping and poultry husbandry. The former are the disadvantaged group in which less labour is required, while the later requires intensive care, labour payment and treatment, which for the poor is difficult to afford. Urban agriculture tends to supplement incomes of unemployed or under-employed men and women. For example, 68% of respondents *i.e.* smallholder farmers are involving in other social economic activities in their settlements to supplement the produce. The hot climate, the seasonality of the activity depending on rainfall and water from river streams seems to affect the production in the three study settlements.

4. Conclusions

4.1. Legal and Policy Environment

The agriculture policy environment in Tanzania is earmarked by a common dichotomy between urban and rural development administration. It leaves little scope for acknowledging the specific characteristics and needs of agriculture implementation in urban areas. Agricultural policies and programmes are primarily designed for rural areas, and are therefore not always compatible with the needs of urban agriculture. To bridge this gap, there are opportunities for linking activities and programmes in respect to land development including land use planning and National development programmes. For example, conducting workshops with smallholder farmers, linking with researchers, extension officers and policy makers having the capacity to affect agriculture is crucial. This therefore calls for involvement of different stakeholders and policy enforcement to effect urban development programmes as supported in legal and policy documents in the. This helps to create awareness among different stakeholders and therefore meets the interests of smallholder farmers and other land developers.

Urban planning professionals in practise use a traditional master plan approach stipulated under [8] in urban land use planning processes, which seems to be rigid and inflexible to changed circumstances in urban areas. Principally, the Act restricts urban agriculture. Despite of good-intentions in policy and legislations reforms intentions to ensure the community's involvement in land use planning, there still exists a vacuum regarding its operation at the grass-root level. For example, under the land regularization scheme as per [13], the power to determine whether or not to declare the regularization scheme and its implementation falls under the Minister for lands. The

Minister is also empowered to delegate functions to the Commissioner for Lands for the scheme execution. However, the substantive involvement of private sectors, and popular sector institutions to initiate and implement seems not to be emphasised in regularization process. Thus, in one way the Act, places smallholder farmers and their local authorities including village council who may want to formalise their land in a difficult position to directly and promptly intervene in an area, which has to be upgraded through land regularisation. The situation points out the need to decentralise land regularisation (*i.e.* land use planning) powers from central to local government, and to give the local community the power to initiate, implement and monitor land use planning outputs. With this respect, local authority and central government may wish to facilitate in terms of expertise, resources and plan endorsement as well as in implementation and monitoring modalities. This calls for a private public partnership in improving urban land governance.

4.2. Social Capital Ties, Linkages and Community Assets

Collaboration is about linkages and ensuring that all stakeholders perceive themselves as part of one group and not as outsiders facing "the others". Inadequate linkages and flow of information between smallholder farmers, the Local Authority, and the Ministry responsible for Local Government who endorse and approve policies and by-laws, and who may have a real functional mandate in coordinating and controlling urban agriculture livelihoods is common. Likewise, Ward Councillors who are political representatives of landholders including smallholder farmers are hardly visiting smallholder farmers to understand their problems. It was noted that even the Ward Councillors' decisions in practise are based on personal vested interest rather than technical and smallholder farmers' needs. Politics were also observed in decision making to interfere in spatial land use planning, land allocation and delivery systems in the study settlements, the city and the country as a whole. One smallholder farmer in Chang'ombe "A" remarked:

"*…if the person is noted to belong to the opposing party, even if he presents the issue and that issue looks genuine for settlement development the issue is hardly taken into account.*"

The statement principally shows that policy making process is influenced by politics in land use planning. Similarly, smallholder farmers view the government (both central and local) as neglecting them as their needs are not taken into account. This discourages smallholder farmers from considering farming seriously as employment rather than as a temporary activity. No one wants to address this issue in public. This is creating an unfriendly environment between the two particularly when you consider food insecurity, employment and urban poverty in the development processes. Thus, linking politicians, government and private institutions through public/private partnership in land use development is rational in the case study areas and in the city as a whole for improving urban land governance in spatial land use planning processes.

Some investments have so far been made in the three study settlements. These include multi-million Tanzanian shillings worth of vertical and horizontal residential building construction. Others include agricultural production, commercial activities and other micro-economic activeties. The available investments are however still too few to foster rapid economic development of the areas and much has to be done to attract investments in these settlements. Urban agriculture forms formidable investments. In connection to agriculture production where at least every household uses agriculture products each day, land availability and management is important in maintaining this investment. Land as a community asset therefore provides potential livelihoods enabling residents living in the case study settlement and the city. A main finding was the weakness of extension services and marketing alternatives for smallholder farmers. This suggests a need for urgent design of new market areas in periurban zones of the city of Dar es Salaam to serve the urban poor. To avoid institutional conflicts over land use, deregistration of village land after declaration is vital in the city of Dar es Salaam for enhancing this community asset.

4.3. Involvement

The concept of community involvement in planning for sustainable ecological conservation is highly insisted upon in country urban planning and land management policy, programmes and legal documents. The changing of urban planning approaches from master plan and structure plan (*i.e.* unconventional approaches) to more inclusion of actors (*i.e.* conventional approaches) is one of the strategy to ensure involvement. The later includes land regularisation, which earmarks urban planning transition approaches towards sustainability through actors' involvement. The case study shows that the involvement of landholders including smallholder farmers in land use planning is critical in integrating urban agriculture in urban planning processes. Likewise, the case shows lack of review of the 1978 Dar es Salaam Master Plan has led to increased unorganised urban farming in the city. The review may be important for improving land and including urban agriculture through negotiations with actors and experts in the stages involved in land use planning practise.

4.4. Financial Mobilisation

Financial mobilisation for implementing land use plans forms a remarkable need for involving the local community in their settlements to upgrade and plan their neighbourhoods. The Ubungo Darajani case shows the power of local community resource mobilisation in improving urban land governance through involvement in land regularisation processes. The case of Goba landholders contributed funds to find lawyers when the government was implementing the land use plans without involving them. One smallholder farmer remarked:

"···*we contributed funds meriting TSh.* 300,000/= (*i.e* 240 US$) *to fetch the lawyer to assist in interpreting the land Law towards understanding what is our right and what belongs to the government. It was in this base where we understood our land rights and our right in involvement in land use plan making, making alternative plans and rejecting plans.*"

The statement indicates how different actors including the Lawyers and Ardhi university by then UCLAS as consultants help enable landholders to fight for their land rights in land use planning processes. The two examples show the willingness of landholders in their settlement towards improving their settlements. This shows the power of local communities when they are facing the same felt problems. The government may consider this readiness to improve the urban settlements and include urban agriculture within urban plans where possible to exploit this potential.

4.5. Conclusions

The paper contributes different issues and associated factors towards improving urban land governance through urban agriculture integration in spatial land use planning practice. Urban agriculture implementation continues to face challenges. Among the challenges demonstrated in this chapter we pinpoint the mindset of planners, urban managers and decision makers who do not see and value it the way the smallholder farmers and researchers do. The social subsistence of urban farmers and the economic contribution to income including employment are important. Contributions to nutrition, environmental conservation are also benefits. Planners, urban managers and decision makers think that urban agriculture is less valuable and makes a smaller contribution than rural agriculture. It is the authors' view that small contributions of urban agriculture to the poor sector of the society matters a lot as in future may result in crime if no proper production systems, forms and scale are less considered in urban planning. Thus, mainstreaming the sector in a proper framework in urban planning and management for apt city growth and development is rational. Likewise, ex-

Urbanization and Its Impacts to Food Systems and Environmental Sustainability in Urban Space: Evidence
from Urban Agriculture Livelihoods in Dar es Salaam, Tanzania

73

clusion of urban agriculture land use plans and inadequate water infrastructure provision, fear of health impacts, weak institutional coordination and legal enforcement are other challenges. Besides decision making, inadequate access to resources including information, communication, funds and lands are other constraining factors.

Power for policy making, initiation, declaration, acquisition, control and land use change for spatial land use planning belongs to the central government in most developing countries including Tanzania. Strong facts are needed to convince the government to address and promote urban agriculture. Its financial returns, health aspects, and potential to green the city are facts that could convince the government to make urban agriculture priority. For example, the country is implementing its strategy for poverty reduction of 2005; urban agriculture is not included in this important development vision document. The author is not of the view that community needs should overstate the country policy and legal setting, but defining critical roles of each actor, conduct monitoring, empowering local community to determine their development path and linking the legal setting to the reality is important. This helps break through the theory and practice gap and meet the community needs where the policy and legislation can change. The Policy and Acts are there to change under real challenges in the development process. It is only through addressing the issue that affects the local community and putting in place its importance in national development, that it can be listened to and taken into account rationally. Otherwise, subjective decision making will likely occur. To achieve this takes time and dedication from planners and other stakeholders. This suggests that in Dar es Salaam, where population is expected to expand, a clear spatial land use planning and management strategy is required to overcome these challenges. However, rejecting any systematic role of economics of urban agriculture livelihoods growth, especially in cases of urbanization process in developing countries is inevitable as poverty persist in African cities. The results suggests that, where population is expected to expand, a clear spatial land use planning and management strategy is required to overcome these challenges towards reducing negative externalities for enhanced food systems and urban environmental sustainability in Dar es Salaam and other Sub-Saharan Africa Cities of the same context.

5. Acknowledgements

I would like to thank the community members of the study areas in Dar es Salaam (Goba, Chang'ombe, Ubungo and Mbezi settlements) involved in providing information during this research implementation. I thank DAAD of Germany for providing funds for this research implementation.

REFERENCES

[1] W. J. Kombe, "Land Use Planning Challenges in Peri-Urban Areas in Tanzania," Spring Research Series, Vol. 29, Dortmund, 2002.

[2] United Nations Human Settlements Development Report, New York, 2009.

[3] Worldwatch Institute, "The Renewables 2010 Global Status Report Provides an Integrated Perspective on the Global Renewable Energy Situation," Washington DC, 2010.

[4] L. J. A. Mougeot, "Urban Agriculture, Definition, Presence, Potentials and Risks," In B. Niko, Eds., Feldafing, German Foundation for International Development (DSE), 2000.

[5] Un High Level Task Force on the Global Food Crisis, New York, 2008.

[6] World Health Organization (WHO), "Guidelines for the Safe Use of Wastewater, Excreta and Grey Water: Policy and Regulation Aspects, Vol. 1," WHO France, 2006.

[7] S. Dongus, D. Nyika, K. Kannady, D. Mtasiwa, H. Mshinda, U. Fillinger, A. W. Drescher, M. Tanner, M. C Castro and G. F. Killeen, "Participatory Mapping of Target Areas to Enable Operational Larval Source Management to Supress Malaria Vector Mosquito in Dar es Salaam, Tanzania," International Journal of Health Geographics, Vol. 6, No. 1, 2007, p. 37.

[8] A. Inocensio, H. Sally and D. J. Merrey, "Innovative Approaches to Agriculture Water Use for Improving Food Security in Sub-Saharan Africa," Sri-Lanka, IWA, Colombo, 2003.

[9] United Republic of Tanzania, "Town and Country Planning Ordinance-Cap 378 Revised in 1961," Government Printer, Dar es Salaam, 1956.

[10] V. Kreibich and W. H. A. Olima, "Urban Land Management in Africa," Spring Research Series, Vol. 40, Dortmund, 2002.

[11] United Republic of Tanzania, "Land Use Planning Act," Government Printer, Dar es Salaam, 2006.

[12] R. Nugent, "The Impact of Urban Agriculture on the Householdss and Local Economics," 2000.

[13] United Republic of Tanzania, "Land Act," Government Printer, Dar es Salaam, 1999.

7

For a Holistic View of Biotechnology in West and Central Africa: What Can Integrated Development Approaches Contribute?

Francis Rosillon

University of Liège, Water, Environment, Development Unit, Arlon Campus, Arlon, Belgium.

ABSTRACT

Africa, ever on the lookout for development levers that will allow its economy to take off, is turning more and more towards technology. This is one of the possible modern avenues to success, especially the use of the biotechnologies that are so touted by Western countries. However, the hope placed in these new technologies must not hide the long-proven fact that technology alone is not enough to solve development problems. Biotechnologies do not escape this rule. Biotechnologies can be the best and the worst things for the people of Africa. Beyond their technical contributions, we must be wary of their boomerang effects and collateral damage. A country's development is actually more complex than simply implementing technology, and in the current global environmental context a holistic vision is necessary to ensure sustainable development. In the area of water, this integrated vision emerged on the international scene during the Dublin Conference in 1992, which consecrated the principles of Integrated Water Resources Management (IWRM). More recently, the Eco-Health concept strives to combine human health and ecosystem health while incorporating a socioeconomic dimension into the health and environmental spheres. The concern to mesh human activities better with environmental protection was materialized previously, in the 1970s already, through impact studies. After presenting this set of tools in the service of a holistic approach to the environment and development, we shall see that these approaches can inspire the players when it comes to the ways they implement biotechnologies. At the end of the day, a holistic approach to biotechnologies in Africa will be facilitated by enhanced information and communication and reliance on peasant farmers' expertise. It will have to be rooted in broader participation of the players concerned. This integration will also concern environmental and land-owning aspects, without forgetting socio-cultural acceptance of the projects and the links with health. Ultimately, it will also mean putting the human at the heart of development by taking all the richness and particularities of African society into account.

Keywords: Biotechnology; Africa; Development; Integration; Holistic Approach

1. Introduction

Africa, ever on the lookout for the levers of development that will get its economy to take off, is turning more and more towards technology. It is one of the possible modern avenues, especially that of biotechnology, which is so touted by Western countries. The Economic Community of West African States (ECOWAS) has held a spate of conferences and meetings on the subject, e.g., in Dakar in 2004, Ouagadougou in 2004, Bamako in 2005, Accra in 2007, and so on. However, the hope placed in these new technologies must not hide the fact that it has long been known that technology alone is not enough to solve development problems. Biotechnology is no exception.

Biotechnologies can be the best and the worst things for the people of Africa. However, beyond their purely technical contributions, we should be wary of their boomerang effects and collateral damage.

Biotechnology is sometimes presented as the "new means of the new green revolution that will usher humankind into a new era of abundance" (Pallante, 2011) [1], with reference to the green revolution that stroves to increase agricultural yields through the use of fertilizers and pesticides. However, according to an International Labor Organization (ILO) report quoted by Pallante (2011), "hunger and malnutrition are increasing very quickly, especially where the green revolution occurred"

[1]. Let us hope that the same will not happen with the application of agricultural biotechnologies.

Developing a country is actually more complex than simply applying techniques and, given the current global environmental context, a holistic vision is necessary to achieve sustainable development. Let us recall that the neologism "holism" is coined in 1926 and comes from the ancient Greek *holos* meaning "the whole, the entirety." It was proposed by the South African statesman Jan Christiaan Smuts in his book "Holism and Evolution". According to Smuts, holism is the tendency in nature to form wholes that are greater than the sum of their parts through what he calls "creative evolution". It is thus the opposite of narrow, sectorial approaches, but it aims to integrate the various interacting components within a complex system.

While various integrating concepts of development existed, that of sustainable development, which was inherited from the Brundtland report and spread during the Rio Conference in 1992, conquered the entire planet. However, well before then, in the early 1970s, the desire to mesh human activities better with environmental protection had already materialized in the form of environmental impact studies. The integrated, holistic view emerged on the international water management scene during the Dublin Conference of 1992, which consecrated the principles of Integrated Water Resources Management (IWRM). More recently, the Eco-Health concept is aimed at combining human health and ecosystems' health while integrating a socioeconomic dimension into the health-environment sphere.

The complex and multifold nature of biotechnologies calls for this type of integrated approach. After reviewing biotechnology's complexity, we shall recall the series of integrating concepts mentioned above. We shall also see how these approaches, which make it possible to go from a sectorial approach to a holistic vision of development, can inspire the biotechnology players in implementing their biotechnology choices. The idea will also be to put the human at the heart of development processes by taking all the richness and particularities of African society into account. However, much uncertainty exists about the consequences of implementing biotechnologies on human and environmental health and the economy of a country. We shall thus try to show here the advantages of taking a holistic approach as far upstream as possible from the projects concerned, an approach that considers the environmental, societal, economic, and other externalities, so as to circumscribe this uncertainty as best one can.

2. Biotechnologies at the Intersection of Several Fields

The term "biotechnology" already contains a binomial in which two different worlds meet, namely, the world of living things and the world of production processes. The word is often used in the plural, adding an additional layer that refers to the diversity of processes developed. The Organization for Economic Cooperation and Development (OECD) defines biotechnologies as "the application of scientific principles and engineering to the transformation of materials by biological agents to produce goods and services." The term can also be broken down into the following three elements: *bio* for life, *techno* for the tools that are developed, and *logis* for mastery of the process. This adds up to biotechnology for mastering the tools of the living. The various types of biotechnology also take on different colors: the green biotechnology of agricultural value, red biotechnology of medical value, and white biotechnology, which EuropaBio defined in 2003 as the application of natural processes to industrial production, especially that of biological engineering in the service of chemistry. Biotechnologies are thus situated at the crossroads of three areas of expertise: health, agrifood, and the environment.

According to the West and Central African Council for Agricultural Research and Development (WECARD), biotechnologies are taking root in the world market, with a turnover of \$9.2 billion in 2009. They account for more than 40% of the seed market and 30% of the pesticide market, while 14 million peasant farmers use them [2]. The developing countries do not want to be excluded from this new market.

The developing countries, especially those in Sub-Saharan Africa, are thus placing great hope in biotechnology as a way to solve the problems of hunger and poverty in particular. So, according to A. Traoré (2005), the strategic choice of promoting the planting of transgenic Bt (*Bacillus turingensis*) cotton should be a powerful way to fight poverty [3]. According to Jacques Diouf, director of the UN's Food and Agriculture Organization (FAO), as quoted by A. Traoré (2005), "It is now widely recognized that we have entered a post-Green Revolution era and conventional selection has reached a ceiling in terms of crop yields. Biotechnologies and genetic engineering should be able to help solve this problem by increasing yields significantly".

A Sahel and West Africa Club (SWAC) secretariat consultation of the players concerned by the introduction of biotechnologies showed the many advantages that the spread of these biotechnologies were expected to have in Africa. However, caution is still necessary, as shown by the consultation report's chapter on perceptions of the risks that can be linked to the use of biotechnologies and are feared by farmers and civil society organizations [4]. These concerns belong to a variety of spheres, namely,

- the socioeconomic sphere;

- regulatory, political, and strategic constraints;
- the environment;
- human health;
- and ethical considerations.

The diversity of fields affected by biotechnologies thus gives us cause to envision their implementation within an integrated approach. Concepts and tools exist to frame the integrated development of these biotechnologies.

3. A Few Integrating Concepts in the Service of African Biotechnologies

3.1. The Sustainable Development Concept

The various spheres listed above recall the multidimensionality of sustainable development, which can be applied to biotechnology so as to meet the people's needs without jeopardizing those of future generations. Sustainable development's multidimensionality is actually a reflection of all the complexity that characterizes the human being and society alike. As Edgard Morin pointed out in 1999, the human being is biological, mental, social, emotional, and rational all at the same time, while society contains historical, sociological, economic, religious, and ethical dimensions [5]. Sustainable development does not escape this fundamental complexity, which is traditionally situated at the intersection of three main fields (ethics, economics, and the environment) and crisscrossed by anthropocentric or ecocentric currents.

In the years after the Rio Conference, and even today, implementing sustainable development in actual fact was fraught with many difficulties. One such difficulty was a lack of criteria or characteristics to help guide the players' strategic choices and use of appropriate management tools guaranteeing that all these characteristics are taken into account. So, as part of a research program on sustainable development and water we compiled a list of criteria in order to judge whether the strategy adopted strayed from or was close to the concept of sustainable development (**Table 1**) [6].

The criteria proposed above can be used as a grid for analyzing all development projects, including biotechnology development projects. However, besides this chief notion of sustainable development, other, more hands-on, approaches, such as impact assessments, can also be mobilized.

3.2. Impact Assessments

Environmental and social impact assessments are part of the project evaluation processes that donors and international bodies already require. The aim is to estimate and predict the consequences of a development action on the environment and society. The assessment can concern projects, policies, technologies, and consumption choices

Table 1. A few sustainable development criteria (from Rosillon, 2011).

A few sustainable development criteria
Social criteria
Agreement on a philosophy and an outlook
Guaranteeing democratic operation
Taking all needs into account
Promoting inter- and intra-generational management
Acting on the long term
Promoting education and training
Environmental criteria
Overall ecosystemic approach
Good knowledge of the problems and things at stake, along with reliance on expertise
Heeding the biogeochemical cycles
Heeding ecosystems' carrying capacities
Promoting socio-cultural and biological diversity
Incorporating environmental policies
Precautionary principle
Principle of prevention
Economic criteria
Economic efficiency
Productivity of the resource
The "polluter pays" principle
Sustainable financing
Asset management

[7]. The idea is to analyze their effects on the environment and society, measure their degree of acceptability, propose remedial actions, and provide assistance in deciding whether or not developing a project is opportune. This is an interdisciplinary process that also pulls in the local population by making the study—as a rule supplemented by a non-technical summary for the public at large—available and organizing information sessions to ensure the project's transparency.

Allusions to impact assessments can already be found in Burkina Faso's legislation in 1977. Article 17 of law 0052/97/ADP of January 30, 1977, concerning the Environment Code of Burkina Faso stipulates, "...activities likely to have significant impacts on the environment are subject to the Environment Minister's prior opinion. This opinion is made on the basis of an environmental impact assessment or notice." This provision was confirmed more recently by law 005/97/ADP of January 30, 1997, concerning the Environment Code of Burkina Faso. This text refers to the environmental impact assessment (EIA) and environmental impact notice (EIN) as tools for incorporating environmental concerns in development projects and plans.

While these impact studies are excellent integrating tools, they sometimes come up against a lack of data and quantifiable indicators in the countries of the South. Consequently, it is necessary to bolster expertise regarding the generation of knowledge allowing the best possible characterization of the project and its possible impacts on the environment and health. Shouldn't this expertise cover the development of assessment methods that are particularly suitable for biotechnologies while including the notion of time as well, since it can take a very long time for these impacts to arise?

3.3. Integrated Water Resources Management (IWRM)

After this succinct presentation of two generalist approaches to integration, let us examine the IWRM concept regarding the water sector.

Water management—basically access to water and drainage systems—was long considered to be mainly a technical problem. Thousands of miles of pipes were laid down by hydraulic engineers around the world to convey water to the places where it was needed or remove it where it was bothersome. Catchment works, dams, manmade reservoirs, hydroelectric power plants, and drainage works were built and are still being built today [8]. However, this technical approach overlooked the fact that water had many facets beyond being the simple molecule "H_2O". Today, water is the subject of a wide range of interests expressed by numerous users. Finally, all the inhabitants of the planet feel concerned by this familiar asset that is indispensable for life. Along with its technical dimension, water thus has various other dimensions: environmental, social, economic, legal, educational, cultural, and spiritual.

The definition of IWRM, which was historically attached to the four principles set during the International Conference on Water and the Environment in Dublin in 1992, was clarified later on. The definition given by the Global Water Partnership (GWP) is regularly quoted, to wit, "Integrated Water Resources Management is a process which promotes the coordinated development and management of water, land and related resources in order to maximise economic and social welfare in an equitable manner without compromising the sustainability of vital ecosystems and the environment". IWRM thus takes water resources' dynamics in the natural areas of watersheds and aquifers into account, with all of the players in

the water sector being involved in a new management framework that strikes the best balance among all of water's uses.

However, what does integration mean in the field of water? The term "integration" meets the following two definitions [8]:

1) Integration of an element in a larger set:
- integration in the hydrosphere and the global water cycle;
- integration in the development concepts or programs that are deemed important and/or priorities (sustainable development and climate change, for example).

2) Integration of the parts with each other:
- integration of all resources: water in all its forms, quantitative and qualitative resources, an ecosystemic approach that respects the ecosystem's integrity, the quality of aquatic environments, which results from three components (physical, chemical, and biological), regenerative potentials, and ecosystems' carrying capacities versus users' ecological footprints;
- integration of knowledge and expertise;
- social integration: integration of needs and functions and harmonization of uses;
- integration of the players: participation and policy integration;
- economic integration; and
- spatial integration: the catchment area.

In actual practice, this integration, which is so strongly desired, still often finds itself hobbled by strong sectorial management patterns, while in developing countries IWRM is hemmed in by organizational and socio-cultural barriers [9]. Yet IWRM is characterized by a holistic, crosscutting, multidisciplinary approach that can also be applied in the field of biotechnology.

3.4. Eco-Health: Human Health and Ecosystems' Health

Here is another example of an integrating concept, this time in the field of health. Like the concept of sustainable development, the Eco-Health approach puts health at the intersection of environmental, economic, and social factors. This concept is widely defended by the International Development Research Centre (IDRC), a Canadian state entity that puts its faith in researchers from the South to solve their countries' development problems. According to Jean Lebel (2003), no less than human beings' places in their environment are riding on the ecosystemic approach to human health [10].

The ecosystemic approach thus gives as much importance to good environmental management as to economic factors and the community's aspirations when it comes to resource management [10]. An Eco-Health approach also takes care to get the local communities directly involved in research programs and initiatives that are implemented with them in mind. It does so through monitoring committees that supervise the process and in which representatives of users and civil society take part.

It is thus necessary to go beyond the purely health and environmental perspectives in order to solve, with all the players, the sources of environmental degradation that affect health. The Eco Health approach aims both to improve human beings' health but also to enable ecosystems to function well so that they can carry out their natural functions in terms of maintaining biodiversity and potential resources.

This holistic approach can also be a way to combat poverty, which can be translated as a lack of satisfactory responses to all human beings' vital needs, *i.e.*, access to resources (water, food, energy, etc.), the availability of a safe living space, accommodations, and so on. However, poverty also drives the poor to take risks to survive. This is how Kofi Annan, then Secretary-General of the UN, put it in 1999: "…poverty and demographic pressure are driving more and more poor people to live exposed to danger, on plains that are subject to flooding, in areas exposed to earthquakes and on unstable hillsides". (Kofi Annan, 1999, in USAID, 2006) [11].

The situation of a degraded environment and chronic poverty is often compounded by bad governance. Here we have all the macabre elements of the vicious circle of human and environmental distress (**Figure 1**) [12].

Human health results from environmental conditions and chronic poverty, but also from problems of organization and governance in the country. The main challenge that must be taken in poor countries is to break the vicious circle that is maintained by sectorial approaches and inappropriate policy choices, be they national initiatives or imposed on the country by international bodies.

We have dared to draw a parallel between the Eco-Health approach and biotechnologies, since the latter are also mentioned as providing possible ways to fight poverty in Sub-Saharan Africa [3]. Will biotechnologies be able to break this vicious circle of human and environmental distress, or do we run the risk of causing a shift to other forms of environmental and social degradations, health problems, and a new type of poverty?

4. Some Elements for the Integrated Development of Biotechnologies in West and Central Africa

What can we take from these integrated approaches when we look at the implementation of biotechnologies in West and Central Africa? Below we have set out a few of the elements that can be taken into account in favor of a holistic vision of biotechnologies, especially in the area of agriculture, based on the four concepts outlined previously. This list is far from exhaustive.

4.1. Information and Communication

The information and communication duo is definitely one of the conditions for integrated development. In the field of biotechnologies, various voices have been raised to denounce the all-too-frequent absence of an impartial information and communication policy, especially when it comes to genetically modified organisms, which trigger fear of the unknown and open the door to all sorts of misunderstandings and assumptions [4]. One cannot content oneself with the unilateral information provided by seed-producing multinationals. Cross-examination of the subject and confrontational debate must be accepted. Before anything else, it appears to be vital that the information circulate among the players directly involved but also to the public at large. The biotechnology firms and researchers brought together by the IDRC in Ouagadougou in November 2004, it should be added, voiced the need to develop a public information system on biotechnology [4].

To trigger a broad democratic countrywide debate, might it not be opportune to organize a national biotechnology conference? The Academy of Science and Technology of Senegal (ASTS) had already proposed a national conference for Senegal in 2004 [13]. This would be an opportunity to launch a broad information campaign on the subject and promote initiatives to explain biotechnologies to the public at large in a language that they can understand while offering an area for exchange and debate on the issue. Care would be taken to let minorities and modest peasants, who are sometimes sidelined or exploited by the new biotechnology bosses, have their say. This conference could also be the crucible in which the policy line to follow in this area could be cast. It could also be a forum for drawing up an appropriate legal framework in which the general conditions of development and implementation, property rights, envi-

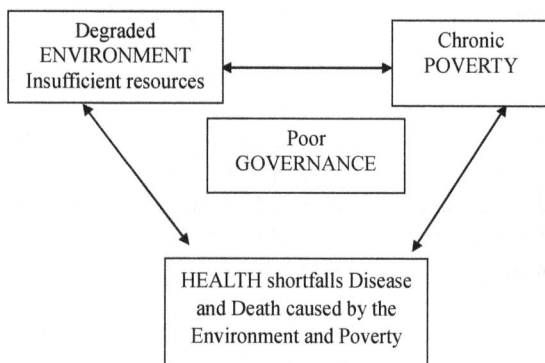

Figure 1. The vicious circle of human and environmental distress [12].

ronmental and health responsibilities, bio-safety, inspection and monitoring schemes, and so on could be spelled out in a manner that would avoid a "cut-and-paste" of Western models and, on the contrary, specify what is good for West and Central Africa.

4.2. Participation

However, informing, even consulting, the people on this topic could also lead to a more intense form of participation. Indeed, when it examined the prospects for implementing biotechnologies in Africa, the ASTS (2004) called for the participation of all the entities involved and the setting of a determined strategy [13]. What is more, the West African Network Farmers' and Producers' Organizations and the West African Network Chambers of Agriculture demanded, back in 2005, "the institution of broad debates in the population in order to enable them to participate in the decision-making" up to and including an at least five-year moratorium in the Economic Community of West African States (ECOWAS) area to give the people time to be informed and participate in making the decision to go with biotechnologies or not [14]. More recently, the FAO made a case for having small farmers participate more in decisions regarding biotechnologies during the Guadalajara Conference in Mexico in 2010 [15].

This participation could also be materialized by the creation of a national biotechnology council as an independent biotechnology development watchdog or observatory. Locally, on the project level, a community steering committee assembling the main forces of the communities concerned might also be created.

4.3. Territorial Integration and the Land

The aim of an important biotechnological subset is to increase crop yields by using genetically improved seeds to replace other plant varieties. As arable land is not infinite, might not certain projects run the risk of being seen as competing with traditional crops? For example, there are possible conflicts of land use between subsistence crops and major cash crops for export, or between local food crops and new crops for biofuel production.

In all cases, land is the object of increasingly stiff competition among various players and not just rural inhabitants. Small farmers, indigenous peoples, immigrants, the private sector, development projects, agribusiness, and other groups have all entered the fray. In developing countries the problem of land use is often complex and divided among a large number of players governed by either customary or modern law.

So, we see a layer cake of land use rights over, even title to, the same area. According to Le Roy et al. (1996),

quoted in Rosillon & Bado-Sama (2008), the surface area gives rise to a right of access and the resource to a right of harvesting, operation, or possession [16]. However, let's not forget that an area's value is contingent on the resources that it harbors or is likely to contain (Weler, 1998, in Rosillon & Bado-Sama, 2008) [16]. Biotechnologies generate new resources that must thus find their places in the available area, but integrated development cannot ignore the land ownership issues mentioned above.

4.4. Environmental Integration

The ecosystemic approach is used to take all of the biophysical environment's compartments, i.e., the ground, air, water, biodiversity, and energy, into account. It is also necessary to examine the effects that are produced locally, in the sub-region, and even internationally, both in the short and longer term, and even in the very long term. The saying "Act locally, think globally" will also apply. Externalities and side or collateral effects absolutely must be taken into account. (Example: Using smaller amounts of pesticides improves the quality of the environment and reduces the production costs linked to purchasing such inputs, but are the treated plants more vulnerable to other parasites not covered by the genes of resistance that have been incorporated in the new seeds' DNA, and what will the effects on health and biodiversity be?) Preserving biodiversity will thus be vital, and there is the risk of "genetic contamination by transfer of the modified genes to local primary strains" that can lead to "a disorganization of the ecosystem and, ultimately, the disappearance of the local genetic heritage" [4].

The new products' whole life cycles and overall ecological footprints both upstream (in the research and development phases) and downstream (in the production, consumption, and elimination phases) must be analyzed. This means factoring in all the indirect costs and environmental and social externalities.

4.5. Health and Biotechnologies

What are the links between human health and biotechnological food production? The problem is complex. What are the links between the resurgence of cancers and GMOs? What are the links between cases of allergies and the consumption of GMOs? What about the development of resistance to antibiotics, the production by transgenic plants of toxins that are harmful to human health [4]? All these questions have yet to be answered.

Today, as Professor Séralini's team (2012) is disseminating the results of their research on the impact of GMOs on rats fed Round Up-tolerant transgenic corn, the debate about using these GMOs correctly has been reopened. According to their studies, rats fed for two years with transgenic NK 603 corn and non-genetically engineered corn treated with Round Up died earlier than their fellow rats fed with untreated corn. The rats in the first two groups developed kidney, liver, and breast tumors, revealing the herbicide's long-term chronic toxicity [17]. However, Séralini's research has been criticized for a lack of scientific method and refuted by France's High Council of Biotechnologies [18] and the National Food Safety Agency [19]. The seeds of doubt have been planted and the international scientific community is split, reflecting once again all the complexity that surrounds the use of biotechnologies.

We are operating in an increasingly complex world in which yesterday's truths are challenged and swell the ranks of our uncertainties and doubts. The wake-up call, when it comes, will be all the more of a shock if we develop policies and projects sectorally, by looking at only a small part of the problem, without imagining the consequences of a policy or project on elements that are sometimes very remote from the subject. Given these uncertainties, shouldn't we give priority to the precautionary principle inherited from the foundation of sustainable development? And shouldn't the health field be tackled using the previously described Eco-Health approach?

4.6. Socio-Cultural Acceptance

While technology alone cannot solve the problems of development, it nevertheless must first be accepted by society and taken up by the populations that are concerned.

Today's biotechnologies come for the most part from research programs carried out in Western countries. It is very often a matter of exporting technologies to developing countries in contexts that are very different from those in which they arose. According to Pallante (2011), simply adjusting production patterns to territories, know-how, and forms of social organization must be avoided. He continues by mentioning the risks of agro-industrial excesses: "…getting people to grow exportable commodities by supplying the seeds and inputs, complete with user's manual, and after the harvest shipping the products to the West, steps that the small farmer has no control over, so that he is dependent on the global economy and its fluctuations" [1].

Leading-edge biotechnologies for cash crops are not sustainable in the long run if the rural populations are excluded, with the risk of creating a social divide affecting the most vulnerable family farmers, and if poverty is not overcome. The idea is not to eradicate subsistence farming in favor of agro-industry. On the contrary, agribusiness must leave room for crop diversity, local marketing and processing of harvests, and respect for the peasants' loyalty to their own culture, while guaranteeing their self-sufficiency and independence [1]. Finally, sociocultural acceptance of biotechnologies will also entail the peasants' freely given acceptance of changes in production patterns regarding the fact, for example, that farmers will no longer be able to harvest their own seeds.

So, we must bear in mind that genetically modified crops can also run aground on religious and cultural sand bars, for example, due to new varieties' potential impacts on traditional seed systems [20] or the destruction of inter-specific barriers between species, especially given the matter of gene mobility between the animal and plant kingdoms [4].

4.7. Integration of Local Expertise

Even if the development of biotechnologies calls for high-level human skills and high-tech material means of research, with the creation of centers of excellence and research clusters in particular, local expertise must not be overlooked. That is the message that François Traoré (2006) wanted to put across in his acceptance speech of the title of Doctor *honoris causa* from the Agricultural College of Gembloux (Belgium) [21]. To avoid these new technologies' being perceived as "impositions from on high," farmers must be involved in the processes of developing new varieties [20].

In fact, when it comes to intellectual property rights, the FAO proposes recognizing the rights of farmers as the holders of local genetic assets within the International Union for the Protection of New Plant Varieties [4].

4.8. New rules for the World Market

According to an OECD report from 2009, "…between now and 2015 around half of global production of the major food and feed crops will probably come from plant varieties developed with the help of biotechnology" and by 2030 biotechnology could contribute up to 2.7% of gross domestic product (GDP) in industrialized countries and even more in developing countries, given that the latter's economies are more heavily dependent on agriculture [22]. Moreover, the European Union is investing 1.9 billion euros in the creation of a European bioeconomy under the banner of "food, agriculture, fisheries, and biotechnology" as part of the 7th R & D Framework Program (FP7) [22].

The economic stakes riding on biotechnology are thus enormous, but for developing countries the problem is to cope with the new masters of biotechnologies without locking themselves into new relationships of dependence on Western countries and the seed-producing multinationals that are taking up monopoly positions on the global market. Even more crucially, how can food sovereignty be guaranteed if the countries have no control over the production and distribution of the seeds that are used?

From a global standpoint, we can also wonder how Africa is going to position itself in the geopolitics of biotechnology, caught as it is between the US-EU pincers' jaws and given Asia and Latin America's increasing power [23]. Rather than having each country react individually to developments, the African position should be strengthened if the countries of West and Central Africa band together and pool their capacities, expertise, and experience while making sure that the countries coordinate their policies.

5. Conclusions

It is understandable that Africa does not want to miss out on this new biotechnology market in which the countries of the North dominate. There is even the feeling that biotechnologies are the door through which the countries of Sub-Saharan Africa will enter globalization [13]. However, these techniques and technologies can also engender new types of dependence, notably on the seed suppliers' monopolies, for these multinationals are flooding the market with their selected seeds, against which the peasants' seeds have little say. It will also be necessary to oversee the spread of biotechnologies in Africa through new regulatory frameworks that spell out the rules of biosafety, intellectual property, and patentability, among other things, while safeguarding local interests.

To claim to benefit from the fallout of these biotechnologies as levers of development, the countries of West and Central Africa's scientific and technological lag will have to be overcome by bolstering their capacities and stepping up collaboration between research teams from the North and the South. Along with WECARD [2], we think that technical and financial means will have to be freed up for researchers in developing countries. However, this expertise will have to take specifically African environmental and socio-cultural aspects into account as well.

It is also a good idea for the African development model to differentiate itself from Western models by taking a holistic approach that puts the human back at the heart of the debate and factors in all aspects of life in society. Great pains must be taken not to dissect the living and pick out the favorable elements that could lead to more profitable production and consumption processes. Without wanting to teach anyone a lesson, this plea for a holistic approach to biotechnology has recalled a few integrating concepts that the African biotechnology strategy could mobilize.

A vast field of research and adaptation is opening up

to turn biotechnologies into a genuine lever of human development in which all the peoples of Africa will come out winners. Many types of ability will be required, and besides engineers, multidisciplinary teams of environmentalists, doctors, sociologists, etc., and local farmers' representatives will have to be set up in order to grasp all of biotechnology's intrinsic complexity. This broadening of competences will be facilitated by the creation of forums for dialog and knowledge sharing along the lines of the regional colloquium organized by Ouagadougou University in November 2012 [24].

May biotechnologies help to break the vicious circle of poverty by bolstering food self-sufficiency and improving human and environmental health in West and Central Africa!

REFERENCES

[1] M. Pallante, "La Décroissance Heureuse, la Qualité de vie ne Dépend pas du PIB," Nature et Progrès, Namur, 2011, p. 222.

[2] D. Zarour Medang, "West Africa: The Biotechnology at the Heart of the Development, WECARD Makes the Plea," 2010. http://fr.allafrica.com/stories/printable/201012220555.html

[3] A. Traoré, "La Biotechnologie, un Outil Efficace Dans la Lutte Contre la Faim et la Pauvreté en Afrique Subsaharienne: Cas du Burkina Faso," *Actes Maîtrise des Procédés en vue d'améliorer la Qualité et la Sécurité des Aliments*, 8-11 November 2005, Ouagadougou, p. 8.

[4] J. S. Zoundi, L. Hitimana and K. Hussein, "Agricultural Biotechnology and the Transformation of West African Agriculture: Synthesis of the Regional Consultation with African Actors," Sahel and West Africa Club Secretariat (SWAC/OECD), 2006, p. 27.

[5] E. Morin, "Les Sept Savoirs Nécessaires à l'éducation du Futur," *Ecole Nationale Supérieure des Mines*, Saint-Etienne, 1999. http://www.agora21.org/menu.html

[6] F. Rosillon, "Gestion Participative de L'eau et Développement Durable, Application à la Gestion de L'eau en Région Wallonne à travers L'expérience des Contrats de Rivière," Editions Universitaires Européennes, Saarbrücken, 2011, p. 360.

[7] M. Tchindjang, "Les Études D'impacts Environnementaux," University of Yaoundé 1, Yaoundé, 2009. http://web.cm.refer.org/eie/chapitre2_1.html#2.1.1

[8] F. Rosillon, "La GIRE Décryptée, Éléments Pour un Renforcement de la GIRE en Haïti et Dans les Pays en Développement," PROTOS-ULG, 2010, p. 144.

[9] F. Rosillon, "Quelques clés de Succès et Verrous Pour des Portes Résistantes à la GIRE," *Revue Liaison Energie-Francophonie*, Vol. 92, 2012, pp. 16-21.

[10] J. Lebel, "La Santé: Une Approche Écosystémique," IDRC, Ottawa, 2003, p. 84.

[11] USAID, "Vulnérabilité Environnementale en Haïti,"

USAID, Haiti, 2006, p. 146.

[12] F. Rosillon, "La Gestion Intégrée de L'eau en Haïti, Levier de Santé Environnementale et Humaine," In: E. Emmanuel, *et al.*, Eds., *Santé Humaine et Équilibre Biologique des Écosystèmes à Port-au-Prince*: *Analyse et gestion des Risques Urbains et Environnementaux*, Quisqueya University, accepted 2013, p. 38.

[13] Academy of Science and Technologies of Senegal (AS TS), "The Biotechnologies, the Potential, the Stakes and the Perspectives: The Case of the Senegal," *ASTS Report*, Dakar, 2004.

[14] S. Traoré, "Produire Plus, Sans Mettre en Danger L'homme et Son Environnement?" *Initiatives*, *Grain de sel* No 32, 2005, pp. 29-30.

[15] FAO, "Biotechnologies Should Benefit Poor Farmers in Poor Countries, Mexico Conference Takes Stock of Conventional Biotechnology Applications in Food and Agriculture," 2010.
http://www.fao.org/news/story/en/itm/40390/icode/

[16] F. Rosillon and H. Bado-Sama, "Contribution à la Gestion Intégrée des Eaux et des Sols à Travers L'application du Contrat de Rivière Sourou au Burkina Faso," *Proceedings of Journées Scientifiques du Réseau de Chercheurs en Développement Durable de L'AUF*, IRD-AUF, Hanoï, 2008, p. 12.

[17] G.-E. Séralini, E. Clair, R. Mesnage, S. Gress, N. Defarge, M. Malatesta, D. Hennequin and J. Spiroux de Vendômois, "Long Term Toxicity of a Roundup Herbicide and a Roundup-Tolerant Genetically Modified Maize," *Food and Chemical Toxicology*, Vol. 50, No. 11, 2012, pp. 4221-4231.

[18] Haut Conseil des Biotechnologies, "Avis en Réponse à la Saisine du 24 Septembre 2012 Relative à L'article de Séralini *et al.* (*Food and Chemical Toxicology*, 2012)," Comité Scientifique, 2012, p. 37.
http://www.hautconseildesbiotechnologies.fr

[19] ANSES, "Avis de l'Agence Nationale de Sécurité Sanitaire de L'alimentation, de L'environnement et du Travail Relatif à L'analyse de L'étude de Séralini *et al.* (2012), Long Term Toxicity of a ROUNDUP Herbicide and a ROUNDUP-Tolerant Genetically Modified Maize," 2012, p. 51.
http://www.anses.fr/Documents/BIOT2012sa0227.pdf

[20] O. Ezezika and J. Mabeya, "Comment Collaborer Avec les Agriculteurs sur les Cultures GM?" *Le Réseau Sciences et Développement*, 2012,
http://www.scidev.net/fr/agriculture-and-environment/

[21] F. Traoré, "Les Chroniques de François Traoré: La Crise du Coton Vue par un Agriculteur Burkinabé," *Biotechnology, Agronomy*, *Society and Environment*, Vol. 10, No. 4, 2006, pp. 295-297.
http://popups.ulg.ac.be/Base/document.php?id=536

[22] CORDIS, "OECD Sets Out Opportunities and Challenges for Future Bioeconomy," 2009.
http://cordis.europa.eu/fetch?CALLER=EN_NEWS_FP7 &ACTION=D&DOC=3&CAT=NEWS&QUERY=013d3 21c507d:b165:23801e16&RCN=30881

[23] T. Raffin, "Les Plantes Génétiquement Modifiées dans les PVD: Entre Discours et Réalité," *Revue Tiers Monde*, ND, pp. 705-720.
http://www.cairn.info/article_p.php?ID_ARTICLE=RTM _188_0705

[24] CRSBAN, University of Ouagadougou, "The Biotechnologies in Front of Challenges of the Sustainable Development in West and Central Africa," Center of Research in Biological Sciences Food and Nutritional, CRSBAN, 2012, p. 60.

8

A Study of Phytoplankton Communities and Related Environmental Factors in Euphrates River (between Two Cities: Al-Musayyab and Hindiya), Iraq

Jasim M. Salman[1], Hassan J. Jawad[2], Ahmmed J. Nassar[1], Fikrat M. Hassan[3*]

[1]College of Science, University of Babylon, Hilla, Iraq; [2]College of Science, University of Karbala, Holly Karbala, Iraq; [3]College of Science for Woman, University of Baghdad, Baghdad, Iraq.

ABSTRACT

The phytoplankton communities and related physical-chemical features of the Euphrates River at its middle region inside Iraqi territory were studied during the study period from October 2011 to September 2012. Samples were taken from Al-Musayab district extending to Al-Hindia district. The phytoplankton community (quantitative, qualitative and Chlorophyll-a) have been studied, in addition to many environmental parameters such as temperature, pH, electric conductivity (EC), Salinity (‰), TDS, TSS, dissolved oxygen, BOD_5. A total of 105 phytoplankton taxa belonging to Bacillariophyta (69), Chlorophyta (19), Cynophyta (12), Euglenophyta (3), and Dinophyta (2) were recorded within the present study period. Some algal genera dominated mostly in the study period and sites such as *Scendesmus*, *Melosira*, *Cymbella*, *Diatoma*, *Navicula*, *Nitiazschia* and *Syndera*. A statistical analysis was done using the canonical correspondence analysis (CCA).

Keywords: Phytoplankton; Water Quality; Quantitative and Qualitative Study; Euphrates River; Iraq

1. Introduction

Phytoplankton, considered as the basic component of an aquatic food chain, is the source of oxygen and the main autochthonous primary producers [1]. The floristic variation in phytoplankton community depends on the availability of nutrients, temperature, light intensity and on other limnological factors [2]. Phytoplankton is one of the major biological elements used for the assessment of the ecological status of surface water bodies, and the variation in the biotic parameters provides a good indication of energy turnover in aquatic environments, due to its sensitivity to any change in the environment [3,4]. Many authors emphasized the importance of phytoplankton as bioindictors in different aquatic systems [5-7].

Many previous researches on the Euphrates River showed the phytoplankton composition and the effect of lotic characteristics especially in the south of Iraq [8-10]. Few studies worked on the middle region of the Euphrates River inside Iraq [11-15].

The present study aims at filling the gap of information on phytoplankton communities and water quality of the studied area.

2. Materials and Methods

The present study area was chosen along the Euphrates River, middle of Iraq, between al-Musayab city (near the northern of Al-Hindiya barrage) to Al-Hindiya city (formerly Twareej: near the southern holy city of Karbala). The three sites were chosen in this area (**Figure 1**).

The current study was carried out from October 2011 to September 2012. Physical and chemical properties of the river water (Temperature, PH, electrical conductivity, salinity, TDS, TSS, dissolved oxygen, BOD5) were measured according to [16], chlorophyll-a study site was measured [17].

Phytoplankton was collected from the sampling sites with plankton net [18,19] for qualitative study, while a sedimentation technique was used for quantitative study. The micro transect methods were used for counting diatoms, and hemocytometer methods for other groups [20]. Identification of the phytoplankton was done by following references [21-26]. The statistic analyses were done

*Corresponding author.

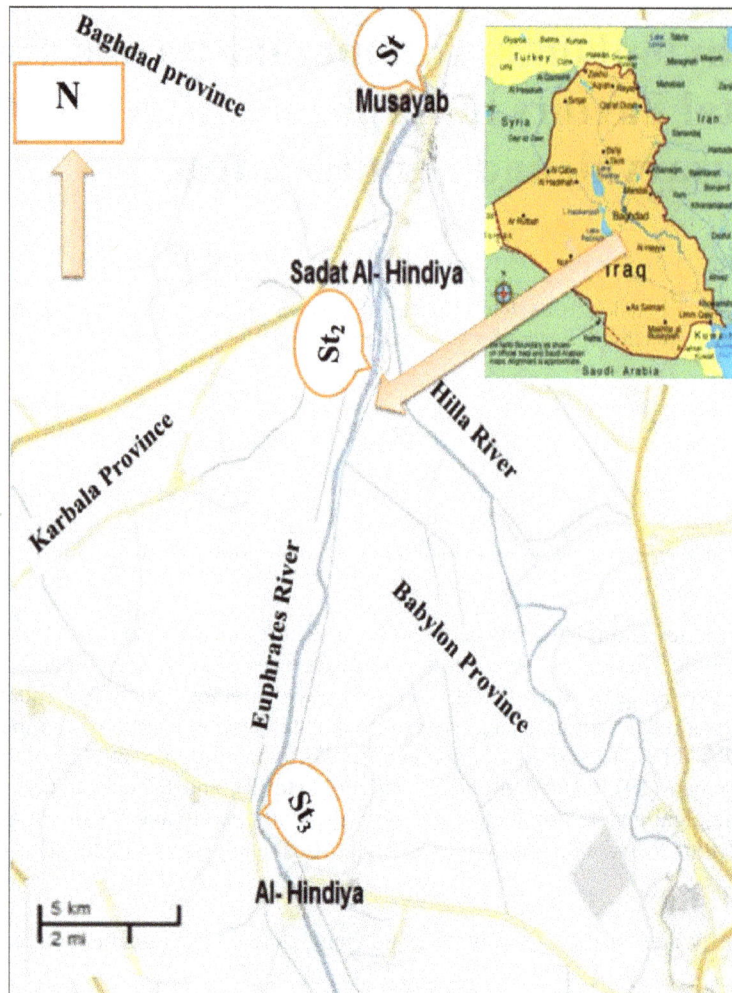

Figure 1. Map of the study area in Euphrates River at its mid region, Iraq.

by using correlation coefficient (r), and canonical correspondence analysis (CCA). CCA method was conducted by using the computer program CANOCO, version 4.5 in order to clarify the relationships between water quality parameters and phytoplankton species.

3. Results and Discussion

The environmental characteristic of the water in the study area is shown in **Table 1**. The air temperature reached its high value (43.5°C) in Site 3 and its lowest value (9°C) in Site 2, while the water temperature ranged between 9.06 to 31. 43 in sites 2 and 3 respectively. Narrow fluctuation of pH was observed during the study period, with the highest average value of 8.7 and a low of 7.3; this observed variation was statics significant between month (p $\leq$ 0.05) but not between the sites (p $\leq$ 0.05). pH variation might be caused by discharge of waste water, photosynthesis and other metabolic process [27], and may be attributed to introduction of silt into the river by rain water

or due to the mixing of the fast flowing water as it moves down stream [28]. These results match with many other studies of [11,15,28-32].

The results showed high conductivity values ranged between (798.7 $\mu s \cdot cm^{-1}$) in site 1 as lower value and (1168.6 $\mu s \cdot cm^{-1}$) in site 2 as high value, while water salinity in study area ranged between (0.5‰ - 0.7‰). The increasing values of conductivity and salinity in Euphrates River may be of the discharge of agricultural and industrial wastewater [13].

The total dissolved solid (TDS) followed the trend as conductivity temporarily and spatial. It ranged from 502. 33 to 789.3 mg/l, while the value of TSS ranged between 0.01 to 0.3 mg/l. Many factors affecting the transparency of water such as silting, microscopic organisms and suspended organic matter [32] and this variation can be explained by a higher nutrient dynamics in the water column [33].

High concentration of dissolved oxygen was recorded in the present study. This concentration ranged from 6.2

A Study of Phytoplankton Communities and Related Environmental Factors in Euphrates River (between Two Cities: Al-Musayyab and Hindiya), Iraq

85

Table 1. Some physical & chemical properties of water in Euphrates river from October 2011 to September 2012 (First line: average, Second line: mean ± SD).

Parameters	Sites		
	1	2	3
Air temp (°C)	11.7 - 39.27 22.27 ± 7.76	9 - 38.2 23 ± 9.34	13.33 - 43.5 26.53 ± 8.86
Water temp (°C)	11.54 - 30.37 19.73 ± 6.69	9.06 - 29.96 20.08 ± 6.85	13.03 - 31.43 21.09 ± 6.23
pH	7.6 - 8.7 8.26 ± 0.34	8.7 - 7.3 8.18 ± 057	7.6 - 8.7 8.29 ± 0.34
E.C (μ·S/cm)	798.7 - 1167.6 961.17 ± 127.16	811.1 - 1168.6 974.6 ± 114.36	903 - 1149 1011.4 ± 95.33
Salinity (o%)	0.511 - 0.746 0.614 ± 0.081	0.521 - 0.747 0.621 ± 0.073	0.577 - 0.736 0.646 ± 0.062
TDS (Mg/L)	540.22 - 758.44 641.1 ± 68.71	502.33 - 739 624 ± 64.81	572 - 789.3 651 ± 76
TSS (Mg/L)	0.0104 - 0.0443 0.0226 ± 0.0209	0.06 - 0.323 0.0172 ± 0.0069	0.063 - 0.031 0.0176 ± 0.00754
DO (Mg/L)	6.9 - 11.38 8.69 ± 1.4	6.27 - 11.59 8.96 ± 1.52	6.63 - 10.78 8.44 ± 1.3
BOD_5 (mg/L)	1.28 - 4.216 2.33 ± 1.1	0.94 - 4.04 2.9 ± 1	1.05 - 3.86 2.25 ± 0.94
Chlorophyll-a (μg/L)	1.22 - 18.84 6.75 ± 5.57	0 - 13.29 5.6 ± 4.84	0 - 24.37 7.99 ± 8.24
Total algae (cell/cm³)	19214	15327	17652

mg/l to 11.59 mg/l. There was an increase in dissolved oxygen in cold months at all the study sites. The concentration of dissolved oxygen was affected by many factors especially biological activities such as photosynthetic, respiration and decomposition process at the river bottom in addition to the rainfall effects [34-36]. The dissolved oxygen concentration was found to be within the (5 - 9) mg/l, which was limited for drinking water [37].

The concentrations of BOD_5 ranged from 0.94 mg/l to 4.22 mg/l in site 2 and site 1 respectively, these results showed acceptance values of BOD5 according to (APHA, 2003). These concentrations may be attributed to the observed human activities such as washing, dumping of refuse and discharge of sewage into the river [38]. The lower concentrations of BOD5 may be due to the self-purification of river [39]. The recorded BOD5 in this study indicated that the studied river is classified as un-polluted to moderately polluted, Adakole *et al.* [40] revealed that the BOD_5 values ranged between >1.0 to <10.0 mg/l was considered as unpolluted to moderately polluted water. BOD is referred to the amount of biodegradable organic materials [41], and its values refer to associated with wastewater concentration [37].

Phytoplankton biomass, as chlorophyll-a, has been measured at least once monthly in all study sites, chlorophyll-a concentrations ranged between N.D in site 2 as lower value and 24.37 mg/l in site 3 as higher value. The species composition has a big effect on chlorophyll-a

concentration in an aquatic systems and low concentration of chlorophyll-a was recorded (**Table 2**), might be due to dominate of diatoms in high number compared with other algal groups [42], and the changes in biomass of phytoplankton have been associated with increased temperature and decreased water discharge [43]. The variation of chlorophyll-a concentrations was highly correlated with phytoplankton density and water quality in the present study sites [44]. Clear peaks were observed in late spring and autumn. Similar results were found in many Iraqi aquatic system studies [12,30,45].

Phytoplankton community of Euphrates River in the present study sites from October 2011 to September 2012 consisted of a total of (105) taxes (**Table 3**) belonging to five taxonomical division; Bacillariophyta (69), Chlorophyta (19), Cyanophyta (12), Euglenophyta (3) and Dinophyta (2). The total numbers of phytoplankton ranged between 216 cell/cm³ to 4565 cell/cm³ in sites 2 and 1 during March 2012 and October 2011 respectively (**Table 2**). High yearly density of Phytoplankton recorder in site 1 (19,214 cell/cm³) and low yearly density in site 2 (15,327 cell/cm³) during the study period, might be due to availability of environmental condition for phytoplankton growth [46]; phytoplankton density which can be explained by the increase of nutrients in the environment [42], suggests that a high density of phytoplankton recorded for the first sampled point may be associated with nutrients and low water current before Al-Hindia

Table 2. Total numbers of phytoplankton (cell/cm^3) in study sites from October 2011 to September 2012.

Sites	→2011←						→2012←						Total
	Oct	Nov	Dec	Jan	Feb	Mar	Apr	May	Jun	Jul	Aug	Sep	
1	4565	2393	819	514	1115	318	395	785	1944	613	1300	4453	19,214
2	3900	2221	712	450	816	216	320	514	955	711	1112	3400	15,327
3	4210	2111	701	510	1022	413	385	716	1714	610	1150	4100	17,642

Table 3. List of identified phytoplankton species in Euphrates River-Middle of Iraq (2011-2012).

Taxa	S1	S2	S3
Cyanophyceae			
AnabeanabergiiOstenf	+	−	-
Chroococcusminuteus (*Ktz.*) *Naegeli*	+	+	+
Chroococcusturgidus (*Ktz.*) *Naegeli*	+	+	+
Gloeocapsa Sp.	−	−	+
Merismopediaelegans (*A.*) *Braum*	+	+	+
Merismopediaglauca (*Ehr.*) *Naegeli*	+	+	+
MicrocystisaeruginosaKuetzing	+	+	−
Nostoc SP.	+	−	−
Oscillatoria articulate Gardnar	+	−	−
OscillatoriachalybeaMertens	−	−	+
Oscillatoria sancta (*Ktz.*) *Gomont*	+	+	+
SpirolinalaxaG.M.Smith	+	−	−
Chlorophyceae			
ActinostrumhantizschiiLagerhein	+	−	−
Ankistrodesmusfalcatus (*Corda*) *Ralfs*	+	−	+
Carteriaklebsii (*Dang*) *Dill*	−	−	+
Chlamydomonasangulosa Dill	−	+	−
Chlorella vulgaris Bejerinck	+	+	+
Oedogonium SP.	−	−	+
Pediastrum duplex Meyen	+	+	+
Pediastrum Simplex Meyen	+	+	−
Scenedesmusdimorphus (*Turb.*) *Lagher*	+	+	−
ScenedesmuslongusMeyen	−	−	+
Scenedesmusquadricauda (*Turb.*) *de Brebisson*	+	+	+
Scenedesmusbijuga (*Turb.*) *Laghere*	−	+	−
Spirogera SP.	+	−	−
Spirogyra fluviatilis Hila	−	−	+
Staurastrumalternans	+	−	−
StaurastrumSP.	+	−	−
Tetraedronhastatum (*Reisch*) *Hansg*	+	−	+
Tetraedron regular Ktz.	+	−	+
Ulothrixzonata (*Webre and Mohr*) *Ktz.*	−	+	−
Euglenophyceae			
Euglena acusEhernberg	+	−	+
EuglengracilisKlebs	+	+	−
Trachelomonas SP.	−	−	+
Dinophyceae			
Ceratiumhirundrnella (*Muell,*) *Du Jardin*	+	+	+
Peridiniumcinctum (*Muell,*) *Ehernberg*	−	−	+

A Study of Phytoplankton Communities and Related Environmental Factors in Euphrates River (between Two Cities: Al-Musayyab and Hindiya), Iraq

87

Continued

Bacillariophyceae		
Order Centrales			
CyclotellaatomusGrunow	+	−	−
Cyclotellacomta (Ehr.) Kuetzing	+	−	+
CyclotellameneghinianaKuetzing	+	+	+
CyclotellaocellataPantocsek	+	−	+
Melosira granulate (Ehr.) Ralfs	+	+	+
MelosiravariansAgradh	−	−	+
Order Pennales			
Amphora ovalis(Ktz.) Kuetzing	−	−	+
Amphora SP.	+	−	−
Amphora venetaKuetzing	+	−	+
Asterionella Formosa	+	+	+
Bacillariafaxillifer (Muell,) Hendey	+	+	+
Caloneis amphisbaena (Bory) Cleve	−	−	+
Caloneispermagna (Bail) Cleve	+	−	−
CocconeispediculusEhernberg	+	+	+
CocconeisplacentulaEhernberg	+	+	+
Cymatopleurasolea (Berb.) W. Smith	+	−	−
Cymbellalanceolata (Ehr.)	+	−	+
CymbellaaffinisKuetzing	+	+	+
Cymbellaaspera (Ehr.) H.Paragallo	+	−	+
Cymbellacistula (Ehr.) Kirchn	+	+	+
Cymbellagracilis	−	−	+
CymbellahelveticaKuetzing	+	−	−
Cymbellaparva (W.Smith) Kirchn	−	+	+
Cymbellatumida (Berb.) Van Heurck	+	+	+
Cymbellaturgid (Greg) Cleve	+	+	+
Diatomahiemale (Roth.) Heiberg	+	−	−
DiatomavulgareBory	+	+	+
Diploneisovalis (Hilse) Cleve	−	+	+
DiploneissmithiiBory	+	+	+
Eutoniacurvata	+	−	−
EutoniamonodonEhr	−	−	+
FlagilariacapucinaDesmazieres	+	+	−
FlagilariacrotonensisKitton	+	+	+
FlagilariaintermediaGrunow	+	−	−
Gomphonema. constricumEhernberg	+	+	+
G. fanensisMaillard	+	+	+
G. gracileEhernberg	+	−	−
G.angustatum (Ktz.) Rabenhorst	+	+	+
G.parvulum (Ktz.) Kuetzing	−	−	+
Gomphoneisolivaceum (Horne) P.Dawson ex Ross et Smith	+	+	+
GomphonemaacuminatumEhernberg	+	+	+
Gyrosigmaacuminatum (Ktz.) Rabenhorst	+	−	−
MstogloiasmithiiThw.Ex.W.Sm	+	+	+
Naviculacincta (Ehr.)	+	−	−
N.anglicaRalfs	+	+	+
N.gracile (Ehr.)	+	+	+

Continued

N.pupulaKuetzing	+	–	–
N.pygmaeaKuetzing	+	–	–
N.viridulaKuetzing	–	–	+
Neidium affine (Ehr.) Pfitz	+	+	+
Ni. CommutateGrunow	–	+	+
Ni.amphibiaaGrunow	+	+	+
Ni.dissipata (Ktz.) Grunow	+	+	+
Hantzsch Ni.gracilis	+	+	+
Ni.hantizschianaRabenhorst	+	+	+
Ni.hungaricaGrunow	+	+	+
Ni.linearisW.Smith	+	+	+
Ni.palea (Ktz.) W.Smith	+	+	+
Ni.rectaHantzsch ex Rabenh	+	+	+
Ni.sigma (Ktz.) W.Smith	+	–	–
Nitiazschiaacicularis (Ktz.) W.Smith	+	+	+
P.viridis (Nitzsch.) Ehernberg	–	–	+
PinnulariadiverginsEhr	+	+	+
Rhoicospheniacurvata (Ktz.) Grunow	+	+	+
S. tenera Gregory	+	–	–
S.capitataEhernberg	+	–	–
S.ulnaNitzsch. Ehernberg	+	+	+
SurirellaelegansEhr	+	–	–
SynedraacusKuetzing	+	+	+

barrage; moreover, the effect of the barrage on phytoplankton density before (site 1) and after (sites 2 and 3) is clear [45].

Bacillariophyta were found in high percentage (65.7%) of total organisms and they were dominating species among the algal groups (**Figure 2**). Bacillariophyta was dominant followed by Chlorophyta, Cynophyta, Euglenophyta, and Dinophyta (**Table 3**). In the present study the river water showed a higher population of diatoms coincided with the higher dissolved oxygen through the study period, in general, the requirement of dissolved oxygen for growth of many diatom species is well documented [43]. Pennales diatom was the dominated group of diatoms (63 species) in the present study, this might be due to high tolerance to wide environmental changes [1,32,47].

The present study results showed the dominate of some genera of phytoplankton on a long study period, such as *Scendesmus, Melosira, Cymbella, Diatoma, Navicula, Nitzschia, Synedra*, these results were recorded by other studies [1,2,36,38]. Phytoplankton densities tended to increase during the months of spring and summer, which can be explained by the increase of nutrients in the environment, especially nitrogen and phosphorus [42]. Euglenophyta and Dinophyta species were very view in a number (3 and 2 species respectively), and their percentage composition value was low: 2.85% and 1.9% respectively. In general, the existence of Euglenophyta species

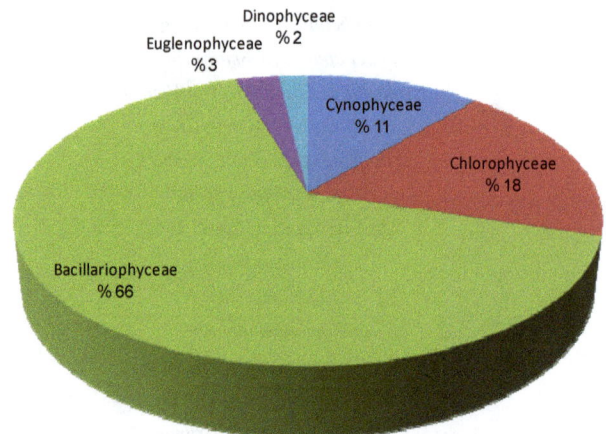

Figure 2. Species abundance of phytoplankton class in Euphrates river.

refers to organic pollution of aquatic system [2].

CCA for water quality and phytoplankton in the present study (**Figure 3**) indicated that negative relationships found between air and water temperature, salinity, TDS, BOD5 and phytoplankton, while, positive relationships were observed between phytoplankton and DO, chlorophyll-a, PH, and TDS. The interaction between various physical, chemical and biological factors is the causative regulator for seasonal variation and standing crop of phytoplankton [48]. The positive correlation of

A Study of Phytoplankton Communities and Related Environmental Factors in Euphrates River (between Two Cities: Al-Musayyab and Hindiya), Iraq

89

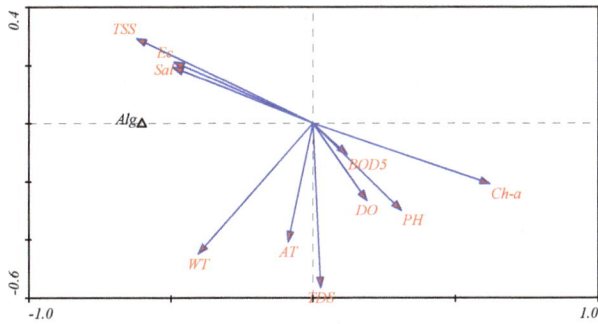

Figure 3. Correlations between water quality parameters and phytoplankton according to Canoco (CCA).

some water properties with phytoplankton density may be due to playing a pivotal role in regulation various biological activities and growth [2].

4. Conclusion

The present study results revealed that temperature, salinity, DO and BOD5 were playing important roles as limited factors to phytoplankton in this study. The variation in the functional groups of phytoplankton reflects the seasonal dynamics of revering Phytoplankton and the impact of water quality.

5. Acknowledgements

We are grateful to Department of Biology, College of Science, University of Babylon for their support to this research.

REFERENCES

[1] M. Shams, S. Afsharzadeh and T. Atici, "Seasonal Variations in Phytoplankton Communities in Zayandeh-Rood Dam lake (Isfahan, Iran)," *Turkish Journal of Botany*, Vol. 36, 2012, pp. 715-726.

[2] S. Ghosh, S. Barinora and J. P. Kesh, "Diversity and Seasonal Variation of Phytoplankton Community in the Santragachi Lake, West Bengal, India," Qscience Connect, 2012:3

[3] C. Forsberg, "Limnological Research Can Improve and Reduce the Cost of Monitoring and Control of Water Quality," *Hydrobiologia*, Vol. 86, No. 1-2, 1982, pp. 143-146.

[4] C. Reynolds, "Ecology of Phytoplankton," Cambridge University Press, Cambridge, 2006.

[5] F. H. Aziz, D. G. A. Ganjo and Y. A. Shekha, "Observation on the Limnology of Polluted Pond in Erbil City, Iraq," *ZANCO*, Vol. 5, No. 4, 2003, pp. 23-30.

[6] F. H. Aziz, "Algae Assemblages as Biological Indictors for Freshwater Quality Assessment," *Journal of University of Duhok, Pure and Engineering Sciences*, Vol. 12, No. 1, 2009, pp. 15-20.

[7] E. G. Bellinger and D. C. Sigee, "Freshwater Algae: Identification and Use as Bioindicators," Johan Wiley & Sons, Ltd., Hoboken, 2010.

[8] R. A. M. Hadi, A. Al-Sabonchi and A. K. Y. Haroon, "Diatoms of the Shatt Al-Arab River Iraq," *Nova Hedwigia*, Vol. 39, 1984, pp. 513-557.

[9] A. Y. Al-Handal, A. R. M. Mobdhamad and D. S. Abdulla, "The Diatom Flora of the Shatt Al-Arab Canal, South Iraq," *Marina Mesopotamica*, Vol. 6 No. 2, 1991, pp. 169-181.

[10] A. A. Al-Lami, H. A. Al-Saadi, T. I. Kassim and R. K. Farhan, "Seasonal Change Epipelic Algae Communities in North Part of Euphrates River, Iraq," *Journal of College of Education For Women, University of Baghdad*, Vol. 10, No. 2, 1999, pp. 236-247.

[11] F. M. Hassan, "A Limnological Study on Hilla River," *Al-Mustansiriya Journal of Science*, Vol. 8, 1997, pp. 22-30.

[12] F. M. Hassan, N. F. Kathim and F. A. Hussein, "Effect of Chemical and Chemical Properties of River Water in Shatt Al-Hilla on Phytoplankton Communities," *E-Journal of Chemistry*, Vol. 5, No. 2, 2008, pp. 323-330.

[13] F. M. Hassan, M. M. Saleh and J. M. Salman, "A Study of Physic Chemical Parameters and Nine Heavy Metals in Euphrates River, Iraq," *E-Journal of Chemistry*, Vol. 7, No. 3, 2010, pp. 685-692.

[14] H. J. J. Al-Fatlawi, "Environmental Study of Algal Community in Euphrates River between Al-Hindia City to Al-Manathere City Region, Iraq," Ph.D. Thesis, College of science, University of Babylon Iraq, 2011.

[15] F. M. Hassan, W. D. Talyor, M. S. Al-Taee and H. J. J. Al-Fatlawi, "Phytoplankton Composition of Euphrates River in Al-Hindiya Barrage and Kifil City Region of Iraq," *Journal of Environmental Biology*, Vol. 31, 2010, pp. 343-350.

[16] APHA (American Public Health Association), "Standard Methods for the Examination of Water and Waste Water," 20th Edition, APHA, Washington DC, 2003.

[17] A. Aminto and F. Rey, "Standard Procedure for the Determination of Chlorophyll a by Spectroscopic Method," *International Council for the Exploration of the Sea*, Techniques in Marine Environmental Science, 2000, 16 p.

[18] T. R. Parsons, Y. Mait and C. M. Laulli, "A Manual of Chemical and Biological Methods for Seawater Analysis," Pergamone Press, Oxford, 1984.

[19] G. Chopra, A. K. Tyor and S. Kumari, "Assessment of Seasonal Density Variation of Phytoplanktons in Shallow Lake of Sultanpur National Park, Gurgoaon, Haryana, India," *The Journal of Biodiversity*, Vol. 112, 2013, pp. 227-232.

[20] R. A. Vollenweider, "A Manual on Methods for Measuring Primary Production Aquatic Environments," Blackwell Scientific Publication Ltd., Oxford, 1974, 225 p.

[21] G. W. Prescott, "Algae of Western Great Lake Area," William, C. Brown Co. Publ., Dubuque, 1982.

[22] N. Foged, "Fresh Water Diatoms in Serilanka (Ceylon) Bibliotheca Phycologi," Herausgeben Von J. Cramer Bond, 1976.

[23] F. Hustedt, "The Pinnate Diatoms," An English Translation of Husted F. *Dickiselal genteliz*, Jensen Iv. Kocwingstein, Gylcoeltz, Sci., Books, 1985.

[24] T. V. Desikachary, "Cynophyta," Indian Council of Agriculture Research, New Delhi, 1959, 686 p.

[25] A. Y. Al-Handal, "Contribution to the Knowledge of Diatoms of Sawa Lake, Iraq," *Nova Hedwiia*, Vol. 59, No. 1-2, 1994, pp. 22-254.

[26] F. M. Hassan, R. W. Hadi, T. I. Kassim and J. S. Al-Hassany, "Systematic Study of Epiphytic Algal after Restoration of Al-Hawizah Marshes, Southern of Iraq," *International Journal of Aquatic Science*, Vol. 3, No. 1, 2012, pp. 37-57.

[27] Y. Tanimu, S. P. Bako, J. A. Adakole and J. Tanimu, "Phytoplankton as Bioindicators of Water Quality in Saminaka Reservoir Northern Nigeria," *Proceeding of International Symposium on Environmental Science and Technology*, Dongguan, 2011.

[28] A. M. Zakariya, M. A. Adelanwa and Y. Tanimu, "Physico-Chemical Characteristics and Phytoplankton Abundance of the Lower Niger River, Kogi State, Nigeria," *IOSR Journal of Environmental Science, Toxicology and food Technology*, Vol. 2, No. 4, 2013, pp. 31-37.

[29] H. A. Al-Saadi, F. M. Hassan and F. M. Alkam, "Phytoplankton and Related Nutrients in Sawa Lake, Iraq," *Journal of Dohuk University*, Vol. 11, No. 1, 2008, pp. 67-76.

[30] J. M. Salman, F. M. AlKam and H. J. Al-Fatlawi, "A Biodiversity of Phytoplankton in Euphrates River, Middle of Iraq," *Iraqi Journal of Science, Special Issue 1st Conference of Biology*, University of Baghdad, 6-7 March 2012, pp. 277-293.

[31] J. M. Salman and H. A. Hussain, "Water Quality and Some Heavy Metals in Water and Sediment of Euphrates river, Iraq," *Journal of Environmental Science and Engineering*, Vol. 1, No. 9, 2012, pp. 1088-1095.

[32] A. A. Ayoade, N. K. Agarwal and A. Chandola-Saklani, "Changes in Physicochemical Features and Plankton of Two Regulated High Altitude Rivers Garhwal Himalaya, India," *European Journal of Scientific Research*, Vol. 27, No. 1, 2009, pp.77-92.

[33] A. P. P. Carralho, T. Zhonghua, M. M. F. Correia and J. P. Neto, "Study of Physical-Chemical Variables and Primary productivity in Bacanga River Estuary Dam, Saoluis, Maranhao, Brazil," *Researcher*, Vol. 2, No. 2, 2012, pp. 15-24.

[34] N. F. Olele and J. K. Ekelemu, "Physicochemical and Periphyton/Phytoplankton Study of Onah Lake, Asaba, Nigeria," *African Journal of General Agriculture*, Vol. 4, No. 3, 2008, pp. 183-193.

[35] A. Campanelli, A. Bulatoric and M. Cabrini, "Spatial Distribution of Physical, Chemical and Biological Oceanographic Properties, Phytoplankton, Nutrients and Coloured Dissolved Organic Matter (CDOM) in the Boka Ko-torska Bay (Adriatic Sea)," *GEDFIZIKA*, Vol. 26, No. 2, 2009, pp. 215-228.

[36] A. S. Amer and H. A. Abd El-Gawad, "Rapid Bio-Indicators Assessment of Macrobiotic Pollution on Aquatic Environment," *International Water Technology Journal*, Vol. 2, No. 3, 2012, pp. 196-206.

[37] UNESCO-WHO-UNEP, "Water Quality Assessment—A Guide to the Use of Biota, Sediments and Water in Environmental Monitoring," E and FN Spon, Cambridge, 1996.

[38] M. A. Essien-Ibok and I. A. Umoh, "Seasonal Association of Physic-Chemical Parameters and Phytoplankton Density in Mboriver, Akwa Ibom State, Nigeria," *IACS TT International Journal of Engineering and Technology*, Vol. 5, No. 1, 2013, pp. 146-148.

[39] K. Piirsoo, P. Pall, A. Tuvikene, M. Viik and S. Vilbaste, "Assessment of Water Quality in a Large Lowland River (Narva, Estonia, Russia) Using a New Hungarian Potamo Planktonic Method," *Estonian Journal of Ecology*, Vol. 59, No. 4, 2010, pp. 243-258.

[40] J. A. Adakole, J. K. Balogun and A. K. Haroon, "Water Quality Impacts Assessment Associated with an Urban Stream in Zaria, Nigeria," *Journal for the Nigerian Society for Experimental Biology*, Vol. 2, No. 3, 1984, pp. 195-203.

[41] A. N. P. I. Amadi, E. A. Olasehinde, E. A. Okosun and J. Yisa, "Assessment of the Water Quality Index of Otamiri and Oramiriukwa Rivers," *Physics International*, Vol. 1, No. 2, 2010, pp. 116-123.

[42] A. P. P. Carvalho, T. Zhonghua, M. M. F. Correia and J. P. C. Neto, "Study of Physical-Chemical Variables and Primary Productivity in Bacanga River Estuary Dam, Sao Luis, Maranhao, Brazil," *Researcher*, Vol. 2, No. 2, 2010, pp. 15-24.

[43] M. M. Ramesha and S. Sophia, "Species Composition and Diversity of Plankton in the River Seata at Seetanadi, the Western Ghats, India," *Advanced BioTech*, Vol. 12, No. 8, 2013, pp. 20-27.

[44] S. Matijevic, G. Kuspilic, M. Morovic, B. Grbec, D. Bogner, S. Skejic and J. Veza, "Physical and Chemical Properties of Water Column and Sediments at Sea Bass/Sea Bream Farm in the Middle Adriatic (Maslinora Bay)," *Acta Adriatica*, Vol. 50, No. 1, 2009, pp. 59-76.

[45] F. M. Hassan, J. M. Salman and A. S. Naji, "Water Quality and Phytoplankton Composition in Al-Hilla River, Iraq," *Proceeding of 4th Conference of Environmental Science*, University of Babylon, Babylon, 5-6 December 2012, pp. 144-160.

[46] T. T. Rajagopal, I. A. Thangamani and G. Archunanl, "Comparison of Physic-Chemical Parameters and Phytoplankton Species Diversity of Two Perennial Ponds in Sattur Area, Tamil Nadu," *Journal of Environmental Biology*, Vol. 31, No. 5, 2010, pp. 787-794.

[47] C. A. Journey, K. M. Beaulieu and P. M. Bradley, "Environmental Factors that Influence Cyanobacteria and Geosmin Occurrence in Reservoirs. Current Perspectives in Contaminant Hydrology and Water Resources Sustainability," 2013.

[48] A. A. H. El-Gindy and M. M. Dorham, "Interaction of Phytoplankton Chlorophyll and Physic-Chemical Factors in Arabian Gulf and Gulf of Oman during Summer," *Indian Journal of Marine Sciences*, Vol. 21, No. 1, 1992, pp. 233-306.

9

Effects of Sea Level Variation on Biological and Chemical Concentrations in a Coastal Upwelling Ecosystem

Marilia M. F. de Oliveira[1*], Gilberto C. Pereira[1], Jorge L. F. de Oliveira[2], Nelson F. F. Ebecken[1]

[1]Civil Engineering Program, Federal University of Rio de Janeiro-UFRJ, Center of Technology, Fundão Island, Rio de Janeiro, Brazil; [2]Geography Postgraduate Program, Fluminense Federal University-UFF, Geoscience Institute, Niterói, Brazil.

ABSTRACT

Oscillations in sea level due to meteorological forces related to wind and pressure affect the regular tides and modify the sea level conditions, mainly in restricted waters such as bays. Investigations surrounding these variations and the biological and chemical response are important for monitoring coastal regions mainly where upwelling shelf systems occur. A spatial and temporal database from Quick Scatterometer satellite vector wind, surface stations from the Southeast coast of Brazil and surface seawater data collected in Anjos Bay, Arraial do Cabo city, northeast of Rio de Janeiro State were used to investigate the meteorological influences in the variability of the dissolved oxygen, nutrients, meroplankton larvae and chlorophyll-*a* concentrations. Multivariate statistical approaches such as Principal Component Analysis (PCA) and Clustering Analysis (CA) were applied to verify spatial and temporal variances. A correlation matrix was also verified for different water masses in order to identify the relationship between the above parameters. A seasonal variability of the meteorological residual presents a well-defined pattern with maximum peaks in autumn/winter and minimum during spring/summer with negative values, period of occurrence of upwelling in this region. This lowering of the sea level is in accordance with the increasing of nutrients and meroplankton larvae for the same period. CA showed six groups and an importance of the zonal and meridional wind variability, including these variables in a single cluster. PCA retained eight components, explaining 64.10% of the total variance of data set. Some clusters and loadings have the same variables, showing the importance of the sea-air interaction.

Keywords: Meroplankton Larvae; Nutrient Concentrations; Coastal Waters; Brazilian Upwelling

1. Introduction

The upwelling shelf systems as reservoirs for biological and chemical concentrations depend as much on the scales and rates of phytoplankton new production as on the dynamical processes that govern heterogeneity of particle sedimentation within the shelf [1]. Tropical seas occurs greater light penetration in the water column, varying little throughout the year which makes it an optimum condition for phytoplankton primary production and so, increasing the concentration of chlorophyll *a*.

In coastal regions subject to the influence of coastal upwelling, phytoplankton production can be greatly affected locally. The transport to the surface of deep water cooler, nutrient rich favors the occurrence of major phytoplankton production peaks that can determine the occurrence of maximum abundance of zooplankton populations [2]. Internal tides and coastal upwelling have important roles in marine ecosystems due to the mixing of water masses with potential redistribution of heat, salt and nutrients to the biological system. There are numerous mechanisms whereby tides might influence shoreline biological and chemical concentrations. Flooding tides can dilute near shore sources and reduce these concentrations. Ebbing tides drain the water from land to sea and spring tides provide a hydrologic connection between the sea and biological and chemical sources at the high water line and upper reaches of the tidal prism in tidal wetlands and subterranean estuaries within the beach aquifer [3].

A flow pattern observed in continental shelf and slope associated with tidal currents is highly modified by factors such as topology of coastline, background bathymetry and climatic conditions of the region. These currents

---

*Corresponding author.

can generate resuspension of sediments and nutrients from the bottom, increasing the mixing layer. Barotropic and baroclinic tides can cause a mixing in the water column, leading an important impact on biological production due to the upward flow of nutrients from the layers below the euphotic zone [4].

Coastal regions are areas with high rates of productivity in the marine environment, and are the areas that suffer most from the impacts of human activities on fish stocks and the pollution from the large coastal urban centers. Human activities such as fishing, shipping and recreation are also centered near the shorelines, where knowledge of the tidal movement is necessary. These coastal regions are areas where tidal ranges are most evident and can, associated with elevated wave energy, cause coastal erosion. Hydrographic mesoscale structures in the South Atlantic subtropical ocean near the Brazilian coastline show a high variability in a environmental conditions, with physical and biological characteristics extremely relevant with mechanisms of ocean-atmosphere (OA) interactions [5]. The upper ocean plays a fundamental role in building a structure of both wind-driven and thermohaline circulation. Large-scale meteo-oceanography patterns are in accordance with the OA interactions through the South Atlantic subtropical high-pressure system, a predominant air mass above the central region of the South Atlantic Ocean basin, centered near 30° latitude that induces the currents of upper ocean due to the wind-driven forces [6]. This air mass modulates the zonal wind field and when easterly winds are dominant in the southeast coast of Brazil, high atmospheric pressure enhances the inflow of Atlantic waters through the Arraial do Cabo region. The water inflow leads to occurrence of coastal upwelling due to the anticyclonic gyre moving Tropical Waters (TW) off the coast followed by the up-flow of the deeper South Atlantic Central Water (SACW) mass, with a decrease in sea surface temperature and a maximum of surface nutrient concentrations located in the coastal area. This pattern is predominant in spring-summer period. On the other hand, in autumn-winter period, when southwesterly winds prevail, lower atmospheric pressure increases the flow parallel to the coastline and the oceanographic and biological structures are different. These physical alterations also induce changes in the distribution of these variables.

Oscillations in sea level due to meteorological driving forces related to wind and pressure occur at different scales and frequencies in all coastal regions. Interactions between meteorological and oceanic variables affect the regular tides and modify the sea level conditions, mainly in restricted waters such as bays. The patterns of tidal waves on the continental shelf are scaled down, being strongly influenced by Kelvin wave dynamics and by basin resonances [7]. Tides on the continental shelf are predicted using the harmonic analysis method which is based in tidal variations represented by N harmonic constituents. These constituents can then be used to provide reliable predictions for future tides at the respective point [8]. Predictions for reference stations are prepared from the astronomical arguments using local constituents determined by previous analysis and do not take into account meteorological influences. Thus, the observed and predicted values of the sea level variations are normally different. Storms are the main cause of these sea level variations known as storm surge and the action of wind stresses on the surface water is the principal factor involved in the generation and modification of the sea level height [7,9,10]. The tendency for surge peaks that occur most often on the rising tide in UK North Sea coastline, using numerical models was confirmed by [11]. They concluded that this pattern occurs independently of the phase between tide and surge. The models made it possible to separate the contribution to interaction from shallow water and bottom friction. [12], see also [13] used an analytical model to show that the shallow water effect becomes dominant for great tidal amplitudes in depths of 10 m or less.

The variability of productivity in upwelling shelf regions must consider the need for a better understanding of the mean distributions and variability of nutrients, as well as the physical—biological coupling, in these regions. Upwelling systems are characterized by the ascension of cold and rich in nutrients water that disturb ecosystem dynamics and increases the environmental heterogeneity [14]. Then, the knowledge about the sea level height variations in shelf regions is very important not only for marine services but also for designing and constructing onshore structures, protection of coastal regions, and mainly for monitoring the changes in marine ecosystems [15].

This study is part of the research developed by the Federal University of Rio de Janeiro—Civil Engineering Program in remote monitoring systems of environmental impacts in coastal regions in order to develop trophic dynamic models to be used in the National Plan and Regional Coastal Management or in any other aquatic system. To continue the study of biological and chemical distribution and meteo-oceanography patterns [16], we have also verified the relations between them and tidal influence in the Arraial do Cabo region. This place is known for its active wind induced shelf upwelling [17]. Seasonal variability of the South Atlantic high pressure system is associated with the occurrence of this phenomenon.

Upwelling is set up in the summer by large-scale, high-speed winds northeasterly blowing over the region off the coast [18]. This point divides Brazilian coast in environments with tropical and subtropical features in a small spatial scale [19]. Moreover this place is very attractive for the tourist and recreational activities, contributing to the local economy, but human activities are significantly affecting the coastal ecosystem [20]. The influences of meteorological patterns in the sea level response at Arraial do Cabo was verified using the Quick Scatterometer satellite vector wind over the South Atlantic Ocean near the coast of the study area, surface stations and seawater harvest sample, applying statistical tools such as multivariate analysis.

The aim of this study is to investigate the influences of the sea level variations in the biological and chemical responses at Arraial do Cabo due to meteorological forcing, considering the main water masses in the southeastern Brazilian shelf.

2. Material and Methods

2.1. Study Area

In order to investigate the sea level dynamic in Arraial do Cabo and its influence in the variability of the studied variables, we verified the spatial and temporal meteorological systems related to the sea level variations. This research is a continuity of [16], in which we used the same study area with the local points from satellite data, surface weather station, tide gauge station and sample water harvest in Anjos Bay.

This study area is under influence by the South Atlan-

tic high-pressure system and is situated on the southeast coast of Brazil near Arraial do Cabo city, (**Figure 1**).

2.2. Climatological Description of the Study Area

This region is influenced by persistent high pressure over the South Atlantic Ocean that enhances northeast flow across the area. This circulation is periodically disturbed by the passage of frontal systems caused by migrating anticyclones that move from the southwest across the northeast in the southeast coast of Brazil.

During the summer (**Figure 2(a)**), the subtropical high, over the continent, becomes weaker than winter (**Figure 2(b)**), moving southerly and on its western side the winds blow northeasterly towards the southeastern coast of South America and they are more intense in the southeastern coast of Brazil. Therefore, in the north region of the Brazil Malvinas Confluence (BMC), at 38° - 40°S, there is an intensification of the northerly winds and a weakening of the southwesterly winds which is difficult for the track of the cold fronts that reach the south and southeast coast of Brazil [21]. This seasonal variability is one of the most important factors related to the occurrence of upwelling in the region of Arraial do Cabo, where the winds blow along the coastline from north to south push the surface waters offshore on summer periods [18].

The mean surface circulation of the South Atlantic

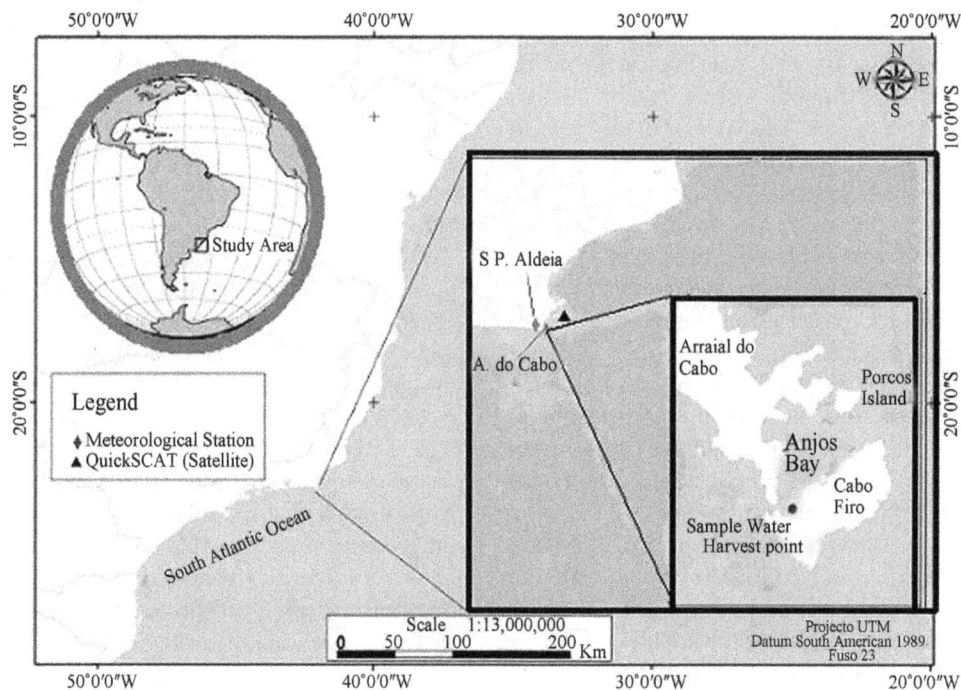

(Adapted from [16]).

Figure 1. Study area with the points representing São Pedro d'Aldeia surface weather station (22°57'S/42°06'W) and Arraial do Cabo (22°57'S/42°14'W) tide gauge station, satellite QuickSCAT winds in 22°52'S/41°52'W and water harvest point in 23°00'S/42°00'W (Anjos Bay).

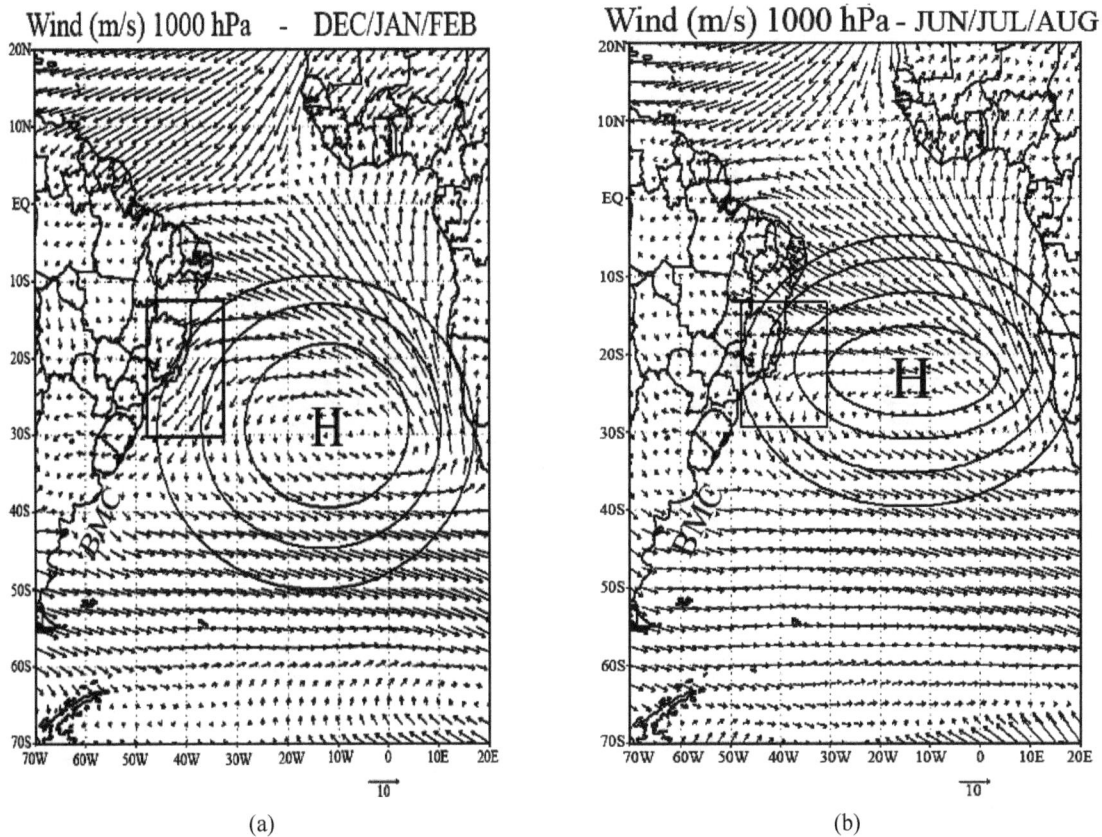

(a)

(b)

(Adapted from [16]).

Figure 2. Surface wind-flow pattern with a scheme of the South Atlantic high-pressure system for summer (a) and for winter (b). The circles represent the isobars and the positions of the high centers moving slightly during the year; the squared shows the winds blowing over the study area; BMC Brazil Malvinas Confluence region.

Ocean is dominated by a closed system known as the South Atlantic subtropical gyre that is composed of several ocean currents and on the western boundary of this gyre it has the Brazil Current (BC) [22]. Thus, the Southwest Brazilian coastline is characterized by the presence of the BC, a warm current of the South Atlantic that goes away inshore from the northern Brazil over a continental shelf toward the south, carrying the TW from the vicinity of the Equator [23]. This current makes this region oligotrophic; therefore, in some areas occur a seasonal wind-driven upwelling of cold, nutrient-rich SACW mass, benefiting on biological productivity. In Arraial do Cabo, the positioning of the Cabo Frio island (23°S, 42°W) forms the small (45 Km^2) and narrow (~10 m deep) Anjos Bay (**Figure 1**).

The latitudinal position of the BC is characterized by a seasonal variation. It means that, on average during the austral summer, the BC extends more southwards. The opposite occurs in winter and BC extends more northwards [22]. The transport of BC that follows the curve of annual variation of wind shear over the subtropical basin with a maximum during the summer (**Figure 3**) and minimum during the winter (**Figure 4**) was suggested by

[24].

The interest in monitoring environmental impacts in this coastal region is because it can be considered yet a pristine area and the hydrologic conditions are strongly influenced by the wind pattern that influences the distribution of water masses. Winds that blowing from northeastern and the Earth's rotation, result in a shunting of the nutrient-depleted surface TW of BC to offshore followed by the up-flow of the deeper (~300 meters) and nutrient-rich of SACW mass [20]. Upwelling events and inorganic nutrients are then supplied to euphotic zone by the exchange of water between nutrient-depleted surface water and nutrient-rich deeper water. When the inverse wind pattern occurs, the winds blow from south and southwestern due to the passage of cold fronts and the oligotrophic TW back to the coast. Thereby, this meteorological event leads to the occurrence of surges which modify the tidal amplitudes due to interactions between winds, shallow water and bottom friction. These processes have a direct impact on the quantity and composition of the phytoplankton communities, modifying the trophic structure [26]. In the other hand, elevated nutrient concentrations are the main origin of coastal eutrophica-

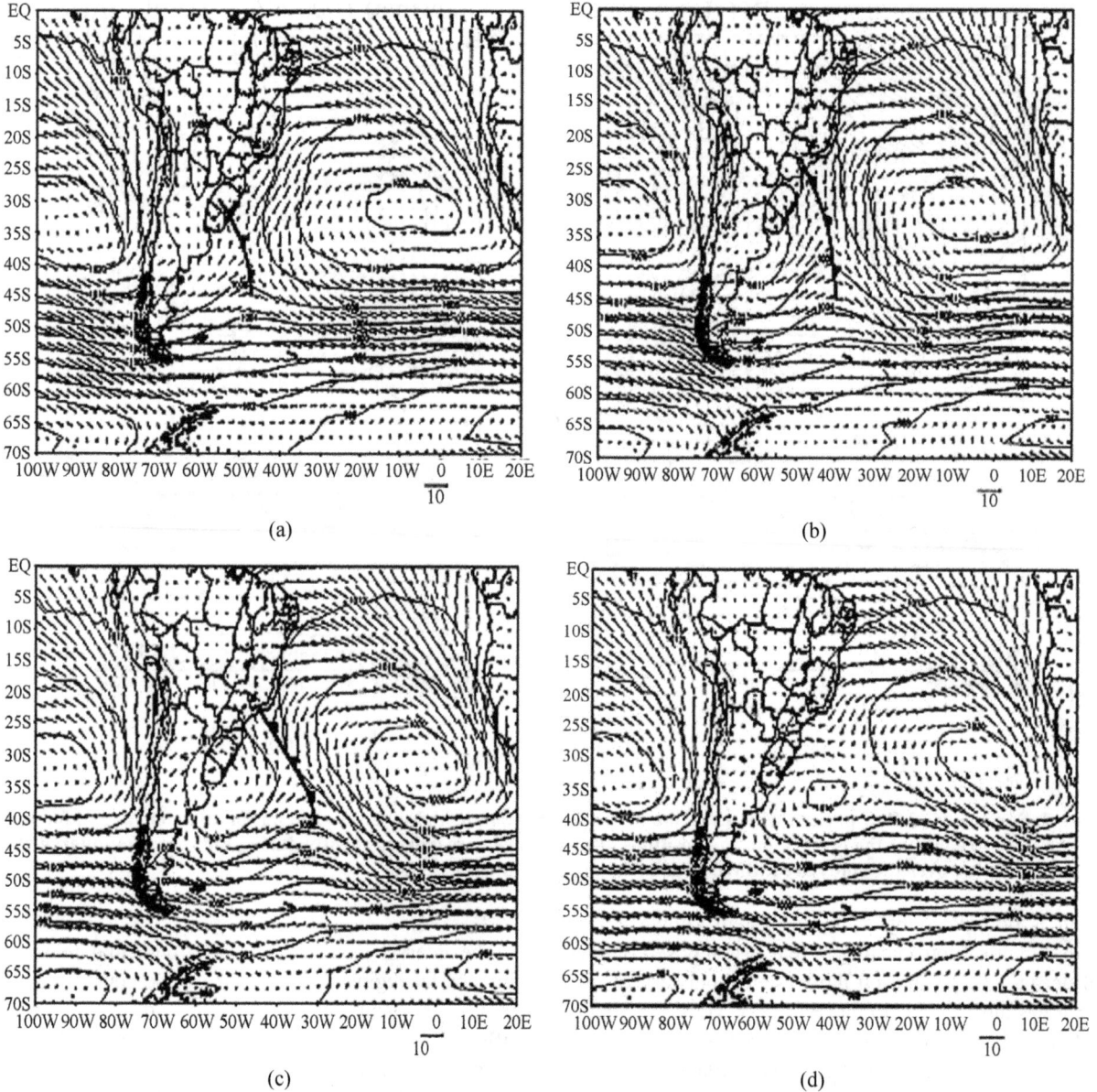

Figure 3. Average fields of sea level pressure (hPa) and 10-m above ground level wind (m/s), for cold fronts passages at (30°S e 47.5°W), in summer months. The contour interval is 4 hPa and the location of surface cold front is also marked.

tion processes and their monitoring allows direct estimates of the degree of contamination, and obviously making possible their management [27,28].

2.3. Astronomical Tides and Meteorological Residual

The astronomical tide pattern at Arraial do Cabo station is mixed tide. The mixed tides are a type of tide in which the diurnal and semidiurnal oscillations are important factors. The tide is characterized by large differences in height between two high tides (HT) or two consecutive low tides (LT). There are usually two HT and two LT

each day, but occasionally the tide can become diurnal. **Figure 5** shows the tidal ranges (vertical distance between high tide and low tide) in Arraial do Cabo that is around 1.20 meter in the largest peaks of low tides due to the diurnal influence, considering periods of spring tide.

The semidiurnal constituents M_2 (principal lunar) and S_2 (principal solar) have the greatest amplitudes. The diurnal constituents O_1 (principal lunar) and K_1 (declinational luni-solar) are also present as well as the shallow water constituents M_4 (quarter diurnal lunar), MNS_2 (semidiurnal harmonic shallow water) and MS_4 (quarter diurnal harmonic shallow water), which indicate the influence of the propagation of the tide wave in the conti-

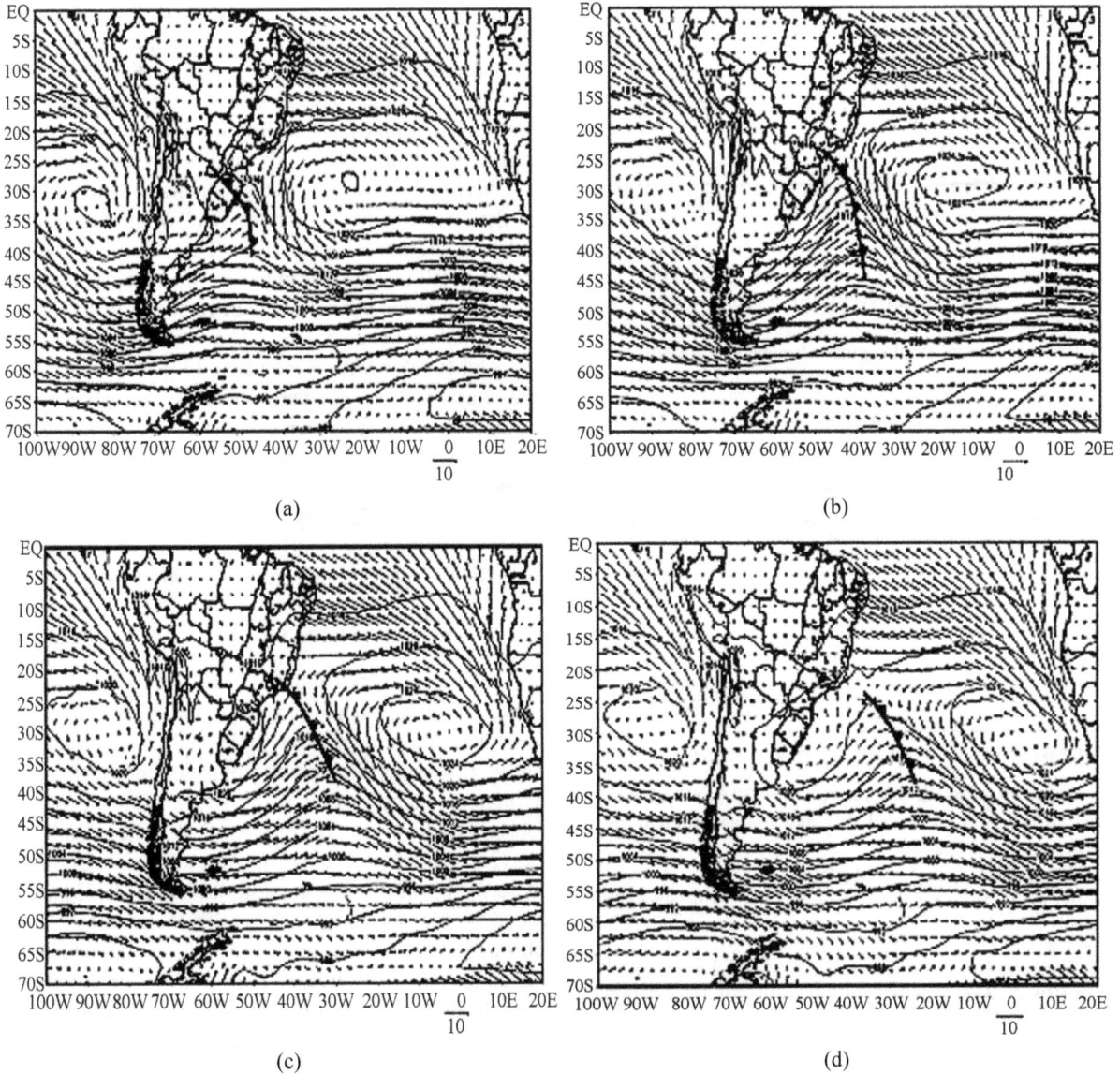

(a)

(b)

(c)

(d)

Source: [25].

Figure 4. Average fields of sea level pressure (hPa) and 10-m above ground level wind (m/s), for cold fronts passages at (30°S e 47.5°W), in the winter months. The contour interval is 4 hPa and the location of surface cold front is also marked.

nental shelf [10]. The regular and predictable oscillation of the tides is modified by meteorological patterns, being the principal ones the atmospheric pressure and winds acting on the sea level. These irregular changes, known as surge are the non-tidal residuals and can be defined as the difference between the observed and predicted levels. The non-tidal residual is alternatively called the surge or storm surge, non-tidal component or meteorological residual [10]. **Figure 6** shows a 30-day hourly height records of the predicted tide at the Arraial do Cabo tide gauge station, showing the tidal regime with distortions by the influence of the shallow water as well as the observed tide and the variation of the meteorological residual.

Recurrent upwelling events provide cold and rich in nutrients water to the coastal zone throughout the year. The semidiurnal tides, ranging between 1.0 and 2.4 m, play a key role in these systems, forcing the upwelled water into the embayment [29]. Many shallow water larvae migrate according to the tidal frequencies from one location to another in synchrony with the tidal cycle. The most common form of synchronized migration with the tides is the intertidal migration that consists of the movement of organisms into and out of the intertidal zone in each oscillation of the sea level on the ebb and flood tide [30]. Decapods are sensitive to the light cycle, and within estuaries, larval release and upward vertical migration occur most frequently on nocturnal high tides. These behaviors occur in a wide range of planktonic animals and

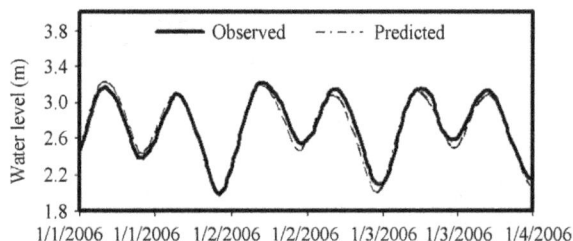

Figure 5. Series of 72 hourly heights of the observed and predicted tides at Arraial do Cabo tide gauge station, starting January 1, 2006. It can verify the mixed regime: semidiurnal with diurnal influence. The smallest peaks of LT characterize this local tidal regime observed on days 1 - 2 at 2300 local time.

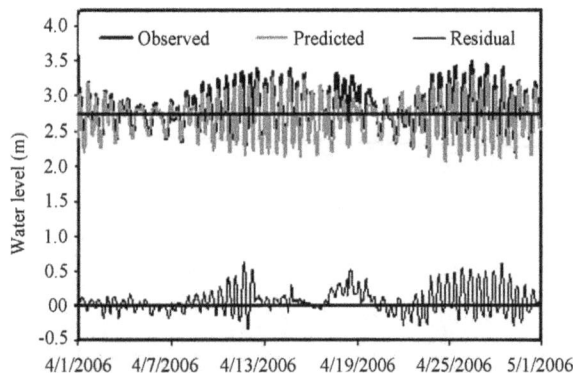

Figure 6. Series of harmonic analysis of 720 hourly heights at Arraial do Cabo, observed tide and meteorological residual starting April 1, 2006.

are considered to be driven in large part by predation [31].

2.4. Water Masses in the Southeastern Brazilian Shelf

A brief description of the main water masses in the southeastern Brazilian shelf is presented in this paper. Traditionally, the classification of water masses is made through temperature/salinity diagrams according to their thermohaline circulation. The water masses can be identified according to the geographic region where they arise, and the depth in which they reach the vertical equilibrium [6].

The TW is a warm and salty South Atlantic surface water mass, which at the western boundary is transported southward by the BC. This surface water is formed as a result of the intense radiation and excess of evaporation in respect to precipitation. This water mass is characterized by temperatures above 20°C and salinities above 36% on Brazilian Southeastern coast [32]. The Sub-Antarctic Water is cold and less saline high-latitude water mass and its western boundary layer reaches northward extensions due to advection by the Malvinas Current. These two water masses mix and form the SACW that

takes place at the Subtropical Convergence Zone (confluence between the SACW and the Antarctic Circumpolar Current) and it extend as far north as 35°S. The SACW is associated with sinking and northward transport and is found flowing into the region of pycnocline, with temperatures above 6°C and below 20°C and salinities between 34.6 and 36 PSS (Practical Salinity Scale). In the Brazilian Southeast the SACW thermohaline circulation is around 20°C and 36.2 PSS [32]. The Coastal Water (CW) has the thermohaline characteristics varying according to the annual cycle of river runoff and mixture with offshore waters [33]. The Environment National Council (CONAMA) Resolution 357/2005 that provides the classification of water bodies and environmental guidelines for its framework determines limits for the levels of several chemicals components, including nitrogenous nutrients [16]. We have used temperature and salinity data provided by the Admiral Paulo Moreira Institute of Sea Studies—IEAPM and the water mass thermohaline indices are presented in **Table 1**.

2.5. Data

2.5.1. *In Situ* Measurements

Samples of physical, chemical, and biological surface seawater (0.5 m deep) were collected with a Nansen bottle with reverse thermometer outside, and in the bottom (water/sediment interface), by scuba diving using a 2-l polyethylene bottle (three samples). The salinity, dissolved oxygen, and nutrients were determined ashore as described in [34]. The method described in [35] was applied to chlorophyll *a*. An inversion thermometer fixed to the outside of a Nansen bottle was employed for temperature. Then the physical and chemical parameters are: Sea Surface Temperature (SST), salinity, dissolved oxygen (DO), nitrogen as ammonium cation (NH_4^+), nitrite (NO_2) and nitrate (NO_3), and ortho-phosphate (PO_4).

The biological variables are composed of chlorophyll *a* (milligrams per cubic meter) measurements as an estimation of microalgal biomass, but probably also contains all free-living autotrophic bacteria of the water column both influenced qualitatively and quantitatively by nutrient entrances that on the other hand, supplies itself as feeding material for meroplankton larvae which are expressed in numbers of organisms per cubic meter of water and were collected by means drag plankton net of 100 mesh and fixed in 10% formalin and then counted under a microscope. These data were collected with weekly frequency from July 21, 1999 to June 28, 2007 in Anjos Bay, Arraial do Cabo city, and the nutrients are in accordance with the CONAMA resolution [16].

2.5.2. Meteo-Oceanographic and Sea Level Data Set

1) Station measurements

Hourly sea level records for the period 1999-2007 ob-

Table 1. Southern Brazilian shelf water mass thermohaline indices in Arraial do Cabo.

Water Mass	Temperature (°C)	Salinity (g/L)
SACW	T < 18	S < 36
SACW/COASTAL	18 < T < 20	35 < S < 36
COASTAL	T > 20	S < 35.4
SACW/TROPICAL	18 < T < 20	S > 36
COASTAL/TROPICAL	T > 20	35.4 < S < 36
TROPICAL	T > 20	S > 36

tained from the tide gauge station installed at Arraial do Cabo near Anjos Bay at latitude 22°58'S and longitude 42°00'W.

The equipment has been operated and maintained by the IEAPM of Brazilian Navy. The tidal predictions were also supplied for the same period. The meteorological residual was obtained from the difference between the observed and predicted levels. 6-hourly (UTC) atmospheric pressure and direction and wind speed from São Pedro d'Aldeia (SPA) meteorological station were also used.

2) Satellite wind measurements

The QuikSCAT is the first satellite-borne scanning radar scatterometer which measures the surface roughness of the ocean, affected by the wind magnitude and direction, by transmitting microwave pulses (13.4 GHz) and receiving the backscatter. Multiple and simultaneous normalized radar cross-section values are obtained from the backscatter power at a single geographical location or wind vector cell and converted to wind speed and direction measurements (10 m neutral winds) using a Geophysical Model Function [36]. High-resolution QuikSCAT vector wind fields suitable for coastal applications and studying of smaller oceanic processes have been produced by combining scatterometer measurements with a regional mesoscale model [37] or by use of "slices" [38].

The QuikSCAT vector wind product of Remote Sensing Systems (RSS) available daily on a 0.25° grid was used for the same period, obtained from the National Aeronautics and Space Administration—Jet Propulsion Laboratory—Physical Oceanography Distributed Active Archive Center.

Direction and wind speed as well as the meteorological residual were calculated from the wind data set and the tide gauge records, respectively. As the surface wind stress provides the most important forcing of the ocean circulation due to the relative motion between the atmosphere and ocean, we used in this work the zonal (ZWS) and meridional wind stress (MWS) calculated by the following equations:

$$T_x = \rho C_d |W| u \qquad (1)$$

$$T_y = \rho C_d |W| v \qquad (2)$$

where: $\rho = 1.22$ kg·m^{-3} (air density);

W = intensity of the wind (m·s^{-1}) calculated from zonal (**u**) and meridional (**v**) wind components; $C_d = 1.1 + 0.053W$ (coefficient of drag for the southeast Brazil coast, [39]). The units used for wind stress are N·m^{-2}, where 1 hPa is equal to 10^2 N·m^{-2}. The meteo-oceanographic time series are then weekly and for the same period as all the others.

In this research the physical, chemical and biological parameters were collected with weekly frequency. Astronomical tides are one of the most evident sea level oscillations, being that the more energetic tides are the semi-diurnal (6-hourly periods) and diurnal (12-hourly periods) and therefore this sample period do not represent the tidal influence when compared with others. Due to the meteorological systems have longer periods of oscillation and influence the variations in sea level near the coastline, we only used the meteorological residual.

2.6. Methodology

2.6.1. Statistical Analysis

The basic statistical analysis was applied in the environmental time series for the period from July 1999 to June 2007. Some outliers were identified and substituted by the average values between the previous and the following weekly data (**Table 2**).

The number of occurrence of NE and SW wind stress directions was verified. These two directions characterize two important meteorological events in the region. Winds that blow from northeastern due the presence of high pressure systems can result in upwelling events and southwestern winds are related to the passage of cold fronts and cause lowing and rising of the sea level, respectively. Meteorological event leads to the occurrence of positive and negative surges which modify the tidal amplitudes due to interactions between winds, shallow water and bottom friction. The relationship between the seasonal patterns of the residual and biological and chemical parameters was verified in order to evaluate the most significant months for monitoring the region during the critical period of upwelling. From the total data set were extracted those relating to each water masses, according to temperature and salinity. Cross-correlation matrix was then applied to the data set of each water masses and then extracted the relationships between these variables and the residual for each marine environment of interest. The knowledge of trophic relationships in the different water masses seems to be crucial for the success of environmental management policies in coastal aquatic systems.

Table 2. Statistical summary of the data set.

Variables	Max	Min	Mean	S. Dev.
Biological				
Chlorophill a (mg/m^3)	11.9	0.0	1.0	1.19
Ascidiacea (Org/m^3)	1115	0.0	11	59.5
Bivalvia (Org/m^3)	1833	0.0	99	194.9
Briozoa (Org/m^3)	101	0.0	2	5.7
Cirripedia (Org/m^3)	3641	0.0	210	362.8
Cypris (Org/m^3)	5192	0.0	22	255.5
Decapoda (Org/m^3)	437	0.0	20	35.9
Isognomon (Org/m^3)	2342	0.0	31	166.5
Mytilidae (Org/m^3)	2636	0.0	93	173.9
Polychaeta (Org/m^3)	1683	0.0	20	91.5
Ostreidae (Org/m^3)	1132	0.0	27	76.9
Sample Water				
Temperature (°C)	26.7	15.9	22.6	1.76
Salinity (g/L)	36.6	33.5	35.7	0.46
Oxigen (DO) (ml/L)	7.0	2.6	5.3	0.45
Phosphate (PO$_4$) (µmol/l)	3.7	0.0	0.3	0.21
Nitrite (NO$_2$) (µmol/l)	0.6	0.0	0.1	0.08
Nitrate (NO$_3$) (µmol/l)	10.2	0.0	0.7	0.97
Ammonium (NH$_4$) (µmol/l)	7.8	0.1	1.2	0.79
Stations				
Pressure_SPA (hPA)	1028.0	1003.0	1016.0	4.65
Wind stress_SPA (N/m^2)	0.3256	0.0	0.0416	0.0512
Tide (cm)	327	178	256.7	27.99
Meteorological residual (cm)	51.5	−48.5	2.15	15.11
Satellite (QuickSCAT)				
Wind stress_Quick (N/m^2)	0.4442	0.0	0.0963	0.0811

2.6.2. Multivariate Statistical Analysis

Cluster analysis is a multivariate technique used to group objects into classes (clusters) on the basis of similarities within a class and dissimilarities between different classes. It is a data classification method. In hierarchical cluster analysis, a dendogram is drawn with samples plotted in clusters on the y axis and linkage distances plotted on the x axis. The linkage distances between the clusters illustrate relative similarities in the characteristics of the samples. Ward's method was used to form the clusters [40]. This method uses an analysis of variance approach to minimize the sum of squares of any two hy-pothetical clusters. Pearson-r distances were used as the similarity measure. CA is also used to verify the spatial distribution data set and discover groups of similar patterns to describe more clearly the structure and composition of the study area ecosystem. The data set was separated seasonally in spring-summer (upwelling period) and autumn-winter (downwelling period) to characterize which variables are more important in these distinct periods relating them with the sea level variation.

PCA is a technique for mapping multidimensional data into lower dimensions with minimal loss of information. This method can take into account the variability of the data set between different locations (spatial scale) and between successive samples (temporal scale). It is used in all forms of analysis because it is a simple, non-parametric method of extracting relevant information from data sets. Numerical analysis procedures, such as PCA, have been developed to interpret large space/time data sets, which can decompose total variance into spatial and temporal variances. The principal axis method was used to extract the components, and this was followed by an orthogonal rotation [41]. In environmental studies with many physical, chemical and biological variables, one way to evaluate an integrated complex data is multivariate statistical methodology where variables can be analyzed together [42,43]. It consists of a linear transformation of all original variables in new variables or components. In such a way, the first computed component is responsible for most of the variance in the observed variables. The second is responsible for the greatest possible variance remaining and so on until all the variance has been explained. PCA is a variable reduction procedure [44]. Then we applied it, which establishes a set of orthogonal factors based on a correlation matrix, providing information about similarities and redundancies of the samples, using Varimax normalized rotation.

3. Results and Discussion in

3.1. Seasonal Variability of the Wind Stress Induced Upwelling

Wind events induce different disturbances in the water mass structure depending on the season. In Arraial do Cabo coastline, when NE-wind events increase the intensity for a period, mixing with the water column, the response is the SACW upwelling with capacity of a redistribution of nutrients with effects on plankton dynamics. Considering the seasonality of the wind events that modulates the zonal and meridional wind field, we also verified the number of occurrence of NE and SW wind stress (**Figure 7**).

The number of occurrence of SW wind direction is similar at SPA and Q, around 11% and 12%, respectively. This result shows the small quantify of frontal systems that reach the region. Although SPA station presents higher percentage than satellite data set for NE direction, the values are very close between them, around 71% and 62%, respectively. These results show that the Quick Scatterometer winds are also a good source of data set for this coastal area. The predominance of NE winds indicates the seasonal variability of the South Atlantic high pressure system that leads upwelling in this region. Therefore the monitoring of this area is important due to the effects that this phenomenon causes in the local biodiversity and economy.

3.2. Seasonal Variability of the Residual and Chemical and Biological Parameters

The seasonality of the mean residual presents a well-defined pattern with maximum peaks in the months of autumn and winter. It can be related to the presence of cold fronts that although they reach the region with no much frequency, they are more intense in this period. Another important aspect in raising the sea level is the internal waves due to the interaction of currents, such as barotropic tides and wind-induced flows, including seabed topography near the continental shelf [45]. Minimum during spring with negative values and summer can be related to the prevailing of the South Atlantic high-pressure with E-N winds blowing offshore the coastline and consequently occurrence of coastal upwelling (**Figure 8**). Seasonal variations were then compared between mean

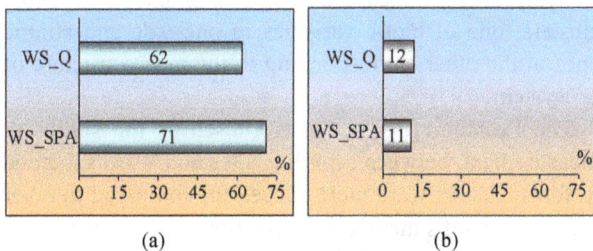

(a)   (b)

Figure 7. Number of occurrence of NE (a) and SW (b) wind direction verified at SPA weather station and vector wind obtained by Quick Scatterometer (Q).

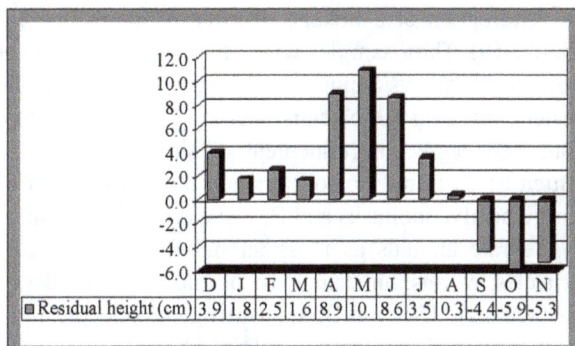

Figure 8. Seasonal distribution of the mean values meteorological residual.

residual, total nitrogen (NO$_2$, NO$_3$, and NH4), DO, PO$_4$, and peaks of maximum values of larvae. **Figure 9** shows an increasing of biological and chemical variables in spring/summer and a minimum of residual which may take place due to the meteorological system previous described. The opposite is verified in autumn/winter with minimum peaks in April and May. However we can verified that Cirripedia and Polychaeta present maximum in July. In this month verifies the presence of the TW mass which at the western boundary is transported southward by the BC [16]. Cirripedia are sessile invertebrate organisms (barnacles) that cover substantial area of the substratum on intertidal and subtidal zones. They live in warm and shallow water and are important in determining and monitoring environmental impacts in coastal areas [46]. Although there are no further reports of economic damage, it is known that the hulls of ships, oil platforms, pipelines and other plant artificial substrates available in the marine environment may be completely covered by the barnacles causing corrosion of metals and an increase in maintenance costs [47]. The Polychaeta is a bristle-worm with species that live in the coldest ocean temperatures and other which tolerate the extreme high temperatures. An inventory of the species found on the beaches of Rio de Janeiro State, Brazil was made by [48].

3.3. Cross-Correlations for Different Water Masses

As shown in **Table 1**, we have used temperature and salinity data (thermohaline indices) to classify the water mass. Thus, the data set were separated according to this criterion and applied cross-correlation to verify the relationship between the biological, chemical and physical parameters. **Table 3** shows these correlations that were verified only for Central Water (SACW), Central/Coastal Water (SACW/CW) and Central/Tropical Water (SACW/TW). These correlations verified in the presence of SACW show the importance of this water mass in the coastal upwelling process. In other water masses these variables did not present correlations. Correlation between the residual and the variables in presence of SACW suggest a sensitivity of microorganisms as Bivalvia (Class of mussels), Mytilidae (Family of mussels), and Ostreidae (oysters) associated, likely, with the decrease of the residual due the presence of the South Atlantic Subtropical High (SASH) for the upwelling period with E-NE wind direction. A direct and inverse correlation is verified for Chlorophyll a (0.90) and Salinity (−0.89), respectively. The residual also presents direct correlation with the nutrients PO$_4$, NO$_2$ and NO$_3$ for the SACW in the presence of CW. This leads to a linear relationship between them, which can suggest a relation to a discharge of water on the coastline. Chlorophyll a, As-

(a)

(b)

(c)

(d)

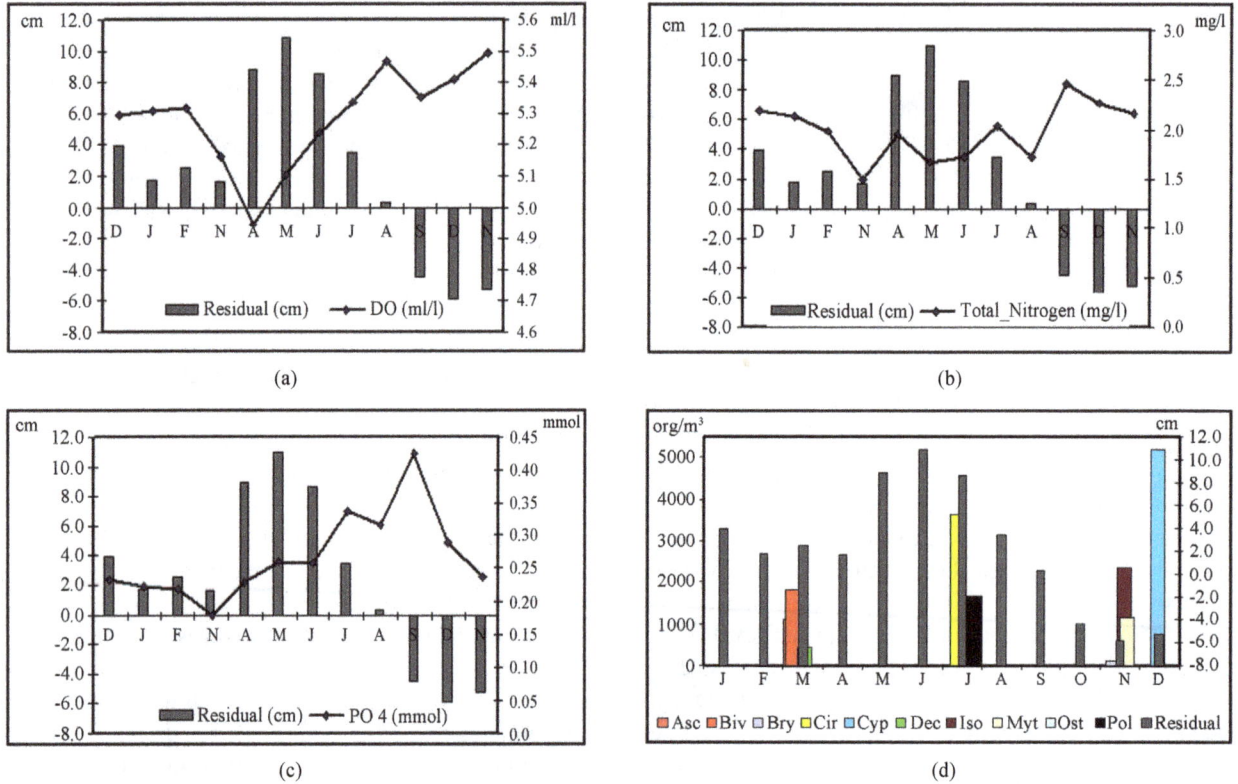

Figure 9. Seasonal variability of (a) DO, (b) total nitrogen (NO_2, NO_3, and NH_4), (c) PO_4, and (d) peaks of maximum values of larvae for 1999-2007 period.

cidiacea, Cypris, Decapoda (−0.77), Polychaeta present a negative correlation with the residual, showing a sensibility of these larvae with sea level variations due to meteorological forcing. For the SACW/TW, the residual presents more correlations with the chemical and biological variables. An inverse correlation with Chlorophyll *a*, DO, nutrients, Cirripedia, Cypris and Decapoda (−0.92), suggest also a relation to the same construct. Ascidiacea, Cirripedia and Cypris present correlations for SACW/CW and SACW/TW and are important in studies on recruitment as well as determining and monitoring environmental impacts in coastal regions.

Bivalve mollusks, particularly marine mussels, correlated in presence of SACW (−0.63) have been used as indicator organisms in environmental monitoring programmers due to their wide distribution, sedentary lifestyle, tolerance to a large range of environmental conditions and because they are filter-feeders with very low metabolism which allows the bioaccumulation of many chemicals in their tissues [49].

3.4. Multivariate Statistical Analysis

Data grouped according to the CA were performed to verify the structure and composition of the regional ecosystem. There are many available algorithms for this task and here we used for the Ward method with Pearson-r

coefficient. **Figure 10** illustrates the results of cluster analysis that grouped the sampling points in two big clusters, one of biotic variables in one side and abiotic ones at the other, corresponding to the macrostructure of ecosystem.

The Pearson-r distance was chosen as the similarity measurement, between sampling sites and Ward's method to form clusters were more successful compared to other methods. Ward's method is distinct from all other linkage rules because it uses an analysis of variance approach to evaluate the distances between clusters, producing the most distinctive groups [50]. The classification of the samples into clusters is based on a visual observation of the dendrogram at a linkage distance of about line 1.2 (**Figure 10**). Thus, samples with a linkage distance lower than line 1.2 allows a division of the dendrogram into six clusters, reaching the objectives of the classification method. The degree of refinement of similarities can be written as C_1 (family of oysters and mollusks, genus of marine bivalve mollusks and Bryosoa (phylum of aquatic invertebrate animals as moss animals)), C_2 (classes of barnacles, worms, tunicate, class of mussels and order of crustaceans), C_3 (wind), C_4 (nitrogen concentration, phosphorus and pressure), C_5 (DO, Chloro *a* and ostracods (a type of crustacean) C_6 (Temperature, Residual, Salinity and ammonium). This CA shows the importance

Table 3. Correlation between the residual and Chemical, Biological and Physical variables (95% confidence interval).

Variable	Residual		
	SACW	SACW/CW	SACW/TW
Sal	−0.89	x	x
Chl a	0.90	−0.38	−0.47
DO	x	x	−0.57
NO_2	x	0.61	−0.34
NO_3	x	0.58	−0.60
NH_4	0.40	x	−0.69
PO_4	x	0.50	−0.44
Asc	x	−0.47	x
Biv	−0.63	x	x
Cirr	x	x	−0.62
Cyp	x	−0.56	−0.47
Deca	x	−0.77	−0.92
Myti	−0.62	x	−0.42
Ost	−0.63	x	x
Poly	x	−0.68	x
WS_X_Q	−0.48	x	x
WS_X_SPA	x	x	−0.46
WS_Y_SPA	x	x	−0.47
WS_Y_Q	x	x	−0.78

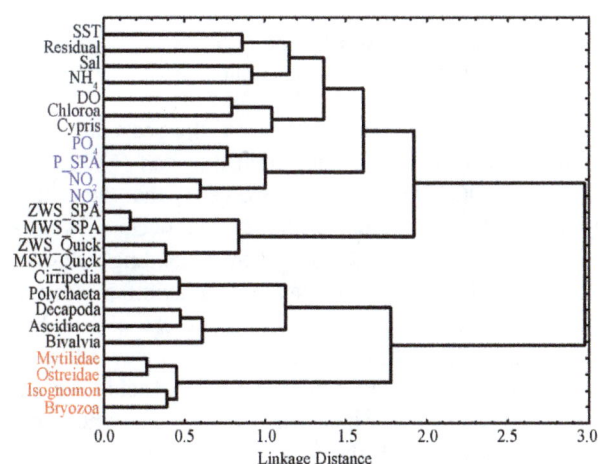

Figure 10. Dendogram for 24-variables illustrating the result of the cluster analysis. Different clusters are indicated by the different colors (C_1-C_6).

of the zonal and meridional wind variability, including these variables in a single cluster as noted in the first two factors of PCA. In reality, the number of components

extracted in a PCA is equal to the number of observed variables being analyzed. This means that an analysis of N variables would actually result in N components. However, in most analyses, only the first few components account for meaningful amounts of variance, so only these first few components are retained and interpreted. In the present study were performed on 24-variables and only eight components were extracted with this criterion. The first component can be expected to have large amount of the total variance. Each succeeding component will account for progressively smaller amounts of variance. Therefore, only the first eight components were retained and they explain 64.10% of the total variance in the data set. The first three components were the most important, with 15.7%, 11.3% and 8.6% of explained variance (**Table 4**). The first component was attributed to larvae that had the greatest variability with scores strongly positively correlated to Mytilidae, Decapoda, Bivalvia, Ostreidae, Isognomon and Bryosoa. This statistical strategy gives greater importance to the larvae in the ordering and feature extraction of the coastal ecosystem, being then denominated here "biotic component" that corresponds to the major larvae economically important to the region as crustaceans, mussels and oysters. The second component was strongly negatively correlated with zonal and meridional wind from SPA and satellite and relatively weak with meteorological residual ($r = −0.40$), showing the interaction sea-air and therefore denominated "wind component". The third component was attributed to temperature and nutrients. It showed a strong positive correlation with SST and relatively weak positive loading on PO_4, NO_2 and NO_3. These abiotic variables characterize the local marine ecosystem and are related to the presence of different water masses. Then this component was denominated here "water masses component" and it can explain the presence of SACW that is very important in upwelling periods. The factor loading of these three components accounts for 35.6% of variance and the other factors are restrict to a few parameters with low variance, indicating that they do not account for great variations. The fourth factor that contributes with a small variance of 7.5% also relates to the biotic components of the classes of zoobenthos (Cirripedia, Polychaeta and Ascidiacea) found in intertidal zones [51]. The ascidiaceas, for example, are larvae bioindicators that react to environmental changes with high filtration, playing a significant role in water purification. They influence the amount of nutrients and pollutants in suspension [52,53]. But they grow rapidly and have a long reproduction period, becoming invasive potential, contributing to incrustation in the port regions [54]. **Figure 11** illustrates the three principal components. The dendrogram produced by this approach is close to the PCA components. Some clusters and loadings have the same

Table 4. Correlation between the residual and chemical, biological and physical variables (95% confidence interval).

Factor loading	Factor 1	Factor 2	Factor 3	Factor 4
SST	0.105	−0.012	**0.716**	0.140
Sal	0.020	0.121	0.259	−0.023
DO	−0.048	−0.064	−0.306	−0.012
PO_4	−0.111	−0.056	**−0.532**	−0.192
NO_2	−0.128	−0.153	**−0.560**	−0.248
NO_3	−0.110	−0.016	**−0.535**	−0.232
NH_4	−0.112	0.096	−0.042	0.014
Chloro a	−0.070	−0.115	−0.158	−0.034
Cirripedia	0.383	−0.054	0.115	**−0.535**
Mytilidae	**0.785**	−0.251	−0.188	0.257
Decapoda	**0.607**	−0.220	0.126	−0.368
Polychaeta	0.326	−0.111	0.066	**−0.583**
Bivalvia	**0.668**	−0.063	0.117	−0.259
Ostreidae	**0.789**	−0.251	−0.112	0.187
Cypris	0.079	−0.045	0.033	−0.105
Ascidiacea	0.387	−0.057	0.240	**−0.519**
Isognomon	**0.715**	−0.197	−0.132	0.376
Bryozoa	**0.595**	−0.274	−0.247	0.479
Residual	−0.194	**−0.400**	0.149	0.073
ZWS_SPA	−0.310	**−0.787**	0.090	−0.064
MWS_SPA	−0.290	**−0.765**	0.095	−0.067
P_SPA	0.010	−0.083	−0.439	−0.150
ZWS_Quick	−0.278	**−0.672**	0.150	0.056
MSW_Quick	−0.281	**−0.718**	0.028	0.015
Eigenvalue	3.773	2.711	2.069	1.794
Variability (%)	15.722	11.296	8.619	7.474
Cumulative (%)	**15.722**	**27.019**	**35.638**	43.112

variables, such as C_1, C_2, C_3 and C_4 with Factor 1, 2, 3 and 4, corresponding to the importance of the biological variables and local environmental structure.

4. Conclusions

Seasonal variability of the mean meteorological residual presents a well-defined pattern with maximum peaks in autumn and winter and minimum during spring and summer, showing the meaningful meteorological patterns

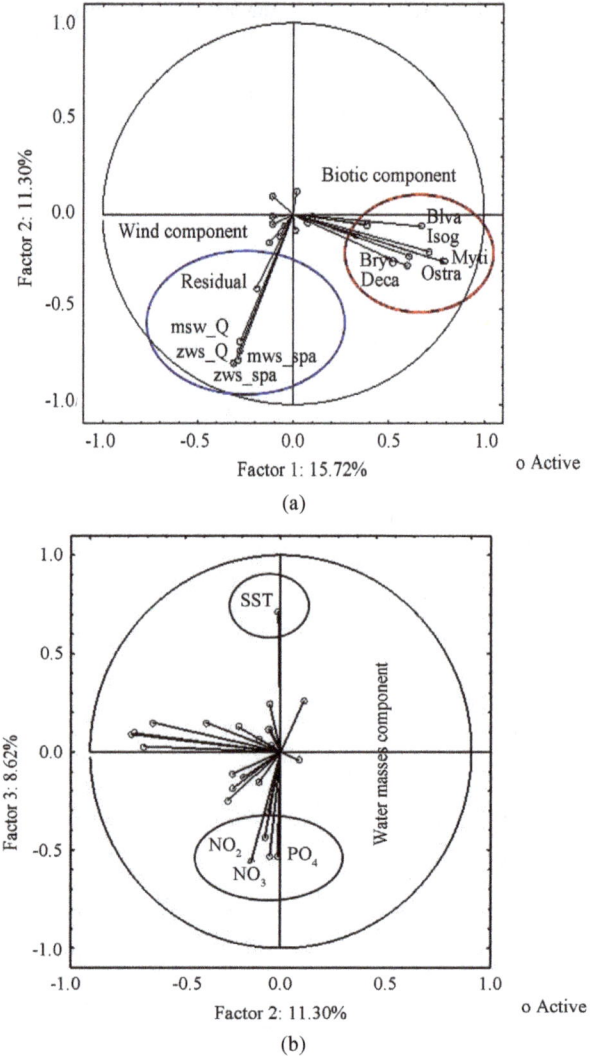

Figure 11. Plot of loadings for the first two components (a), third component (b) with Varimax normalized rotation.

that occur in the region. Comparing the seasonal variations between mean residual, total nitrogen (NO_2, NO_3, and $NH4$), DO, PO_4 and larvae we verify that the critical months with lower values of the residual occur in August, September, October and November. Ascending values of DO are verified from August with maximum peak in November. Total nitrogen and PO_4 present high values from September with maximum peaks. However the larvae present peaks in March to Bivalvia, Ascidiacea and Decapoda; July to Cirripedia and Polychaeta when verifies the presence of the TW; November to Bryosoa, Isognomon, Mytilidae and Ostreidae and December to Cypris, period of occurrence of upwelling.

The cross-correlations between meteorological residual and the biological and chemical variables for different water masses present correlations only in the presence of Central Water (SACW), showing the importance of this water mass in the coastal upwelling process.

CA grouped the sampling in two big clusters, one of biotic variables in one side and abiotic ones at the other, corresponding to the macrostructure of ecosystem. The degree of refinement of similarities allowed a division into six clusters of samples, giving the most satisfactory results at forming distinct clusters, reaching thus the objectives of the classification method. This cluster analysis shows the importance not only the biotic variables but also of the zonal and meridional wind variability, including these variables in a single cluster.

PCA retained and interpreted eight components and they explain 64.10% of the total variance in the data set. The first component was attributed to larvae that had the greatest variability with scores strongly positively correlated, corresponding to the major larvae economically important to the region as crustaceans, mussels and oysters. The second component was strongly negatively correlated with zonal and meridional wind both with the SPA station and with satellite wind data and relatively weak with meteorological residual, showing the sea-air coupling. The third component was attributed to temperature and nutrients, characterizing the local marine ecosystem related to the presence of different water masses. The fourth factor contributes with a small variance and also relates to the biotic components of the classes of zoobenthos (Cirripedia, Polychaeta and Ascidiacea) found in intertidal zones.

The applied methodology shows the relationship between biological and chemical variables and meteorological processes in the main Brazilian region where upwelling occurs, contributing to the improvement of models of coastal management.

5. Acknowledgements

The authors thank the Admiral Paulo Moreira Institute of Marine Studies—IEAPM of Brazilian Navy for data availability and logistical support. The authors also thank the financial support of the Coordination for the Improvement of Higher Level Personnel—Brazilian Research Agency (Capes).

REFERENCES

[1] P. M. S. Monteiro, G. Nelson, A. van der Plas, E, Mabille, G. W. Baileyd and E. Klingelhoeffer, "Internal Tide-Shelf Topography Interactions as a Forcing Factor Governing the Large-Scale Distribution and Burial Fluxes of Particulate Organic Matter (POM) in the Benguela Upwelling System," *Continental Shelf Research*, Vol. 25, No. 15, 2005, pp. 1864-1876.

[2] P. M. A. B. Ré, "Biologia Marinha," Faculdade de Ciências da Universidade de Lisboa, Portugal, 2000.

[3] B. Boehm and S. B. Weisberg, "Tidal Forcing of Entero-

cocci at Marine Recreational Beaches at Fortnightly and Semidiurnal Frequencies," *Environmental Science Technology*, Vol. 39, No. 15, 2006, pp. 5575-5583.

[4] F. Pereira, A. L. Belém, M. C. Belmiro and R. Geremias, "Tide-Topography Interaction along the Eastern Brazilian Shelf," *Continental Shelf Research*, Vol. 25, No. 12-13, 2005, pp. 1521-1539.

[5] S. A. Piontkovski, M. R. Landry, Z. Z. Finenko, A. V. Kovalev, R. Williams, C. P. Gallienne, A. V. Mishonov, V. A. Skryabin, Y. N. Tokarev and V. N. Nikolsky, "Plankton Communities of the South Atlantic Anticyclonic Gyre: Communautés Planctoniques du Tourbillon Anticyclonique de l'Atlantique Sud," *Oceanologica Acta*, Vol. 26, No. 3, 2003, pp. 255-268.

[6] R. H. Stewart, "Introduction to Physical Oceanography," Texas A & M University, 2007.

[7] D. T. Pugh, "Changing Sea Levels: Effects of Tides, Weather and Climate," Cambridge University Press, 2005, 265 p.

[8] S. Franco, "Tides: Fundamentals Analysis and Prediction," IPT, São Paulo, 1981.

[9] N. A. Pore, "The Relation of Wind and Pressure to Extratropical Storm Surge at Atlantic City," *Journal of Applied. Meteorology*, Vol. 3, No. 2, 1964, pp. 155-163.

[10] D. T. Pugh, "Tides, Surges and Mean Sea Level," John Wiley & Sons, 1987, 472 p.

[11] D. Prandle and J. Wolf, "The Interaction of Surge and Tide in the North Sea and River Thames," *Geophysical Journal of the Royal Astronomical Society*, Vol. 55, No. 1, 1978, pp. 203-216.

[12] J. Wolf, "Surge-Tide Interaction in the North Sea and River Thames, in Floods due to High Winds and Tides," Elsevier, New York, 1981, pp. 75-94.

[13] K. J. Horsburgh and C.Wilson, "Tide-Surge Interaction and its Role in the Distribution of Surge Residuals in the North Sea," *Journal of Geophysical Research*, Vol. 112, No. C8, 2007, pp. 1-13.

[14] K. Myrberg, O. Andrejev and A. Lehmann, "Dynamic Features of Successive Upwelling Events in the Baltic Sea—A Numerical Case Study," *Oceanologia*, Vol. 52, No. 1, 2010, pp. 77-99.

[15] M. M. F. de Oliveira, N. F. Ebecken, de J. L. F. Oliveira and I. A. Santos, "Neural Network Model to Predict a Storm Surge," *Journal of Applied Meteorology and Climatology*, Vol. 48, No. 1, 2009, pp. 143-155.

[16] M. M. F. de Oliveira, G. C. Pereira, J. L. F. de Oliveira and N. F. F. Ebecken, "Large and Mesoscale Meteo-Oceanographic Patterns in Local Responses of Biogeochemical Concentrations," *Environmental Monitoring and Assessment*, Vol. 184, No. 1, 2012, pp. 6935-6953.

[17] E. D. Campos, J. L. Miller, T. J. Muller and R. G. Peterson, "Physical Oceanography of the Southwest Atlantic

Ocean," *Oceanography*, Vol. 8, No. 3, 1995, pp. 87-91.

[18] L. P. Pezzi, "Variabilidade do Sistema Oceano-Atmosfera no Oceano Atlântico Sudoeste," I Seminário Sobre Sensoriamento Remoto Aplicado à Pesca, INPE, S: J: dos Campos, 2006.

[19] M. A. Guimaraens and R. Coutinho, "Spatial and Temporal Variation of Benthic Marine Algae at the Cabo Frio Upwelling Region," *Aquatic Botany*, Vol. 52, No. 4, 1996, pp. 283-299.

[20] G. C. Pereira, R. Coutinho and N. F. F. Ebecken, "Data Mining for Environmental Analysis and Diagnostic: A Case Study of Upwelling Ecosystem of Arraial do Cabo," *Brazilian Journal of Oceanography*, Vol. 56, No. 1, 2008, pp. 1-18.

[21] C. D. Ahrens, "Meteorology Today: An Introduction to Weather Climate and the Environment," 6th Edition, Brooks/Cole, London, 2000.

[22] R. B. de. Souza, "Oceanografia por Satélites," 2nd Edition, Oficina de Textos, São Paulo, 2008.

[23] S. A.Gaeta, J. A. Lorenzetti, L. B. Miranda, S. M. M. Susimi-Ribeiro, M. Pompeu and C. E. S. Araújo, "The Vitória Eddy and Its Relation to the Phytoplankton Biomass and Primary Production during the Austral Fall of 1995," *Archive of Fishery and Marine Research*, Vol. 47, No. 2-3, 1999, pp. 253-270.

[24] R. P. Matano, M. Schlax and D. B. Chelton, "Seasonal Variability in the Southwestern Atlantic," *Journal of Geophysical Research*, Vol. 98, No. C10, 1993, pp. 18027- 18035.

[25] M. L. G. Rodrigues, D. Franco and S. Sugahara, "Climatologia de Frentes Frias no Litoral de Santa Catarina," *Revista Brasileira de Geofísica*, Vol. 22, No. 2, 2004, pp. 135-151.

[26] J. L. Valentin, "The Dynamics of Plankton in the Cabo Frio Upwelling," In: F. P. Brandini, Ed., *Memórias do III EBP*, Caiobá-Curitiba, 1989.

[27] D. Topcu, H. Behrendt, U. Brockmann and U. Claussen, "Natural Background Concentrations of Nutrients in the German Bight Area (North Sea)," *Environment Monitoring and Assessment*, Vol. 174, No. 1-4, 2010, pp. 361-388.

[28] R. P. Morgan and K. M. Kline, "Nutrient Concentrations in Maryland Non-Tidal Streams," *Environment Monitoring and Assessment*, Vol. 178, No. 1-4, 2011, pp. 221-235.

[29] M. Ribas-Ribas, J. M. Hernández-Ayón, V. F. Camacho-Ibar, A. Cabello-Pasini, A. Mejia-Trejo, R. Durazo, S. Galindo-Bect, A. J. Souza, J. M. Forja and A. Siqueiros-Valencia, "Effects of Upwelling, Tides and Biological Processes on the Inorganic Carbon System of a Coastal Lagoon in Baja California," *Estuarine, Coastal and Shelf Science*, Vol. 95, No. 4, 2011, pp. 367-376.

[30] R: N. Gibson, "Go with the Flow: Tidal Migration in Marine Animals," *Hydrobiologia*, Vol. 503, No. 1-3, 2003, pp. 153-161.

[31] J. K. Breckenridge and S. M. Bollens, "Vertical Distribution and Migration of Decapod Larvaein Relation to Light and Tides in Willapa Bay," *Estuaries and Coasts*, Vol. 34, No. 6, 2011, pp. 1255-1261.

[32] C. A. Silveira, A. C. K. Schimidt, E. J. D. Campos, S. S. Godoi and Y. Ikeda, "The Brazil Current off the Eastern Brazilian Coast," *Brazilian Journal of Oceanography*, Vol. 48, No. 2, 2000, pp. 171-183.

[33] I. Soares and O. M.oller Jr., "Low-Frequency Currents and Water Mass Spatial Distribution on the Southern Brazilian Shelf," *Continental Shelf Research*, Vol. 21, 2001, pp. 1785-1814.

[34] SCOR, "Protocols for the Joint Global Ocean Flux Study (JGOFS) Core Measurements," Scientific Committee on Ocean Research, International Council of Scientific Unions, Bergen, Vol. 9, 1996, 170 p.

[35] T. A. Richard and T. G. Thompson, "The Estimation and Characterization of Plankton Population by Pigment Analyses. A Spectrophotometric Method for the Estimation of Plankton Pigments," *Journal of Marine Research*, Vol. 11, 1952, pp. 156-172.

[36] N. Sharma and E. D'Sa, "Assessment and Analysis of QuikSCAT Vector Wind Products for the Gulf of Mexico. A Long-Term and Hurricane Analysis," *Sensors*, Vol. 8, No. 3, 2008, pp. 1927-1949.

[37] Y. Chao, Z. Li, J. C. Kindle, J. D. Paduan and F. P. Chavez, "A High-Resolution Surface Vector Wind Product for Coastal Oceans: Blending Satellite Scatterometer Measurements with Regional Mesoscale Atmospheric Model Simulations," *Geophysical Research Letters*, Vol. 30, No. 1, 2003, pp. 1-13.

[38] W. Tang, W. T. Liu and B. W. Stiles, "Evaluation of High-Resolution Ocean Surface Vector Winds Measured by QuikSCAT Scatterometer in Coastal Regions," *IEEE Transactions on Geoscience and Remote Sensing*, Vol. 42, No. 8, 2004, pp. 1762-1769.

[39] J. L Stech and J. A. Lorenzzetti, "The Response of the South Brazil Bight to the Passage of Wintertime Cold Fronts," *Journal of Geophysics Research*, Vol. 97, No. C6, 1992, pp. 9507-9520.

[40] S. Sharma, "Applied Multivariate Techniques," John Wiley and Sons, Inc., New York, 1996.

[41] H. Abdi and L. J. Williams, "Principal Component Analysis," *Wiley Interdisciplinary Reviews*: *Computational Statistics*, Vol. 2, No. 4, 2010, pp. 387-515.

[42] N. Ruggieri, M. Castellano, M. Capello, S. Maggi and P. Povero, "Seasonal and Spatial Variability of Water Quality Parameters in the Port of Genoa, Italy, from 2000 to 2007," *Marine Pollution Bulletin*, Vol. 62, No. 2, 2011, pp. 340-349.

[43] M. Ujević Bošnjak, K. Capak, A. Jazbec, C. Casiot, L. Sipos, V. Poljak and Ž. Dadić, "Hydrochemical Characterization of Arsenic Contaminated Alluvial Aquifers in Eastern Croatia Using Multivariate Statistical Techniques and Arsenic Risk Assessment," *Science of the Total Environment*, Vol. 420, 2012, pp. 100-110.

[44] J. V. E. Bernardi, L. D. Lacerda, J. G. Dórea, P. M. B. Landim, J. P. O. Gomes, R. Almeida, A. G. Manzatto and W. R. Bastos, "Aplicação da Análise das Componentes Principais na Ordenação dos Parâmetros Físico-Quimicos no Alto Rio Madeira e Afluentes, Amazônia Ocidenta," *Geochimica Brasiliensis*, Vol. 23, No. 1, 2009, pp. 79-90.

[45] A. Stigebrandt, "Resistance to Barotropic Tidal Flow in Straits by Baroclinic Wave Drag," *Journal of Physical Oceanography*, Vol. 29, No. 2, 1999, pp. 191-197.

[46] M. Apolinário, "Variation of Populations Densities between Two Species of Barnacles (Cirripedia: Megab- alaninae) at Guanabara Bay and Nearly Islands in Rio de Janeiro/RJ," *Nauplius*, Vol. 9, No. 2, 2001, pp. 21-30.

[47] M A. Champ and F. L. Lowenstein, "TBT—The Dilemma of High-Technology Antifouling Paints," *Oceanus*, Vol. 35, 1987, pp. 69-77.

[48] M. B. Rocha, V. Radashevsky and P. C. Paiva, "Espécies de Scolelepis (Polychaeta, Spionidae) de Praias do Estado do Rio de Janeiro, Brasil," *Biota Neotrópica*, Vol. 9, No. 4, 2009.

[49] S. M. Lima, J. R. Moreira, A. V. Von Osten, M. Soares and L. Guilhermino, "Biochemical Responses of the Marine Mussel *Mytilus galloprovincialis* to Petrochemical Environmental Contamination along the North-Western Coast of Portugal," *Chemosphere*, Vol. 66, No. 7, 2007, pp. 1230-1242.

[50] StatSoft Inc., "*Statistica*," Data Analysis Software System, version 7, 2004.

[51] B. Torrano-Siva, "Fitobentos (Macoalgas) in: Informe Sobre as Espécies Exóticas Invasoras Marinhas no Brasil," Ministério do Meio Ambiente, MMA, 2009.

[52] S. A. Narandio, J. L. Carvalho and J. C. García-Gomes, "Effects of Environmental Stress on Ascidians Populations in Algeciras bay (Southerm Spain)," *Marine Ecology*, Vol. 144, 1996, pp. 119-131.

[53] T. C. C. Lambert and G. Lambert, "Non-Indigenous Ascidians In southerm California Harbors and Marinas," *Marine Biology*, Vol. 130, No. 4, 1998, pp. 675-688.

[54] R. P. M. Bak, M. Joenje, I. De Jong, D. Y. M. Lambrechts and M. L. J. Van Veghel, "Long-Term Changes on Coral Reef in Booming Populations of a Competitive Colonial Ascidian," *Marine Ecology*, Vol. 133, 1996, pp. 303-306.

10

Traffic Impacts on Fine Particulate Matter Air Pollution at the Urban Project Scale: A Quantitative Assessment

Chidsanuphong Chart-asa, Kenneth G. Sexton, Jacqueline MacDonald Gibson

University of North Carolina at Chapel Hill, Chapel Hill, USA.

ABSTRACT

Formal health impact assessment (HIA), currently underused in the United States, is a relatively new process for assisting decision-makers in non-health sectors by estimating the expected public health impacts of policy and planning decisions. In this paper we quantify the expected air quality impacts of increased traffic due to a proposed new university campus extension in Chapel Hill, North Carolina. In so doing, we build the evidence base for quantitative HIA in the United States and develop an improved approach for forecasting traffic effects on exposure to ambient fine particulate matter (PM2.5) in air. Very few previous US HIAs have quantified health impacts and instead have relied on stakeholder intuition to decide whether effects will be positive, negative, or neutral. Our method uses an air dispersion model known as CAL3QHCR to predict changes in exposure to airborne, traffic-related PM2.5 that could occur due to the proposed new campus development. We employ CAL3QHCR in a new way to better represent variability in road grade, vehicle driving patterns (speed, acceleration, deceleration, and idling), and meteorology. In a comparison of model predictions to measured PM2.5 concentrations, we found that the model estimated PM2.5 dispersion to within a factor of two for 75% of data points, which is within the typical benchmark used for model performance evaluation. Applying the model to present-day conditions in the study area, we found that current traffic contributes a relatively small amount to ambient PM2.5 concentrations: about 0.14 $\mu g/m^3$ in the most exposed neighborhood—relatively low in comparison to the current US National Ambient Air Quality Standard of 12 $\mu g/m^3$. Notably, even though the new campus is expected to bring an additional 40,000 daily trips to the study community by the year 2025, vehicle-related PM2.5 emissions are expected to decrease compared to current conditions due to anticipated improvements in vehicle technologies and cleaner fuels.

Keywords: PM2.5; Traffic; Health Impact Assessment

1. Introduction

The World Health Organization and other public health advocates have long stressed the need for formal health impact assessment (HIA) to inform decision-making in sectors outside the health-care industry [1-3]. The rationale is that chronic diseases that pose major health burdens in the post-industrial world are driven largely by policy, program, and planning decisions in transportation, agriculture, urban planning, and other sectors that ordinarily do not include population health as an objective in their decision processes. Commonly cited examples include the effects of government agricultural subsidies on the availability of healthy foods and the effects of trans-

portation plans on population exposure to noise and air pollution. HIA is intended to encourage decision-makers in these and other sectors to make choices that minimize negative and maximize positive impacts on public health, within budgetary and other constraints. The intent of HIA is to prevent the chronic, noninfectious diseases—including heart disease, stroke, and diabetes—that have replaced infectious diseases as the leading health concerns in post-industrialized nations [4]. Health practitioners have long recognized that exposures to risk factors for these chronic diseases are driven by a wide range of policy, planning, and program decisions in multiple sectors and that prevention through better-informed decision-making in all sectors is likely to be less costly than treating the symptoms [2].

While the practice of HIA is well established in the European Union and some other nations, in the United States HIA practice is relatively new [2,5,6]. The first U.S. HIA, which evaluated the health impacts of a proposed policy to increase the minimum wage in San Francisco, was completed in 1999 [2,7]. By the end of 2012, at least 114 additional HIAs had been completed in the United States [8]. However, only 14 of these HIAs provided quantitative estimates of the impacts of alternative choices on health [9]. The rest are qualitative, relying on the judgment of the HIA practitioner to determine whether one choice will be more or less detrimental or beneficial to population health, in comparison with other options. In the US urban planning and transportation sectors, such qualitative HIAs are of little use. In order to prioritize urban planning and transportation projects, state and local planning and transportation agencies employ cost-benefit analysis. To be able to include health impacts in these cost-benefit analyses, quantitative estimates of health impacts—in terms of numbers of illnesses and premature deaths—are essential. Yet, a recent review found that only four HIAs in the transportation and urban planning sectors in the United States had employed quantitative methods, and all of these were conducted in major metropolitan areas in California [9].

In order to expand the evidence base for the use of quantitative HIA to support planning and transportation decisions in the United States, this paper presents an improved approach for quantifying the future air quality effects of increased traffic brought by new urban or suburban development projects. We focus specifically on predicting exposure to airborne fine particulate matter (i.e., particles with diameter less than or equal to 2.5 μm, denoted as PM2.5), which often is used as a marker of near-roadway air pollution to support health effects estimates. We then demonstrate the modeling approach for a case study site: a proposed extension to the campus of the University of North Carolina (UNC) at Chapel Hill, in the United States.

Our modeling approach improves on those in the previous four US transportation-related HIAs in several ways. First, it accounts for the effects of acceleration, deceleration, and idling on all roadway links in the study corridor using an approach recommended by Ritner et al. but not previously employed in an HIA [10]. Second, it compares model predictions to measured pollutant concentrations along the roadway corridor. According to Ritner et al., such a performance evaluation has not been previously completed. Third, it improves on the Ritner et al. approach by developing a new algorithm to incorporate daily temperature variability.

The planned future project used as the case study for demonstrating the new modeling method is known as "Carolina North," which is planned as an extension to the current UNC campus. UNC-Chapel Hill is the oldest public university in the United States and has a current student population of more than 29,000 [11]. The campus is located in the town of Chapel Hill, which has a population just over 57,000 [12]. The planned new campus will be located about 3 km (2 miles) north of the existing campus (**Figure 1**). If constructed, it is expected to increase the number of trips to the area by 10,000 per day by 2015—half of those by private vehicle—and, accordingly, to substantially increase traffic in the surrounding neighborhoods [13]. By 2025, the number of additional daily trips to the campus is expected to increase by as many as 40,000 [13]. The main traffic effects are expected along Martin Luther King Jr. Boulevard, the main thoroughfare connecting the new campus to both the existing campus (to the south) and the nearest highway interchange (to the north).

UNC commissioned a transportation impact analysis in 2009 in order to estimate the anticipated increases in traffic volumes, but the air quality impacts of the increased traffic were not evaluated. Hence, the transportation impact analysis cannot be used directly to support decision-making about whether alternative transportation network designs (including, for example, new or expanded public transit routes) may be needed to prevent traffic-related air quality degradation and associated health impacts. By quantifying the air quality effects of additional traffic generated by the future campus, this paper can support a future quantitative HIA to inform local transportation and planning decisions.

2. Materials and Methods

Our process for modeling population exposure to excess PM2.5 attributable specifically to increased traffic from the Carolina North campus builds on a new approach recommended by Ritner et al. [10], who proposed an algorithm to account for vehicle acceleration, deceleration, and idling at intersections in modeling of near-roadway pollutant concentrations. We improved on the Ritner et al. approach by developing a new algorithm for incorporating hourly temperature variability in the estimation. We then tested our predictions against roadside air quality measurements. We analyzed near-roadway air quality for three different scenarios: 2009 conditions, 2025 conditions assuming the new campus is not built, and 2025 conditions assuming the campus is built. Information on traffic counts for all these scenarios came from the previously completed transportation impact analysis [14]. We modeled air quality effects only for daytime traffic (6 a.m. to 7 p.m.), since we assume that the major impacts will occur during these hours.

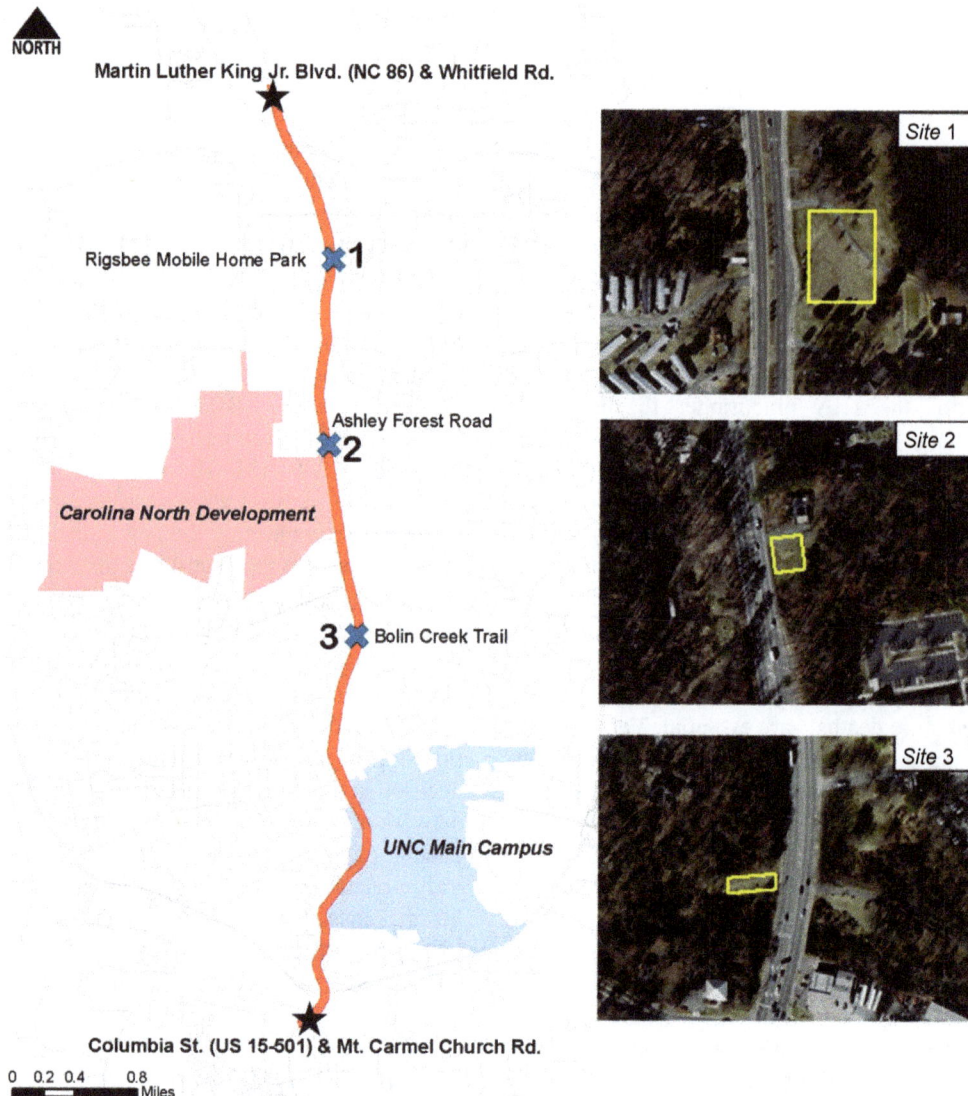

Figure 1. The study corridor runs from the intersection of Martin Luther King Jr. Boulevard and Whitfield Road to the intersection of South Columbia Street and Mt. Carmel Church Road, Chapel Hill, NC. This map also shows the locations of the three selected study sites. Site 1 is on the east side of Martin Luther King Jr. Blvd., opposite the Rigsbee Mobile Home Park. Site 2 is on the east side of Martin Luther King Jr. Blvd. near Ashley Forest Rd. Site 3 is on the west side of Martin Luther King Jr. Blvd., opposite the entrance to Bolin Creek.

We modeled PM2.5 concentrations at each of the 160 census blocks located within 500 m of the study corridor (following guidance from the Health Effects Institute suggesting that key traffic-related pollution impacts occur within 300 - 500 m of major roadways) [15]. Approximately 16,000 people live within these census blocks [16]. In this study, the population exposures in each census block are represented by the estimated 24-hour PM2.5 concentrations at each receptor.

2.1. Modeling Approach

Our modeling framework includes nine Steps (**Figure 2**):
  Step 1: Divide roadway into links for analysis. Air

emissions from any single vehicle depend substantially on the vehicle speed, vehicle acceleration, time spent idling, and road grade. To account for these effects, we followed the approach of Ritner *et al.* by dividing the study corridor roadway into very short links [10]. In total, we modeled 1200 links along the 8.2 km (5.1 mile) study corridor. Each link has a roughly constant road grade; fraction of vehicle time spent decelerating, idling or accelerating; and moving speed. We used ArcGIS 9.3.1 (ESRI, Redlands, CA) and 2010 aerial photos from the Orange County Geographic Information Systems (GIS) Division to draw the series of links [17]. Link-specific traffic activities were determined based on the simulated traffic data for 2009, 2025 no-build, and 2025 build

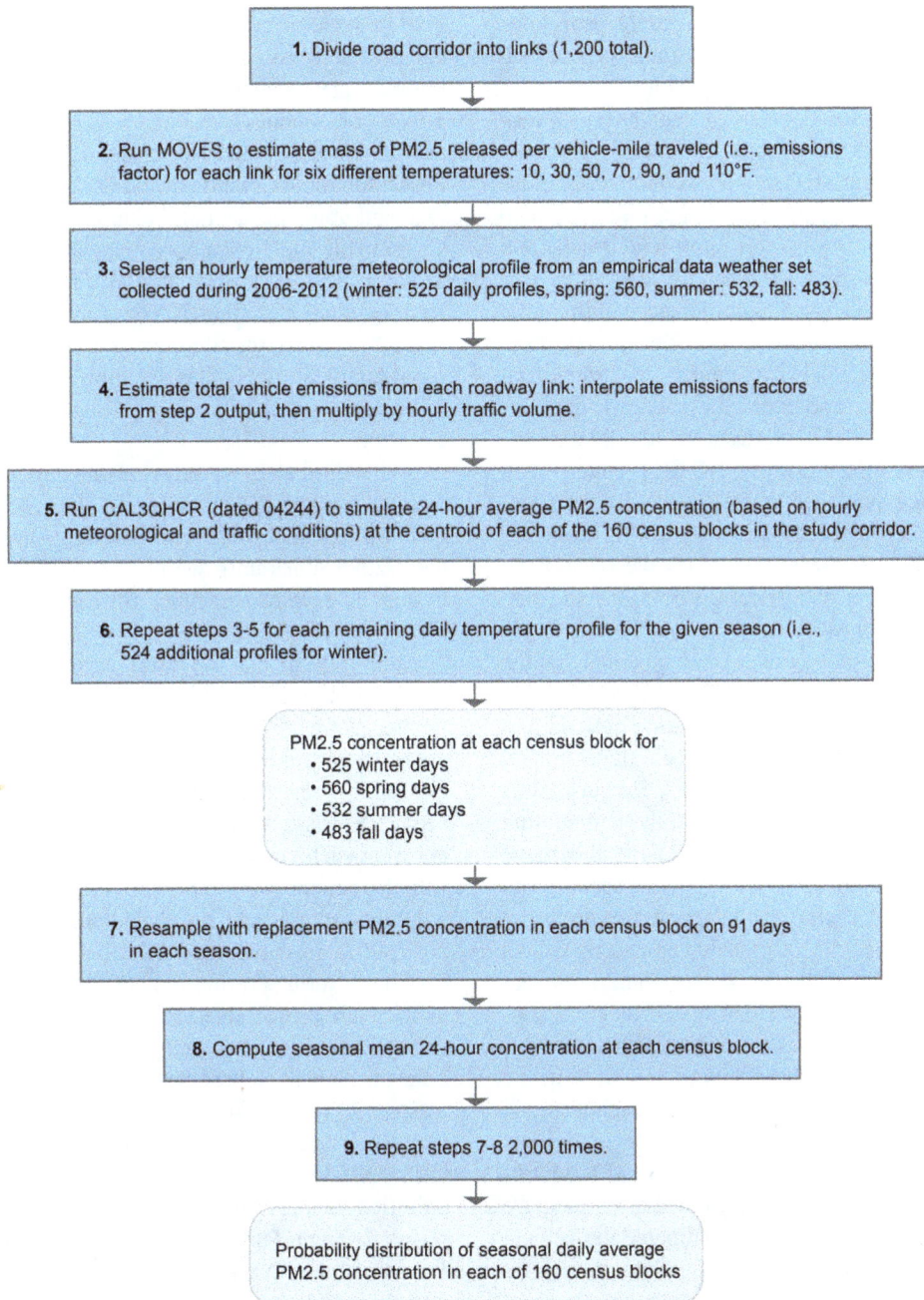

Figure 2. Flowchart showing the nine steps of our modeling framework.

scenarios from the transportation impact analysis [14]. Link-specific average speeds were assumed to be equal to speed limits based on GIS street maps from the Town of Chapel Hill [18]. The speed limit was 25 mph for 17% of the links, 35 mph for 68% of the links, and 45 mph for the remaining 15%. Link-specific grades were derived from GIS contour maps from the Town of Chapel Hill [19] and ranged from 0% - 10%.

Step 2: Estimate vehicle emissions factors for six different temperatures for each link using MOVES. As suggested by both Ritner *et al.* [10] and the US Environ-

mental Protection Agency's (EPA) "Guidance on Quantitative PM Hot-Spot Analyses for Transportation Conformity" [20], we used MOVES 2010b (Motor Vehicle Emission Simulator, EPA, Washington, DC) to develop 2009 and 2025 link-specific emission rates of PM2.5 (grams/vehicle-mile), according to link-specific traffic activities, average speeds, and grades. The MOVES model was developed by the EPA based on laboratory tests that measured emissions from different kinds of vehicles under conditions designed to represent typical driving behaviors. Unlike its predecessor, known as

MOBILE6, MOVES can provide separate emissions factors for different vehicle operation modes: acceleration, deceleration, idling, and cruising [10].

MOVES models emissions for 13 vehicle types: motorcycle, passenger car, passenger truck, light commercial truck, intercity bus, transit bus, school bus, refuse truck, single unit short-haul truck, single unit long-haul truck, motor home, combination short-haul truck, and combination long-haul truck. It also considers three fuel types: gasoline, diesel, and compressed natural gas. Hence, in order for the model to provide accurate estimates for any specific roadway segment, the fraction of vehicles in each class and fuel type category must be estimated. For this analysis, we used vehicle fleet distribution data from Guilford County, NC [21] (county seat: Greensboro), since data specific to Chapel Hill were unavailable. The fuel type distributions as well as fuel supply and formulation in the project areas were based on national defaults. These data (fleet distributions and fuel types) were fixed in all MOVES runs.

The EPA's PM hot-spot guidance recommends that the link-specific emission rates should be prepared based on average temperatures for four different time periods in a day for each season, meaning that each development scenario would require 16 MOVES runs. However, this approach does not fully account for daily temperature variability within a given season. Previous studies have shown that PM emission rates are highlight sensitive to temperature, and hence omitting temperature variability could decrease the accuracy of modeled emissions factors [22,23]. Our new algorithm for representing intra-seasonal variability in temperature and meteorological conditions runs MOVES for six different temperatures: 10, 30°F, 50°F, 70°F, 90°F, and 110°F [24]. Later steps of the algorithm (described below) interpolate between these six estimates to determine temperature-specific emissions factors for each roadway link. For example, if a wintertime simulation of any given hour yielded a temperature of 40 degrees for that hour, we then estimated the vehicle emissions factors to be the average of the emissions factors for 30 and 50 degrees.

Step 3: Select an hourly temperature and meteorological profile from empirical weather data. The meteorological data to estimate probability distributions of the effects of weather on PM2.5 concentrations for each season were obtained from the EPA's Meteorological Processor for Regulatory Models, using 2006-2012 surface and upper air data at the national weather stations in Chapel Hill and Greensboro respectively [24,25]. A total of 2,100 days with complete required data were used in the modeling, including 525 days for winter, 560 days for spring, 532 days for summer, and 483 days for fall. Seasonal temperature profiles are shown in **Figure 3**. **Figure 4** shows the distributions of seasonal wind speed and direction.

In this third step, we selected one day from these 2,100 days to support the modeling in steps 4 - 5 below, and then we repeated this selection (step 6) without replacement 2099 times until we had estimated PM2.5 concentrations in each census block for each day having a complete weather record.

Step 4: Estimate the total emissions from vehicles traveling on each roadway link. The MOVES model estimates average per-vehicle emissions in grams per vehicle-mile, accounting for the specific distribution of vehicle types, ages, and fuel sources at the study site. The next step was to compute the total mass of PM2.5 emitted from each vehicle on each roadway link. For this step, vehicle counts were needed. The link-specific traffic volumes were based on the simulated traffic data for 2009, 2025 no-build, and 2025 build scenarios from the Carolina North Traffic Impact Analysis [14]. For the temperature profile selected in step 3, we estimated emissions factors by interpolating between the outputs of step 2 for the nearest two temperatures.

Step 5: Model dispersion of PM2.5 from roadway emissions into the surrounding neighborhoods using CAL3QHCR. The PM hot-spot guidance suggests two air pollution dispersion models—CAL3QHCR (EPA, Research Triangle Park, NC) or AERMOD (EPA, Research Triangle Park, NC)—for simulating PM2.5 pollution dispersion from roadways. Both models are based on

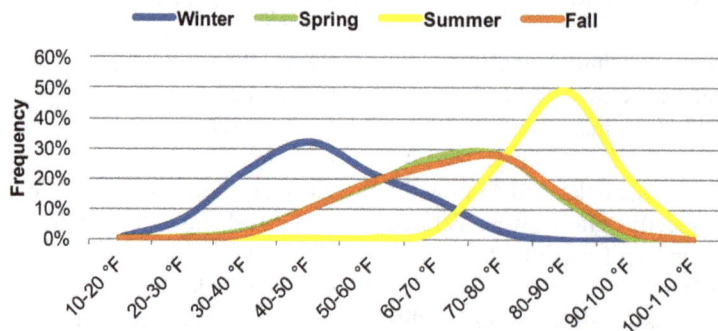

Figure 3. Seasonal temperature profiles from 6 a.m. to 7 p.m., according to the meteorological data used in the CAL3QHCR modeling.

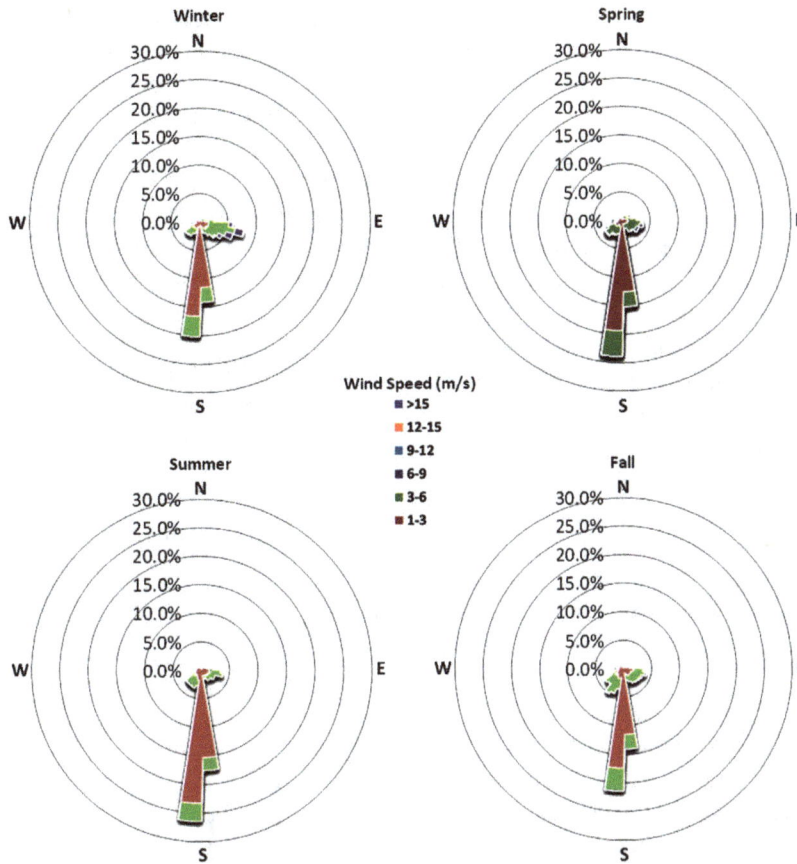

Figure 4. Seasonal wind roses from 6 a.m. to 7 p.m., according to the meteorological data used in the CAL3QHCR modeling.

Gaussian plume dispersion. However, a recent model comparison study suggested that CAL3QHCR requires less meteorological data and user effort and appears to perform better than AERMOD for analyses at the urban project scale [26]. In this study, we tested and used CAL3QHCR for estimating population exposure to PM2.5 ($\mu g/m^3$) from the study corridor. As described below under "model validation approach," we tested two different versions of CAL3QHCR: one dated 13196 and the other dated 04244. We then used the best-performing of the two in subsequent simulations. We ran CAL3QHCR for each roadway link using the meteorological profile from step 3 and the per-link total PM2.5 emissions from step 4. We modeled concentrations at an elevation of 1.5 m, corresponding to the elevation of the adult breathing zone.

Steps 6 - 9: Generate probability distribution of seasonal average 24-hour PM2.5 concentration. As **Figure 2** outlines, we first repeated steps 3-5 for each of the days (2,100 in total) for which historical empirical weather data were available. The result was 2,100 separate daily estimates of the PM2.5 concentration at each of the 160 census block centroids: 525 winter day estimates and 560, 532, and 483 spring, summer, and fall estimates, respectively. We then used a bootstrap technique to estimate a

probability distribution for the average daily PM2.5 concentration in each season. Specifically, for each season, we resampled with replacement 91 days from the simulated daily PM2.5 concentration estimates. We then computed the mean value of these 91 daily estimates for each receptor. Then, we repeated this process of computing a seasonal mean 1999 times, in order to generate a sample of 2000 seasonal mean 24-hour PM2.5 concentrations. This sample then served as the basis for developing a probability distribution of the seasonal mean concentration for each season.

2.2. Model Validation Approach

This study tested the performance of the combined MOVES-CAL3QHCR modeling approach by comparing model predictions against roadside measurements at three selected sites along the study corridor (**Figure 1**). Furthermore, we compared the predictive validity of two versions of CAL3QHCR (dated 04244 and dated 13196) According to the model change bulletin, the mixed mode rounding in the internal calculations of CAL3QHCR dated 04244 was removed from CAL3QHCR dated 13196. Consequently, the simulated concentrations from these two model versions are different in some cases.

We used a DustTrak DRX Aerosol Monitor Model 8534 (TSI, Shoreview, MN) to measure total PM2.5 concentrations at each of the three sites The DustTrak DRX instrument or similar models have been used in roadside measurements in several previous studies [27-29]. The DustTrak can detect concentrations from 1 to 150,000 $\mu g/m^3$ with an error of $\pm 0.1\%$ of the monitored concentration [30]. All of these instruments are calibrated at the factory with a known mass concentration of Arizona Test Dust (ISO 12103-1, A1 test dust) [31]. In addition, in each sampling period, we calibrated the instrument before taking measurements. During all sampling events, the DustTrak was held about 1.5 m above the ground (the adult breathing zone height) and programmed to record the total concentration every five seconds.

We collected samples on two separate days at Site 1 and on one day at Sites 2 and 3 for a total of four sampling days in the study corridor. During three of the four sampling days, we monitored PM2.5 concentrations during the morning and evening peak traffic periods and also in the middle of the day four an hour at a time (roughly 8:00 - 9:00 a.m., noon-1:00 p.m., and 5:00 - 6:00 p.m.). At Site 2, the property owner requested that we not collect samples in the evening, so we only sampled during the morning and noon hours. **Table 1** shows sample collection dates and measured PM2.5 concentrations.

During each sampling event, we drew continuous air samples for three minutes at 10 m from the roadway and then repeated the three-minute sampling at locations of 30 m and 50 m from the roadway (except at Site 2, where

obstructions prevented sampling at 50 m). Then, we repeated this process over the course of about one hour. As a result, at each site and during each sampling event, we collected PM2.5 concentrations for six three-minute intervals at 10 m, 30 m, and 50 m perpendicular distances from the roadway, as **Figure 5** illustrates. For each event, we then computed the average PM2.5 concentration measured during these three-minute intervals; **Table 1** shows the resulting estimated one-hour average concentrations.

During each sampling event, we simultaneously collected traffic counts and meteorological data. Traffic was monitored with a hand-held counter, and the counts were confirmed by viewing digital video recordings from a portable video recorder positioned on a tripod to film the roadway during sampling. We measured wind speed using a Skymate model SM-18 wind meter with accuracy within 3% (Campbell Scientific, Inc, Logan Utah); wind

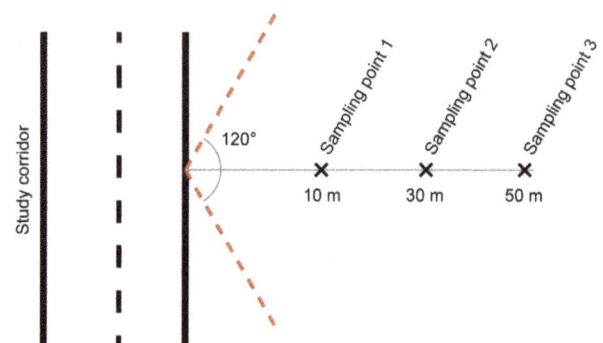

Figure 5. Diagram of sampling points along the study corridor.

Table 1. Measured and modeled PM2.5 concentrations ($\mu g/m^3$).

Site	Date	Time period	Measured concentrations[*]			Measured concentration difference[**]			Predicted concentration differences: CAL3QHCR (04244)			Predicted concentration differences: CAL3QHCR (13196)		
			10 m	30 m	50 m	10 vs. 30 m	10 vs. 50 m	30 vs. 50 m	10 vs. 30 m	10 vs. 50 m	30 vs. 50 m	10 vs. 30 m	10 vs. 50 m	30 vs. 50 m
1	16-May	Morning	14.9	13.8	13.9	1.1	1.0	NEG	0.7	0.9	0.2	0.7	1.0	0.3
		Noon	9.0	8.7	8.3	0.3	0.7	0.4	0.5	0.8	0.3	0.4	0.6	0.1
		Evening	9.7	10.0	9.7	NEG	NEG	0.3	1.0	1.3	0.3	0.9	1.3	0.3
	31-May	Morning	5.1	5.1	4.9	0.0	0.2	0.2	0.7	0.8	0.1	0.7	0.9	0.2
		Noon	2.6	2.2	1.6	0.4	1.0	0.6	0.5	0.8	0.3	0.5	0.6	0.2
		Evening	3.0	2.6	2.4	0.4	0.6	0.2	1.1	1.4	0.3	1.0	1.3	0.4
2	24-Apr	Morning	21.4	20.8	NA	WD	NA	NA	0.5	NA	NA	0.5	NA	NA
		Noon	10.5	10.4	NA	WD	NA	NA	0.7	NA	NA	0.6	NA	NA
3	16-Apr	Morning	10.8	10.8	10.5	NEG	0.3	0.3	0.7	0.9	0.2	0.6	0.8	0.2
		Noon	9.7	9.2	9.0	0.5	0.7	0.2	0.6	1.0	0.4	0.5	0.7	0.2
		Evening	9.2	8.9	8.5	WD	WD	WD	0.0	0.0	0.0	0.1	0.1	0.0

[*]NA indicates PM2.5 could not be measured at this location due to a physical obstruction; [**]Negative values excluded during data cleaning are labeled as "NEG"; those excluded due to unfavorable wind direction are labeled as WD.

direction using a windsock and compass; and temperature, dewpoint, and relative humidity using an Extech model 445814 thermometer-psychrometer with temperature accuracy of ±1.8°F and relative humidity accuracy of ±4%. Data on atmospheric stability class and mixing height were estimated using EPA's Meteorological Processor for Regulatory Models [36]. **Table 2** shows the traffic counts and meteorological conditions for each sampling event.

The measured concentrations at each sampling point represent the sum of background concentrations, PM2.5 contributions from other nearby sources, and traffic-related PM2.5. Therefore, in order to evaluate the performance of the CAL3QHCR model, concentrations of PM2.5 attributable to background and other sources must be subtracted from the monitored concentrations, in order to determine how much of the measured PM2.5 comes from the roadway. In testing model performance, other studies have used background concentrations measured at an upwind location or central air quality monitor [26,32,33]. However, Chapel Hill does not have an active PM2.5 monitor; the nearest PM2.5 monitor is about 45 km away, in Raleigh. Furthermore, due to resource limitations, we were able to use only one DustTrak monitor and hence were unable to capture background concentrations while simultaneously measuring near-road concentrations. Hence, we accounted for the effect of background PM2.5 by characterizing the differentials between the measured concentrations at pairs of sampling points at distances 10 m and 30 m, 10 m and 50 m, and 30 m and 50 m from the roadway. **Table 1** shows these differ-

entials, as computed from the measured concentrations.

A factor-of-two plot has been commonly used to evaluate the performances of the CALINE series of dispersion models (e.g., CALINE3, CAL3QHC/CAL3QCHR, and CALINE4) [26,32-35]. That is, modeled PM concentrations are plotted against measured concentrations to see whether the model estimates are within a factor of two of measured concentrations. Typically, the model is considered valid in predicting the traffic-related concentrations if at least 75% of the comparing pairs are within a factor-of-two envelope. This criterion was also applied in this study. We adopted this approach, comparing measured PM2.5 concentration differences between pairs of points with differences predicted by the two different CAL3QHCR model versions.

2.3. Data Cleaning

In total, the sampling events shown in **Table 1** yielded 29 data points. Of these, five points had to be eliminated because the wind direction was outside of a 120° degree arc from a line drawn perpendicular to the roadway (see **Figure 5**). In such conditions, the monitoring locations were not downwind of the roadway and therefore could not capture roadway contributions to PM2.5 [37]. Four additional data points were eliminated because they indicated negative dispersion (that is, PM2.5 concentrations increased rather than decreased with distance from the roadway). This data cleaning process left 20 data points for comparing measured PM2.5 concentrations to modeled concentrations.

Table 2. Traffic and meteorological data used in CAL3QHCR modeling.

Site	Date	Period	Average Traffic Count (veh/min)	Average Wind Direction (deg)	Average Wind Direction within 120° Arc from Study Corridor?	Average Wind Speed (m/s)	Average Temperature (°F)	Stability Class[*]	Mixing Height (m)[*]
1	16-May	Morning	34	80	Yes	0.8	73.7	Slightly unstable	678
		Noon	26	83[*]	Yes	1.4	85.9	Unstable	1315
		Evening	42	91	Yes	0.7	80.5	Slightly unstable	1395
	31-May	Morning	34	91	Yes	0.9	77.5	Slightly unstable	878
		Noon	29	55	Yes	1.5	88.6	Unstable	1676
		Evening	38	41	Yes	0.8	99.4	Slightly unstable	1776
2	24-Apr	Morning	32	349	No	0.6	56.9	Slightly unstable	670
		Noon	27	37	No	1.1	74.8	Unstable	1360
3	16-Apr	Morning	24	252	Yes	0.2	68.5	Neutral	1869
		Noon	22	264	Yes	0.7	80.0	Very unstable	1939
		Evening	31	20*	No	0.7	77.8	Neutral	1944

NOTE: Wind speeds below 1 m/s were reset to 1 m/s in CAL3QHCR, as suggested by the US EPA [36]. [*]Data obtained from MPRM.

3. Results

3.1. Vehicle Emission Rates

The output from MOVES can provide useful insights about the vehicle classes contributing most to roadside pollution, the effects of meteorological and road characteristics on per-vehicle emissions, and the effects of future vehicle technologies.

To identify the vehicle classes contributing most to roadway emissions, we ran MOVES for a study corridor link with 0% grade, a 35 mph speed limit, and an ambient temperature of 90˚F. **Figure 6** shows the results. This analysis reveals that trucks are the major contributors to roadside emissions for this corridor. In total, trucks of all categories contribute 79% of emissions: 19% from passenger trucks (e.g., sport utility vehicles) and the remaining 60% from various kinds of commercial trucks. Consistent with this result, diesel-fueled vehicles account for nearly two-thirds (64%) of emissions whereas gasoline-fueled vehicles account for 36%. As well, vehicles more than 10 years old account for half of the roadside

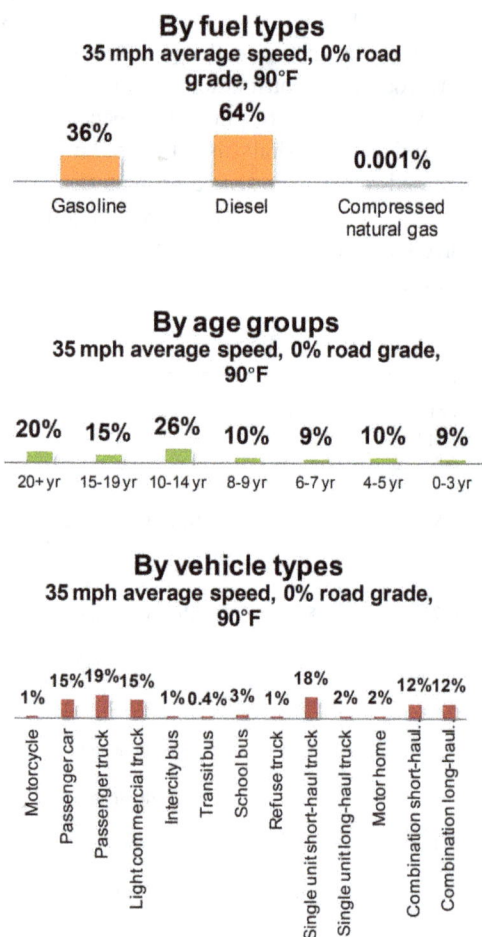

emissions. Hence, improving emissions controls or engine efficiency in diesel-fueled trucks, plus retiring older vehicles, could greatly reduce roadside emissions in the study corridor.

MOVES output also shows the important effects of temperature, road grade, and vehicle speed on roadway emissions. As **Figure 7** shows, emissions decrease as temperature increases, increase as road grade increases, and decrease as vehicle speed increases. These results illustrate the importance for modeling of accurately capturing temperature, vehicle speed, and especially road grade—hence the importance of dividing a study corridor into short links as in our study.

Interestingly, the results show that 2009 link-specific emission rates (ranging from 0.02 - 0.50 g/veh-mile) are *higher* than 2025 link-specific emission rates (ranging from 0.01 - 0.26 g/veh-mile). The differences result from the assumption, built into MOVES, that future vehicles will have more efficient engines that reduce emissions and will use cleaner fuels.

3.2. Model Performance Evaluation

Figure 8 compares the predictions of the two CAL3QHCR model versions to measurements of pollutant dispersion along the roadway corridor. The figure also shows the "factor-of-two envelope:" that is, the range of predictions that are within a factor of two of the measured dispersion. As shown, the models contain both under-predictions of

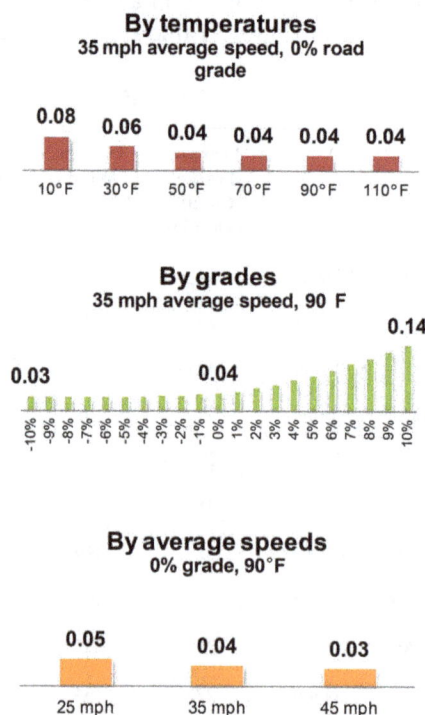

Figure 6. Example of 2009 link-specific emission rate fractions (%) at 35 mph average speed, 0% grade, and 90˚F by fuel types, age groups, and vehicle types.

Figure 7. Examples of 2009 link-specific emission rate (g/veh-mile) changes by average speeds, grades, and temperatures.

the amount of dispersion (*i.e.*, data points below the factor-of-two envelope) and over-predictions (data points more than twice the measured amount). However, both models are more likely to over-predict than to under-predict dispersion: that is, to predict greater concentration differences as one moves away from the roadway than were actually measured. Possible reasons for this prediction error include physical obstacles to dispersion (for example, at site 3, a large rock outcropping may interfere with dispersion) and intermittent winds. Previous model evaluations also have observed that the predecessor to CAL3QHCR did not perform well in the presence of street canyons or other physical obstacles or when winds are intermittent [32].

Of the two models, model 1 (the version dated 04244) performs better than model 2 (the version dated 1196). For model 1, 15 modeled estimates (75%) were within a factor of two of the measured value. Previous studies have suggested that a 75%, factor-of-two prediction capability indicates reasonable model performance, and model 1 achieves this metric [32]. For model 2, 13 observations (65%) were within a factor of two of observed values. Because model 1 better predicted the observed data than model 2, we used model 1 for our exposure predictions.

3.3. Estimated PM2.5 Exposure under Current and Future Scenarios

Our modeling approach can be used to predict the effects of the Carolina North campus on ambient PM2.5 concentrations in census blocks in the study corridor if the campus is built.

Even if the new campus is built, the roadway contribution to ambient PM2.5 levels in the study corridor is predicted to be very low by 2025. The maximum contribution the new campus contributes to any one census block occurs in winter and is predicted to be 0.11 μg/m^3, which is quite low in comparison with the ambient air quality standard (12 μg/m^3 annual average PM2.5 concentration). In comparison, if the new campus is not built, the maximum PM2.5 concentration in any one census block is 0.085 μg/m^3, which is 24% lower than if the campus is built. In both cases, though, the maximum concentration is higher under current conditions than under future conditions, despite the anticipated traffic growth. Under current conditions, the model predicts that the maximum roadway contribution to seasonal PM2.5 in any one census block is 0.14 μg/m^3, which is 24% higher than expected in 2025, even if the new campus is built. These future emissions reductions reflect the built-in assumptions of MOVES that the future vehicle fleet will become more efficient (less polluting) and that fuels will be cleaner. The results thus illustrate the value of ensuring continued improvements in vehicle fuel economy and

emissions standards.

Our modeling approach included a new method for representing meteorological variability. Our results illustrate that variability can be important in some locations. Overall, the daily meteorological variability caused little change in seasonal daily mean PM2.5 concentrations. For example, in the 2025 scenario in which the Carolina North campus is built, the average coefficient of variation (standard deviation of the predicted seasonal mean divided by average of the seasonal mean) is 0.06, meaning that seasonal variability on average has a relatively small effect on model predictions. The maximum coefficient of variation in this scenario was less than 0.5, which means that 95% of the time, meteorological variability will change the predicted seasonal mean by less than a factor of 2. (According to the Central Limit Theorem, the seasonal mean converges to a normal distribution, and hence 95% of the time, the seasonal mean should be within two standard deviations of the actual mean, and in this case the standard deviation is about half the mean.) Thus, this meteorological variability is less important than the model uncertainty shown in **Figure 8**.

The modeling approach can be used to characterize spatial variability in roadway emissions effects on surrounding neighborhoods. **Figure 9** shows the resulting spatial variability for current conditions, and **Figure 10** shows the spatial variability for future conditions. In both instances (because both models rely on the same set of meteorological data), the census block with the maximum concentration is in the same location and also (despite changes in wind directions) does not vary seasonally. The most affected census block (shown with arrows in **Figures 9** and **10**) is located on the east side of Martin

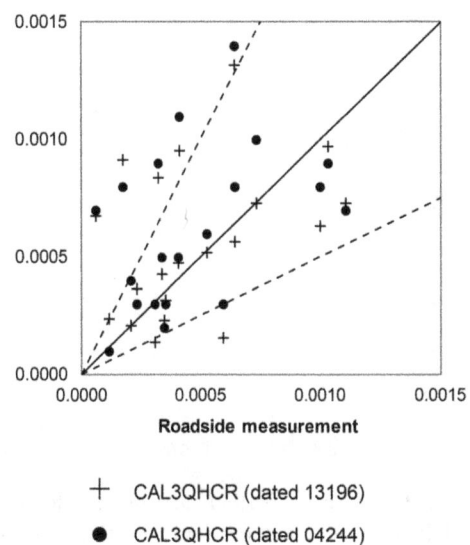

+   CAL3QHCR (dated 13196)

●   CAL3QHCR (dated 04244)

Figure 8. Factor-of-two plots of concentration differences (mg/m^3) observed during roadside measurements and predicted by CAL3QHCR (dated 13196 and dated 04244).

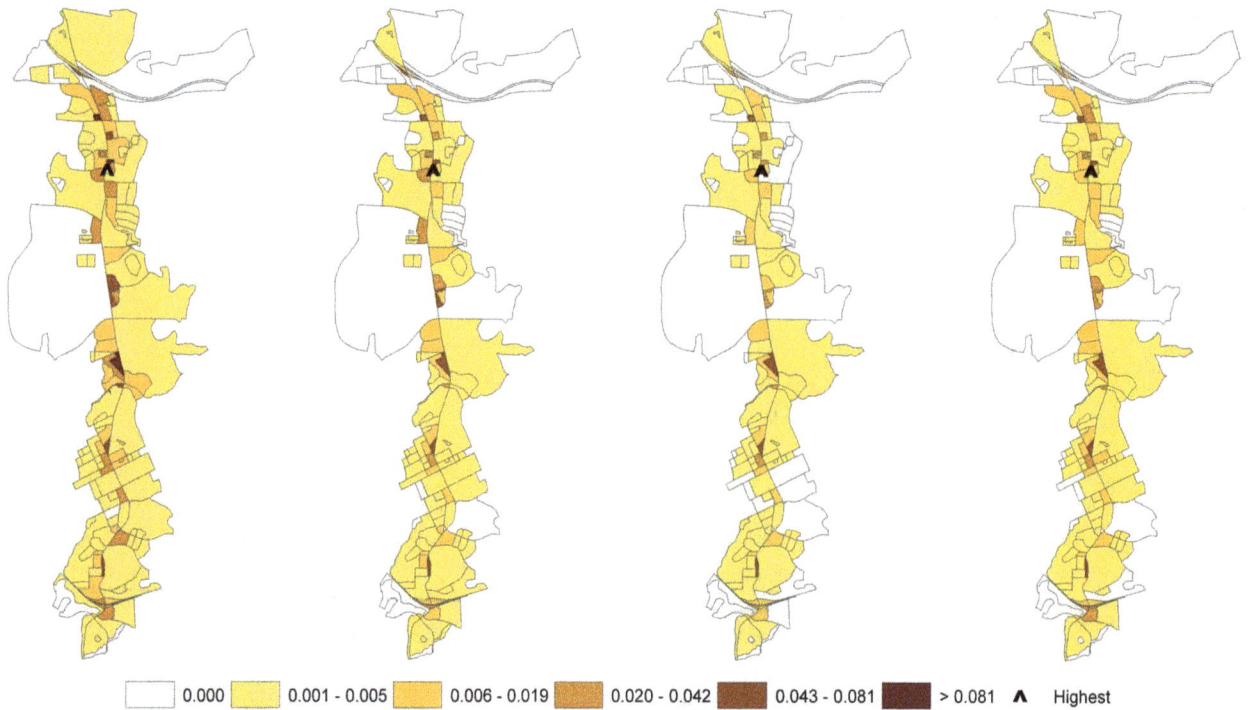

| | 0.000 | | 0.001 - 0.005 | | 0.006 - 0.019 | | 0.020 - 0.042 | | 0.043 - 0.081 | | > 0.081 | ∧ Highest |

Figure 9. PM2.5 concentrations attributable to roadway emissions from the study corridor, as predicted by the combined MOVES-CAL3QHCR approach (μg/m³) by season for the year 2009.

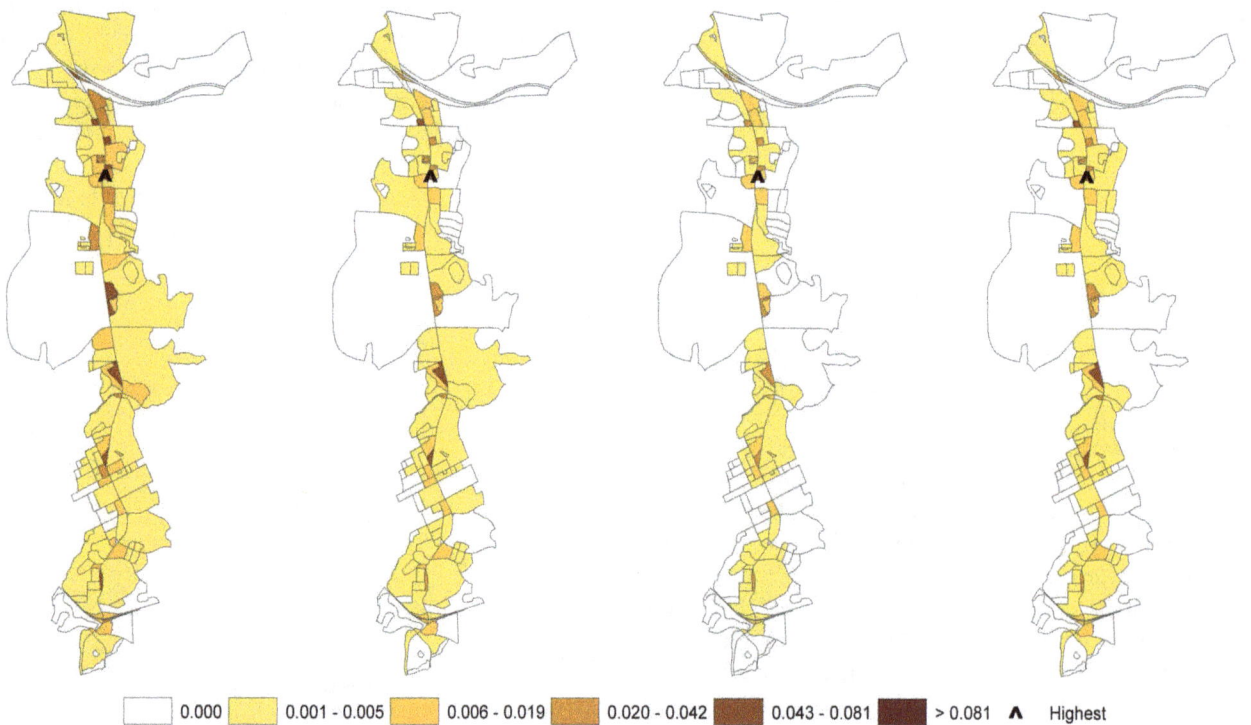

| | 0.000 | | 0.001 - 0.005 | | 0.006 - 0.019 | | 0.020 - 0.042 | | 0.043 - 0.081 | | > 0.081 | ∧ Highest |

Figure 10. PM2.5 concentrations attributable to roadway emissions, as predicted by the combined MOVES-CAL3QHCR approach (μg/m³) by season for the year 2025, assuming the Carolina North Campus is built.

Luther King Jr. Boulevard at Blossom Lane. Such information could be useful for zoning decisions (e.g., decisions about locations for schools, retirement homes, or other land uses attracting sensitive populations).

4. Discussion

Our results are consistent with the few empirical evaluations of the accuracy in predicting roadway PM2.5 concentrations of CAL3QHCR and its predecessor, known as CALINE. Yura et al. compared CALINE predictions of PM2.5 to measured PM2.5 concentrations at a busy intersection in a suburban community in Sacramento, California, and an urban site along a six-lane road in London, England [32]. They found that 80% of model predictions were within a factor-of-two envelope of measured concentrations at the suburban site but that only 56% of predictions were within the factor-of-two envelope for the urban site. They attributed the poor performance at the urban site to limitations of the emissions factors they used (they relied on scaling United Kingdom PM10 emissions factors) and to street canyon effects. Chen et al. extended Yura's work by comparing the performance of the CAL3QHC model to that of the CALINE model for the same two sites (although during a different time period) as Yura used [26]. Chen et al. found that predicted PM2.5 concentrations were within the factor-of-two envelope for 69% of the Sacramento data points and for 59% of the London data points. In both cities, CAL3QHC outperformed CALINE. Gokhale and Raokhande compared the CALINE and CAL3QHC models' ability to predict roadside PM2.5 concentrations at a busy intersection in Guwahati, India [38]. They found that the CAL3QHC model predictions were within a factor-of-two envelope for 65% of (66 of 102) hourly PM2.5 observations during winter and that the CAL3QHC model outperformed the CALINE model (the latter of which produced predictions within the factor-of-two envelope for 46 of 102 data points).

Our findings about the amount of PM2.5 contributed to a given location by a single busy roadway also are consistent with findings of the few modeling studies and quantitative HIAs of local effects of traffic in the United States. In a modeling study, Zhang and Batterman used CALINE along with the predecessor to MOVES, known as MOBILE6.2, to estimate the amount of PM2.5 pollution contributed by a busy roadway in Detroit, Michigan [33]. They found that the local roadway contributed only a small amount of the measured PM2.5: total measured PM2.5 concentrations averaged 16.8 $\mu g/m^3$, but Zhang and Batterman attributed "no more than 0.5 $\mu g/m^3$" to the roadway. They attributed the majority of observed PM2.5 "to long range transport of sulfate and other aerosols from the Ohio River Valley." Chen et al. also found that roadways in Sacramento and London contributed relatively small fractions to observed PM2.5 concentrations at the study sites [26].

Of the four transportation-related quantitative HIAs identified in the comprehensive review by Bhatia and

Seto et al., three predicted PM2.5 concentrations attributable to vehicles on roadways (the fourth predicted PM10 concentrations) [9]. All of these HIAs (including the HIA that estimated PM10 concentrations) focused on proposed new development projects in or near Oakland, California, and all used CAL3QHCR to support their predictions. The first, an HIA of a proposed residential development to be constructed near a highway (with an average daily traffic volume of about 119,000 vehicles) in Pittsburg, California, used CAL3QHCR to estimate that traffic-attributable exposures adjacent to the highway are about 2 $\mu g/m^3$ but that these exposures decline rapidly with distance to about 0.2 $\mu g/m^3$ [39]; this estimate assumed a constant emissions factor of 0.15 g/vehicle-mile travelled, whereas our estimate employed MOVES to estimate link-specific emissions factors, resulting in a range of emissions factors of 0.02 - 0.5 g/vehicle-mile travelled. The second of these three quantitative HIAs considered the potential traffic-related health effects of potential affordable housing sites in Oakland, California; this HIA used CAL3QHCR to estimate that two major roadways with combined annual average daily traffic counts of about 225,000 vehicles would contribute about 0.4 - 0.5 $\mu g/m^3$ to PM2.5 exposures at the locations under consideration, all of which were within meters of the roadways [40]. The third HIA concerned a potential new residential development near a transit station in Oakland; it estimated that alongside a major highway (with daily traffic counts averaging 144,000 vehicles) neighboring the proposed development site, about 0.3 $\mu g/m^3$ of PM2.5 could be attributed to traffic but that this traffic-related contribution decreased to 0.1 $\mu g/m^3$ at a distance of 150 m from the highway [41]. In summary, these HIAs estimate that directly adjacent to highways running through the Oakland area, traffic contributes anywhere from about 0.3 - 2 $\mu g/m^3$. All of these highways have daily traffic counts at least five times as high as the current traffic along the roadway corridor analyzed in the present study. The estimated roadway contributions that our modeling approach yielded (with the maximum roadway-contributed concentration of 0.14 $\mu g/m^3$ under current conditions) hence are quite consistent with these previous estimates when traffic volumes and distances of census block centroids to the roadway are considered. That is, if one multiplies the maximum estimate from our modeling approach by 5, then the estimated maximum predicted concentration in any census block in the study corridor is 0.7 $\mu g/m^3$. This is within the range of concentrations predicted in the California studies.

5. Conclusions

In this study, a new modeling framework to quantify the project traffic growth impacts on population exposure to

PM2.5 air pollution was proposed and then demonstrated by quantifying exposure to roadway PM2.5 emissions that may occur in the future due to the Carolina North development in Chapel Hill, North Carolina. This modeling framework should benefit others conducting quantitative HIAs of the built environment and transportation projects. Whereas previous HIAs employing air dispersion models have used average meteorological data and have assumed that vehicles move at a constant cruising speed along roadway links, our approach considers link-by-link variation in vehicle behavior and hourly meteorological variability.

Our results reveal that improvements in vehicle technologies and fuels will be a key factor in protecting public health from the air pollution generated by increases in traffic expected to occur due to local and regional developments in the future. In fact, the models we employed predict that traffic-related PM2.5 in the study corridor may actually decrease in the future, even if traffic increases, due to improved vehicle technologies and fuels.

Our results also reveal the need for improve models to predict near-road PM2.5 concentrations. While the CAL3QHCR dispersion model was able to predict dispersion reasonably well, about 25% of model predictions over-estimated dispersion. This overestimation bias results in under-estimates of pollutant exposure. Hence, reducing model bias is critical to ensuring that decision-makers are adequately informed about air quality and health risks associated with roadway traffic.

REFERENCES

[1]   J. Kemm, "Perspectives on Health Impact Assessment," *Bulleting of the World Health Organization*, Vol. 81, No. 6, 2003, p. 387.

[2]   National Research Council, "Improving Health in the United States: The Role of Health Impact Assessment," National Academy Press, Washington DC, 2011.

[3]   A. Wernham, "Health Impact Assessments Are Needed in Decision Making about Environmental and Land-Use Policy," *Health Affairs* (*Project Hope*), Vol. 30, No. 5, 2011, pp. 947-956.

[4]   R. Lozano, M. Naghavi, S. Lim, K. Foreman, K. Shibuya, V. Aboyans and C. J. L. Murray, "Global and Regional Mortality from 235 Causes of Death for 20 Age Groups in 1990 and 2010: A Systematic Analysis for the Global Burden of Disease Study 2010," *Lancet*, Vol. 380, No. 9859, 2012, pp. 2095-2128.

[5]   A. L. Dannenberg, R. Bhatia, B. L. Cole, S. K. Heaton, J. D. Feldman and C. D. Rutt, "Use of Health Impact Assessment in the U.S.: 27 Case Studies, 1999-2007," *American Journal of Preventive Medicine*, Vol. 34, No. 3, 2008, pp. 241-256.

[6]   M. Wismar, J. Blau, K. Ernst and J. Figueras, "The Effec-

[7]   R. Bhatia and M. Katz, "Estimation of Health Benefits from a Local Living Wage Ordinance," *American Journal of Public Health*, Vol. 91, No. 9, 2001, pp. 1398-1402.

[8]   L. Singleton-Baldrey, "The Impacts of Health Impact Assessment: A Review of 54 Health Impact Assessments, 2007-2012," University of North Carolina at Chapel Hill, Chapel Hill, 2012.

[9]   R. Bhatia and E. Seto, "Quantitative Estimation in Health Impact Assessment: Opportunities and Challenges," *Environmental Impact Assessment Review*, Vol. 31, No. 3, 2011, pp. 20301-20309.

[10] M. Ritner, K. K. Westerlund, C. D. Cooper and M. Claggett, "Accounting for Acceleration and Deceleration Emissions in Intersection Dispersion Modeling Using MOVES and CAL3QHC," *Journal of the Air and Waste Management Association*, Vol. 63, No. 6, 2013, pp. 724-736.

[11] University of North Carolina at Chapel Hill, "About UNC," 2013. http://www.unc.edu/about/

[12] Town of Chapel Hill, "Snapshot of the Town of Chapel Hill," 2012. http://www.townofchapelhill.org/Modules/ShowDocument.aspx?documentid=12177

[13] K. Ross, "Carolina North Hearing Highlights Traffic Impact," The Carrboro Citizen, 2009. http://www.carrborocitizen.com/main/2009/05/14/carolina-north-hearing-highlights-traffic-impact/

[14] Vanasse Hangen Brustlin, Inc., "Transportation Impact Analysis for the Carolina North Development," Watertown, 2009.

[15] Health Effects Institute, "Traffic-Related Air Pollution: A Critical Review of the Literature on Emissions, Exposure, and Health Effects," HEI Special Report 17, Boston, 2010.

[16] US Census Bureau, "Census Block Shapefiles with 2010 Census Population and Housing Unit Counts," 2011.

[17] Orange County, "Aerials 2010," 2012.

[18] Town of Chapel Hill, "Street Centerline," 2009.

[19] Town of Chapel Hill, "2ft Elevation Contours," 2009.

[20] US Environmental Protection Agency, "Guidance on Quantitative PM Hot-Spot Analyses for Transportation Conformity," 2011. http://www.epa.gov/otaq/stateresources/transconf/policy/pm-hotspot-guide.pdf

[21] N.C. Division of Air Quality, "MOVES Input and Output Giles: Hickory and Triad PM2.5 Redesignation Demonstration and Maintenance Plan," 2011.

[22] P. Mulawa, S. Cadle, H. Knapp, R. Zweidinger, R. Snow, R. Lucas and J. Goldbach, "Effect of Ambient Temperature and E-10 Fuel on Primary Exhaust Particulate Matter Emissions from Light-Duty Vehicles," *Environmental*

tiveness of Health Impact Assessment: Scope and Limitations of Supporting Decision-Making in Europe," Political Science, Copenhagen WHO Regional Office for Europe, 2007.

Science & Technology, Vo. 31, 1997, pp. 1302-1307.

[23] D. Choi, M. Beardsley, D. Brzezinski, J. Koupal and J. Warila, "MOVES Sensitivity Analysis: The Impacts of Temperature and Humidity on Emissions," The MOVES Workshop 2011, US Environmental Protection Agency, Ann Arbor, 2011.

[24] National Climatic Data Center, "Quality Controlled Local Climatological Data (QCLCD)," 2013.

[25] National Oceanic and Atmospheric Administration, "NOAA/ESRL Radiosonde Database," 2013.

[26] H. Chen, S. Bai, D. Eisinger, D. Niemeier and M. Claggett, "Predicting Near-Road PM2.5 Concentrations," *Transportation Research Record: Journal of the Transportation Research Board*, Vol. 2123, 2009, pp. 26-37.

[27] M. G. Boarnet, D. Houston, R. Edwards, M. Princevac, G. Ferguson, H. Pan and C. Bartolome, "Fine Particulate Concentrations on Sidewalks in Five Southern California Cities," *Atmospheric Environment*, Vol. 45, No. 24, 2011, pp. 4025-4033.

[28] J. S. Wang, T. L. Chan, Z. Ning, C. W. Leung, C. S. Cheung and W. T. Hung, "Roadside Measurement and Prediction of CO and PM 2.5 Dispersion from On-Road Vehicles in Hong Kong," *Transportation Research Part D: Transport and Environment*, Vol. 11, No. 4, 2006, pp. 242-249.

[29] Y. Wu, J. Hao, L. Fu, Z. Wang and U. Tang, "Vertical and Horizontal Profiles of Airborne Particulate Matter near Major Roads in Macao, China," *Atmospheric Environment*, Vol. 36, No. 31, 2002, pp. 4907-4918.

[30] TSI, Inc., "Spec Sheet Model 8533/8534 DustTrak DRX Aerosol Monitor, 2012.

[31] TSI, Inc., "DustTrak DRX Aerosol Monitor Theory of Operation (EXPMN-002)," 2012.

[32] E. A. Yura, T. Kear and D. Niemeier, "Using CALINE Dispersion to Assess Vehicular PM 2.5 Emissions," *Atmospheric Environment*, Vol. 41, No. 38, 2007, pp. 8747-8757.

[33] K. Zhang and S. Batterman, "Near-Road Air Pollutant Concentrations of CO and PM2.5: A Comparison of MOBILE6.2/CALINE4 and Generalized Additive Models," *Atmospheric Environment*, Vol. 44, No. 14, 2010, pp. 1740-1748.

[34] P. E. Benson, "CALINE3, A Versatile Dispersion Model for Predicting Air Pollutant Levels Near Highways and Arterial Streets," Sacramento, 1979.

[35] P. E. Benson, "CALINE4, A Dispersion Model for Predicting Air Pollutant Concentrations near Roadways," Sacramento, 1989.

[36] US Environmental Protection Agency, "Meteorological Monitoring Guidance for Regulatory Modeling Applications," EPA-454/R-99-005, Research Triangle Park, 2000.

[37] Battelle, "Detailed Monitoring Protocol for U.S. 95 Settlement Agreement," Columbus, 2006.

[38] S. Gokhale and N. Raokhande, "Performance Evaluation of Air Quality Models for Predicting PM10 and PM2.5 Concentrations at Urban Traffic Intersection during Winter Period," *Science of the Total Environment*, Vol. 394, No. 1, 2008, pp. 9-24.

[39] Human Impact Partners, "Pittsburg Railroad Avenue Specific Plan Health Impact Assessment," Oakland, 2008.

[40] Human Impact Partners, "Pathways to Community Health: Evaluating the Healthfulness of Affordable Housing Opportunity Sites along the San Pablo Avenue Corridor Using Health Impact Assessment," Oakland, 2009.

[41] UC Berkeley Health Impact Group, "MacArthur BART Health Impact Assessment," Berkeley, 2007.

11

Decline of VOC Concentrations with the Aging of Houses in Japan

Motoya Hayashi[1], Haruki Osawa[2]

[1]Department of Life Style and Space Design, Miyagi Gakuin Women's University, Sendai, Japan; [2]Department of Healthy Building and Housing, National Institute of Public Health, Wako, Japan.

ABSTRACT

The purpose of this investigation is to know the long-term characteristics of VOC concentrations in houses built before the building code in 2003 and to clarify the countermeasures against indoor air pollution in the houses already built. For example, the improvements of living habits, ventilation and the remove of building materials. The concentrations of VOCs were measured in these houses in summer and winter from 2000 to 2005. The results showed that the concentration of formaldehyde decreased in the first year. After that the decline of the concentration was not seen and the concentration changed only with the temperature. The characteristics of decline were thought to be caused by two sorts of emission. One is an emission of concealed formaldehyde in the process of material production and the other is an emission with the generation of formaldehyde from adhesives of urea resin and moisture. The concentration of toluene decreased rapidly in the first year. The concentrations of xylene, ethyl-benzene and styrene showed a similar change. But the concentrations of acetaldehyde which were measured from the summer of 2002 did not decrease and its concentration in some houses was higher than the guideline even in the winter of 2005.

Keywords: Indoor Air Quality; VOC; Passive Sampling; Questionnaire Survey; Statistics Analysis

1. Introduction

A traditional Japanese house has a structure of wooden posts and beams. Houses with this structure have wide openings. Therefore the traditional house is beneficial to indoor cooling in summer and to the prevention of the deterioration of wooden materials. Also there are many air infiltration routes through the wall, floor, ceiling, etc. Since 1970s, the structures have been improved to be airtight and insulated using insulation materials and films. However, many infiltration routes lurk in the concealed spaces like spaces inside the walls, the ceilings and the floors [1].

The living style of Japanese dwellers has changed in the sixty years after World War 2. The unoccupied time in houses became longer, because the number of family became smaller. And the time of opening windows became shorter with the spread of air conditioner. New building materials have been produced in factories since 1960s, and prefabricated houses have expanded in Japan

since 1970s. Under these situations, "sick house": the indoor air pollution by chemical compounds from building materials, has been closed up in Japan since 1990s. And many countermeasures against indoor air pollution in houses have been taken by the Japanese government, building companies, building material manufacturers and other related groups. And a new building code with countermeasures against indoor air pollution has been forced by the Ministry of Land, Infrastructure and Transport since 2003. According to this building code the concentration of formaldehyde is expected to be lower than the guideline: 0.08 ppm, which was established by the Ministry of Health, Labour and Welfare of Japanese Government in 1993. This building code requires our consideration about the emission rate of formaldehyde from building materials. The regulated materials are not only building materials which are used for interior finish but also materials which are concealed in walls, ceilings and floors. And the installation of ventilation equipment is required in all residential spaces. The required ventilation time is 0.5 times per hour [2]. Before this building code, the concentrations of formaldehyde were higher than the guideline at the completion of houses in many cases. And some investigations showed that the concentrations of formaldehyde do not decrease with time soon [3-8]. Organic compounds are volatizing from building materials and from furniture and articles which are carried into houses by residents, so the indoor concentrations do not decrease with the aging of houses. Therefore the influences of indoor pollution on the residents' health may continue long and these influences may become a cause of a multiple chemical sensitivity [1]. Under these contexts the decline of concentrations with the aging of houses was investigated.

2. Methods

The concentrations of formaldehyde, toluene, xylene, ethyl-benzene, styrene and acetaldehyde were measured in winter and in summer for five years in about two hundred and fifty houses in which formaldehyde concentration was higher than the guideline: 0.08ppm in 2000. The first investigations in 2000 were carried out from summer to winter continuously in new houses which were built less than a year ago. The number of investigated houses decreased gradually and became eighty-four in

the winter of 2005 as shown in **Figure 1**. The concentrations were measured using passive samplers made by Advanced Chemical Sensor Inc. The samplers were sent to the residents from a laboratory. The residents placed samplers in the rooms for 24 hours and sent them back to the laboratory. The concentrations were specified using gas chromatographs in the laboratory. The characteristics of buildings and residents were checked using questionnaires. The indoor temperatures and the humidity were measured by residents and were checked on the questionnaires. The number of investigated houses changed. But the results of statistical analysis using the eighty-four houses in which the concentrations were obtained during all the periods were quite similar to the results using all the data. The difference between the average formaldehyde concentration of these eighty-four houses and that of all the houses is smaller than 4%. Therefore the results using all the data are reported in this paper.

3. Results

Figure 2 shows classifications of investigated houses. They consist of apartment houses and detached houses. Most apartment houses were built with reinforced concrete structure. And the numbers of stories of 45% of the apartment houses were from one to three. And those of the other apartment houses were from three to ten. The

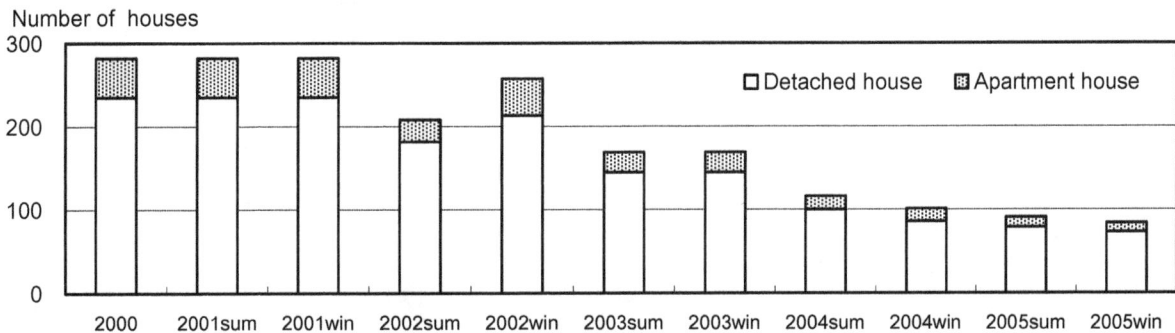

Figure 1. Number of investigated houses.

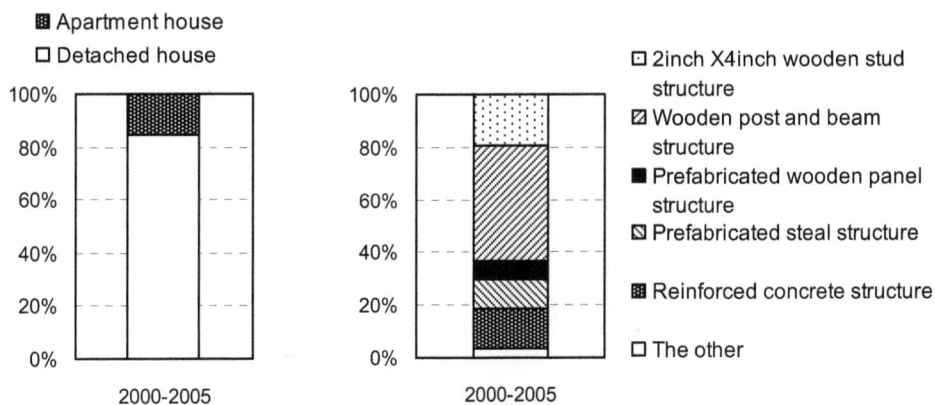

Figure 2. Classification of houses.

structure types of detached houses were various but most detached houses were built with wooden post and beam structure which is common in Japan. Another structure was 2 inch × 4 inch wooden stud structure which has been imported and built generally since 1980s. The others were prefabricated structures. Most of the prefabricated houses are produced in large factories by major housing companies. The percentage of the classification of houses was similar to the percentage of all houses built recently in Japan.

Figure 3 shows the weather conditions and the humidity when the concentrations were measured in houses. In summer, the percentage of "cloudy" was a little smaller than that in winter. The temperatures were shown in **Figure 4**. Naturally the temperatures were high in summer and low in winter. The humidity changed with the temperature; it is hot and humid in summer and it is cold in winter. This annual change is typical in most areas of Japan.

Figure 5 shows how long the residents keep the windows of their houses open in a day. The percentage of open windows does not contain the percentage in the case that the windows are open for less than one hour. Naturally the percentage of open windows is larger in summer than in winter.

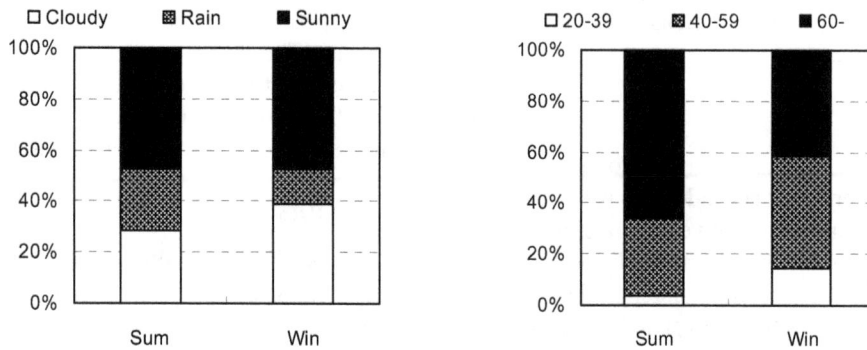

Figure 3. Weather and humidity.

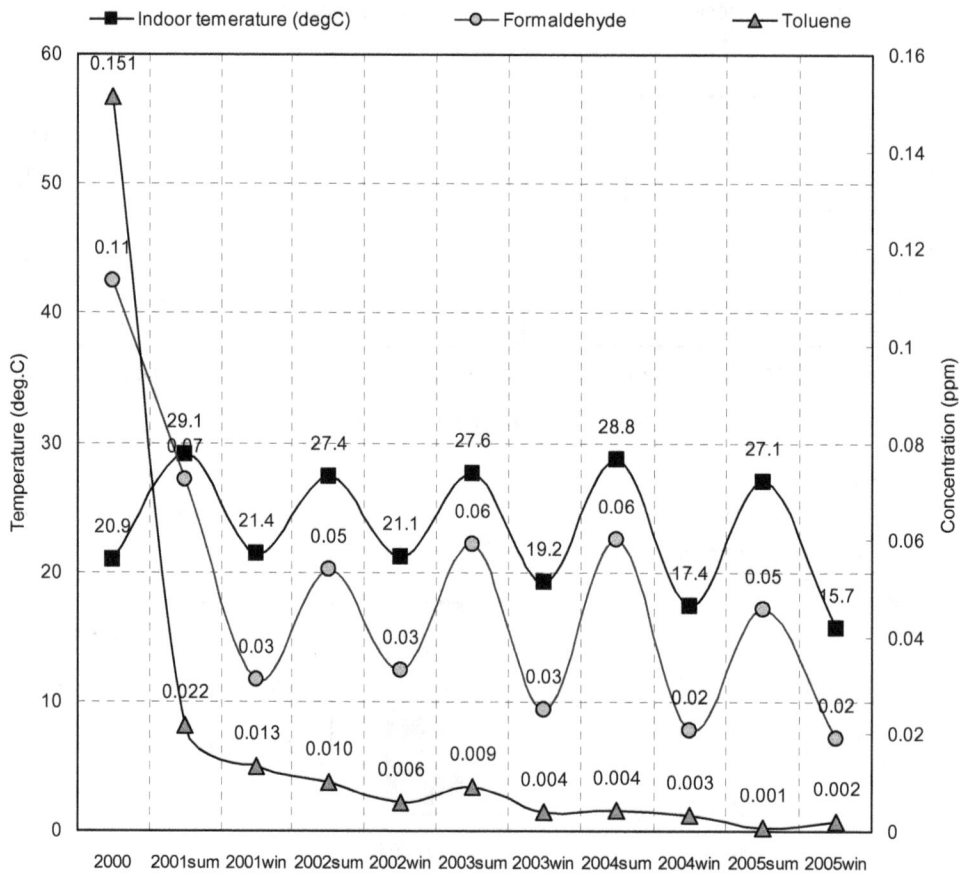

Figure 4. Change of temperature and concentrations.

Figure 6 shows the changes of formaldehyde concentrations. In this figure the standard deviations of concentrations on each measurement period are showed using lines. The average of concentration of formaldehyde was higher than the guideline: 0.08ppm in 2000. This is because the houses were chosen from the houses in which the concentration exceeded the guideline when investigated in 2000. The concentration of formaldehyde decreased from 2000 to the winter of 2001 but the concentration increased in the summer of 2002. After 2003, the concentration increased in summer and decreased in winter. This annual change was repeated during the four years.

Figure 7 shows the concentration of toluene. The average concentration was higher than the guideline: 0.07 ppm in 2000. The concentrations decreased and became lower than the guideline in the summer of 2001. The concentrations did not change with temperature. **Figure**

8 shows the concentration of xylene. The concentration was lower than the guideline: 0.20 ppm in 2000. The concentration decreased during the first year and was low during these five years. In 2003 the concentration increased a little but the reason was not clear.

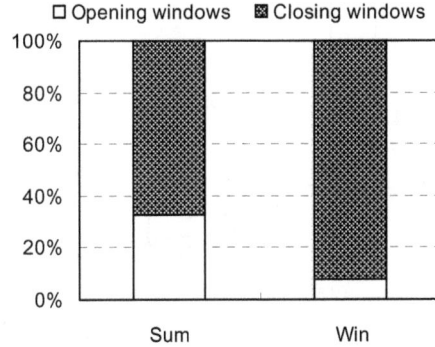

Figure 5. About opening windows.

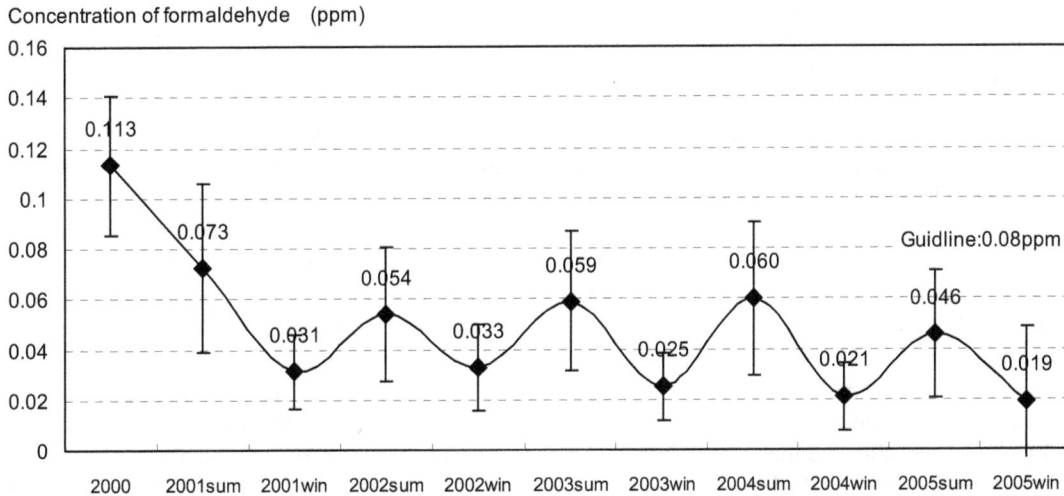

Figure 6. Decline of formaldehyde concentration.

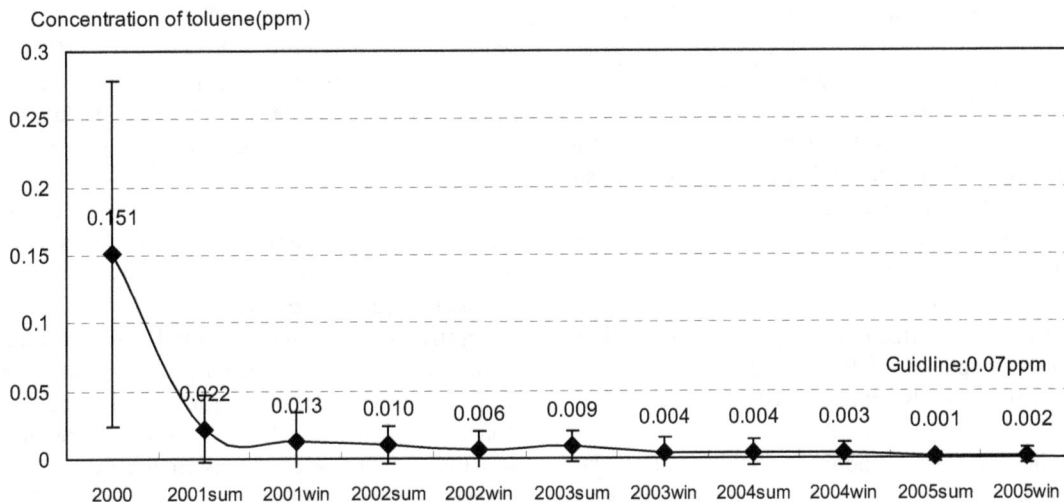

Figure 7. Decline of toluene concentration.

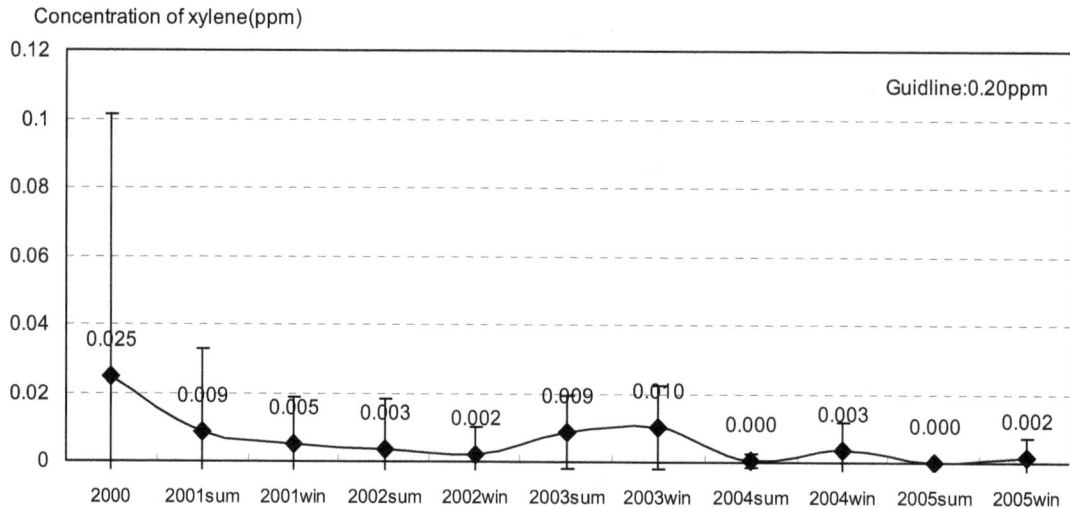

Figure 8. Decline of xylene concentration.

Figure 9 shows the concentration of ethyl-benzene. The concentration was also very much lower than the guideline: 0.88 ppm even in 2000. The concentration decreased during the first year and was low after that. In the winter of 2003 the concentration increased a little but the reason was not clear. These temporary rises are thought to be caused by the pollution sources like furniture or insecticide which are carried in houses by residents.

Figure 10 shows the concentrations of styrene. The average concentration was lower than the guideline: 0.05 ppm in 2000. But in some houses the concentrations were higher than the guideline. The concentration decreased during the first year and was lower during these five years. In 2003 the concentration increased a little.

Figure 11 shows the concentration of acetaldehyde. The concentration was measured after the summer of 2002. The concentration was not enough lower than the guideline: 0.03 ppm in the summer of 2003. And the concentration increased gradually during the four years. The reason of this increase is not clear. One of the emission sources is smoking but the percentage of "smoking" was almost 11% in all measurement periods. The other source is thought to be building materials. Anyway it became clear that the concentration of acetaldehyde exceeded the guideline in some houses.

Figure 12 shows the percentage of houses where the residents felt physical changes in houses. The residents answered the following question: Have you felt a change of physical condition which is thought to be caused by the chemical compound. If the answer was "yes", the following items were also checked: who felt the change, the physical conditions, when the change was felt. And the following physical conditions: nausea, headache, eczema, the pain of eyes, nose and throat were appealed on the questionnaires. The percentage of the residents

who felt the change of their physical conditions was larger in summer and smaller in winter. This change was similar to that of the formaldehyde concentration shown in **Figure 4**. But the percentage decreased gradually and the fluctuation of the percentage also decreased. These results show that it is necessary to investigate the relationship between the indoor air quality and the residents' feeling about the change of physical condition which are checked on the questionnaires. This investigation is expected to be made with their medical researchers.

Figure 4 shows the change of temperature and concentrations. The concentrations of formaldehyde changed with temperature.

Figure 13 shows the relationship between temperatures and concentrations. In the case of formaldehyde the concentrations are distributed on a curve: "Calculation ($C_{25} = 0.046$, a = 1.11)" in **Figure 13**. This curve is calculated using the following equation.

$$C = C_{25}a^{T-25} \qquad (1)$$

where C: concentration, $C25 = E25/(Q + \beta)$, E25: emission rate when temperature is 25deg.C, Q: ventilation rate, β: ratio of sink, a: coefficient of influence of temperature (a = 1.11: a value which was measured in small chambers).

The formaldehyde concentration in 2000 was not on the curve but the other concentrations fitted on this curve generally. This shows that the ability of formaldehyde emission hardly decreased after 2001.

Figure 14 shows the measured concentrations of formaldehyde and the normalized concentrations: C25 which is calculated using equation 1. The normalized concentration is thought to be an indicator of emission possibility in a house. The normalized concentration decreased fast from 2000 to the summer of 2001 and was steady after that. This characteristic of concentration

Concentration of ethyl-benzene （ppm）

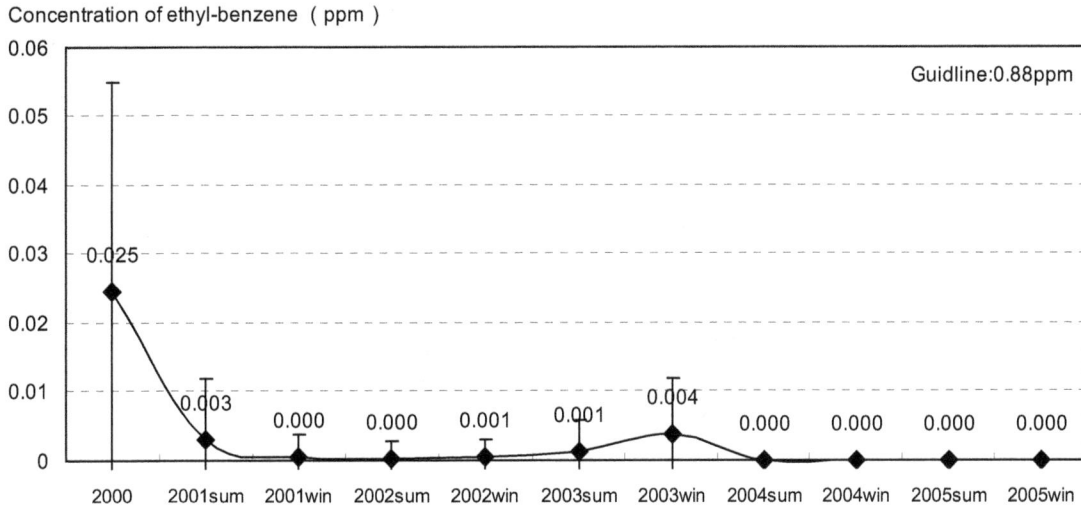

Figure 9. Decline of ethyl-benzene concentration.

Concentration of styrene （ppm）

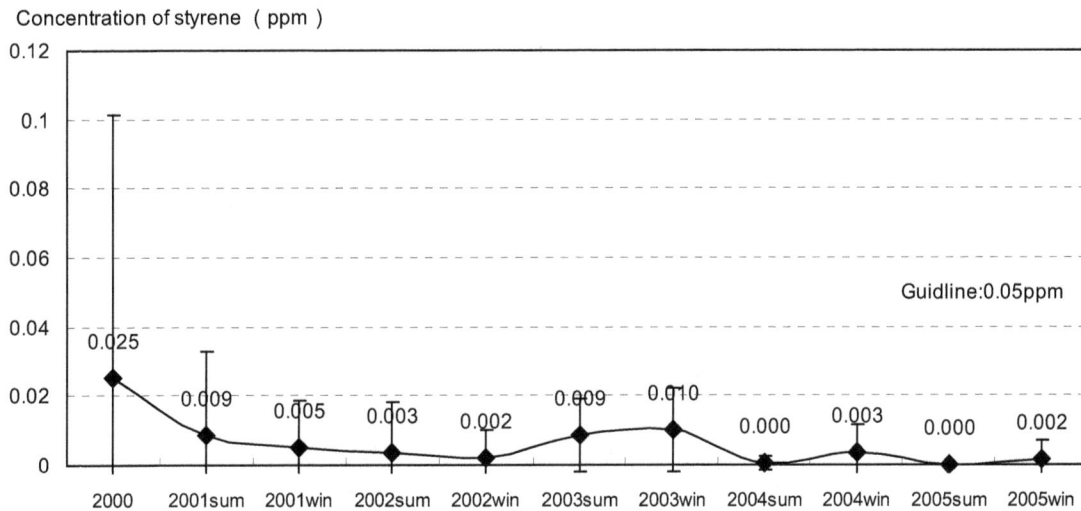

Figure 10. Decline of styrene concentration.

Concentration of acetaldehyde(ppm)

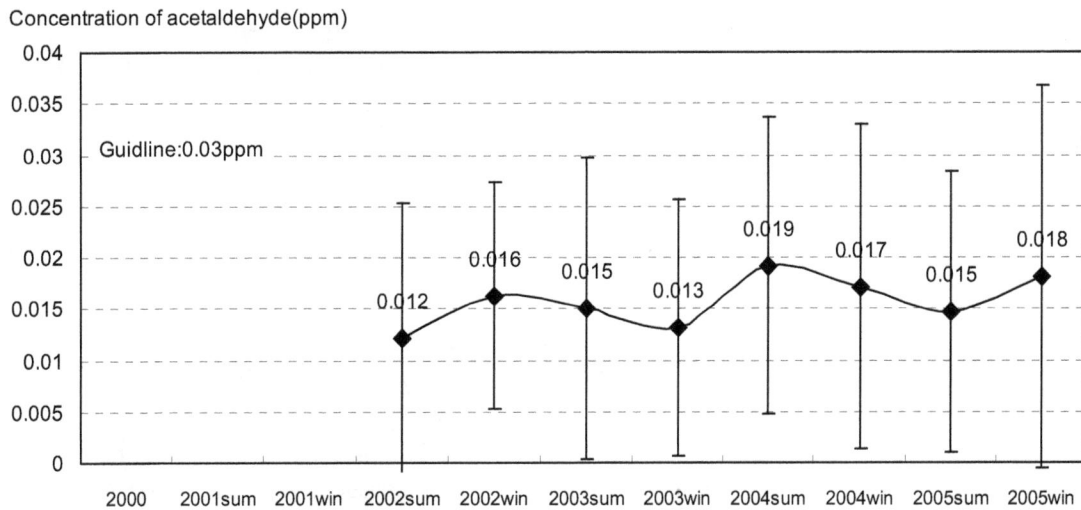

Figure 11. Decline of acetaldehyde concentration.

Percentage of houses where residents felt the change of physical condition(%)

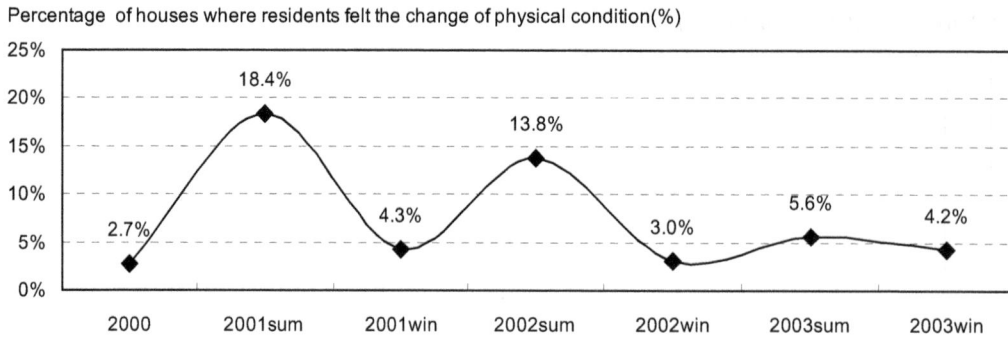

Figure 12. Percentages about change of physical condition with chemical compound.

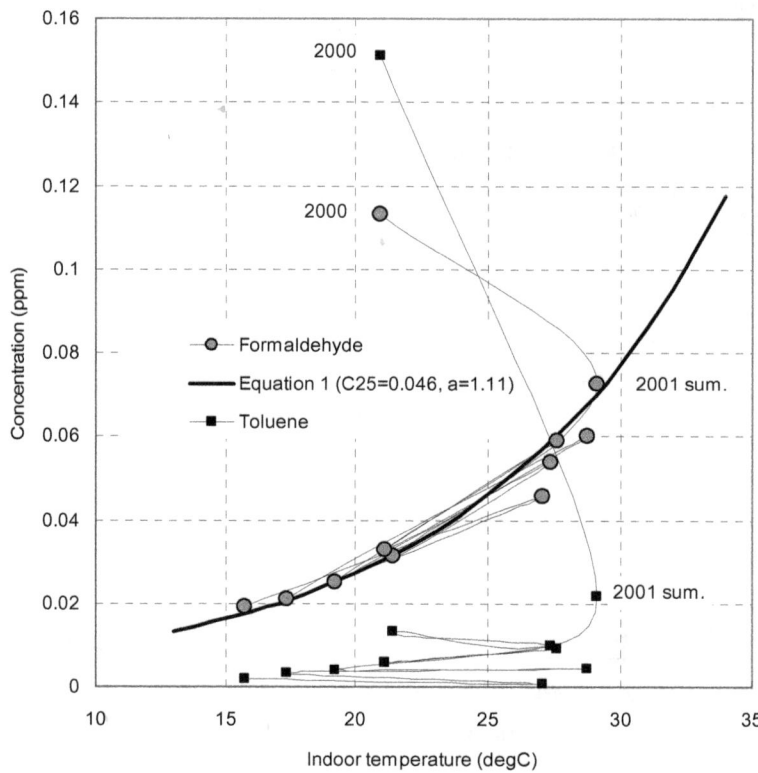

Figure 13. Relationship between temperatures and concentrations.

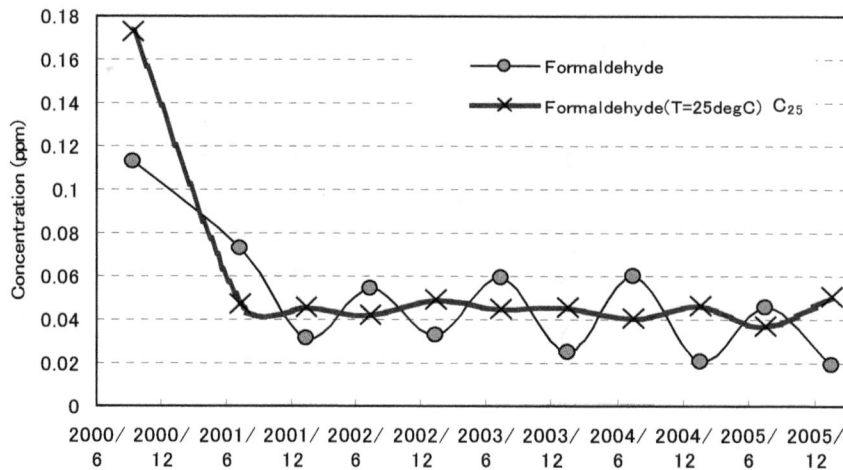

Figure 14. Change of the normalized formaldehyde concentration at a temperature 25°C.

decline may be caused by a mechanism of emission from building materials. In Japanese houses main emission sources of formaldehyde were wooden building materials produced in factories like plywood and particle board. In the case of houses built in 2000, most of these materials were made using adhesives of urea resin which is made with formaldehyde. A part of formaldehyde was concealed in the materials when they were produced in factories. Soon after production these concealed formaldehyde may start to volatilize. On the other hand formaldehyde is generated from inside urea resin and moisture after production. The generation rate is influenced by temperature and humidity. And the generated formaldehyde diffuses to the surface and volatilizes into indoor air. The calculated decline curve of normalized concentration shows that the emission of concealed formaldehyde finished in almost one year and the emission with the generation of formaldehyde continues for a long time at least for five years. The emission will continue until all the sources of formaldehyde are lost.

4. Conclusion

The investigation on the decline of indoor chemical pollution was carried out in Japanese houses. The concentrations of toluene, xylene, ethyl-benzene, styrene decrease fast and the concentrations were lower than the guidelines in most houses at least after one year. But the concentrations of formaldehyde and acetaldehyde were above the safety zone. In the case of formaldehyde the concentration decreased during the first one year but the concentration did not decrease at all during the next four years. The concentration became higher in summer every year. Therefore the long-term countermeasure against chemical pollution in summer is thought to be necessary. If residents have suffered some impairment by indoor chemical pollution, renovations of buildings especially considering indoor air quality will be necessary not only in new houses but also in old houses.

5. Acknowledgements

The study was a part of a national project "Development of Countermeasure Technology on Residential Indoor Air Quality" by National Institute for Land and Infrastructure Management under the Japanese government. The investigations were made with the contributions of many residents. And the investigations were made with the cooperation of Center for Housing Renovation and Dispute Settlement Support and the students of Miyagi-gakuin Women's University. The authors express their gratitude to them and to Dr. Kouichi Ikeda of National Institute of Public Health, Dr. Yasuo Kuwasawa of National Institute for Land and Infrastructure Management, Prof. Hiroshi Yoshino of Tohoku University and the committees of the national projects.

REFERENCES

[1] H. Motoya and O. Haruki, "The Influence of the Concealed Pollution Sources upon the Indoor Air Quality in Houses," *Building and Environment*, Vol. 43, 2008, pp. 329-336.

[2] H. Motoya, E. Masamichi and H. Yoshiko, "Annual Characteristics of Ventilation and Indoor Air Quality in Detached Houses Using a Simulation Method with Japanese Daily Schedule Model," *Building and Environment*, Vol. 36, No. 6, 2001, pp. 721-731.

[3] H. Osawa and M. Hayashi, "Status of the Indoor Air Chemical Pollution in Japanese Houses Based on the Nationwide Field Survey from 2000 to 2005," *The International Journal of Building Science and Its Applications Building and Environment*, Vol. 44, 2009, pp. 1330-1336.

[4] K. Kumagai, K. Ikeda, M. Hori, T. Matsumura, A. Nozaki, H. Kimura, *et al.*, "Field Study on Volatile Organic Compounds in Residences," *Journal of Architecture, Planning and Environmental Engineering: Transactions of AIJ*, No. 522, 1999, pp. 45-52.

[5] Y. Hiroshi, A. Kentaro, M. Mari, N. Koji, I. Koichi, N. Atsuo, *et al.*, "Long-Termed Field Survey of Indoor Air Quality and Health Hazards in Sick House," *Journal of Asian Architecture and Building Engineering*, Vol. 3, No. 2, 2004, pp. 297-303.

[6] Y. P. Zhang, X. X. Luo, X. K. Wang, K. Qian and R. Y. Zhao, "Influence of Temperature on Formaldehyde Emission Parameters of Dry Building Materials," *Atmospheric Environment*, Vol. 41, No. 15, 2007, pp. 3203-3216.

[7] B. Son, P. Breysse and W. Yang, "Volatile Organic Compound Concentrations in Residential Indoor and Outdoor and Its Personal Exposure in Korea," *Environment International*, Vol. 29, No. 1, 2003, pp. 79-85.

[8] I. Aydin, G. Colakoglu, S. Colak and C. Demirkir, "Effects of Moisture Content on Formaldehyde Emission and Mechanical Properties of Plywood," *Building and Environment*, Vol. 41, No. 10, 2006, pp. 1311-1316.

12

Small Mammal Habitat Use within Restored Riparian Habitats Adjacent to Channelized Streams in Mississippi

Peter C. Smiley Jr.[1*], Charles M. Cooper[2]

[1]USDA-ARS Soil Drainage Research Unit, Columbus, Ohio, USA; [2]USDA-ARS National Sedimentation Laboratory, Oxford, Mississippi, USA.

ABSTRACT

Riparian zones of channelized agricultural streams in northwestern Mississippi typically consist of narrow vegetative corridors low in habitat diversity and lacking riparian wetlands. Land clearing practices and stream channelization have led to the development of gully erosion and further fragmentation of these degraded riparian zones. Currently, installation of a gully erosion control structure (drop pipe) at the riparian zone-agricultural field interface leads to the incidental establishment of four riparian habitat types that differ in habitat area, vegetative structure, and pool size. Small mammals were sampled within four sites of each habitat type from June 1994 to July 1995. Small mammal diversity, abundance, and hispid cotton rat (*Sigmodon hispidus*) weight were the least within smallest Type I habitats with the least vegetative structural diversity and were the greatest within the larger Type II, III, or IV habitats having greater vegetative structural diversity and pool size. Small mammal diversity and abundance were the least in the summer 1994, increased in the fall 1994, and then declined later in our study. Hispid cotton rat abundance was the least in summer 1994, winter 1994, and spring 1995 and was the greatest in fall 1994 and summer 1995. Our results suggest that modifying the drop pipe installation design to facilitate the development of larger riparian habitats with greater vegetative structural diversity will provide the greatest benefits for small mammals.

Keywords: Gully Erosion; Channelization; Channel Incision; Erosion Control; Small Mammals

1. Introduction

Within the last 200 years, >80% of riparian zones in North America and Europe have been destroyed and human induced modification of the remaining habitat continues [1]. Intact riparian zones are critical landscape features required for the maintenance of biodiversity [2]. Riparian zones exhibit high levels of plant and animal diversity, and may facilitate ecological and genetic exchange by serving as connectivity areas with other ecosystems [3]. When upland forest or grassland habitat bordering streams is cleared for agriculture, the remaining riparian zones may increase in significance as habitat for small mammals [4]. Riparian zones that differ in vegetative structure and species composition from adjacent upland habitats have greater diversity and abundance of small mammals [5-7].

Riparian zones adjacent to channelized agricultural streams in northwestern Mississippi consist of narrow vegetative corridors low in habitat diversity and lacking riparian wetlands. Agriculture and stream channelization have contributed to the degradation of these riparian zones since the 1830s [8]. The initial reduction of riparian zone width occurred as land adjacent to these streams was cleared and developed for agriculture. Federal channelization projects conducted between 1930 and 1960 initiated severe channel incision that resulted in the destabilization of entire watersheds [9]. Channel incision severs the typical floodplain-stream linkage and frequently results in gully erosion that rapidly migrates perpendicular to the stream through the riparian zone and into the agricultural field. Agricultural land in northwest Mississippi has been the location of some of the most active gully systems in the United States [10,11]. Gully erosion is the most severe form of soil erosion and can result in soil loss rates between 0.1 to 65 $t \cdot ha^{-1} \cdot year^{-1}$ [12]. Fur-

*Corresponding author.

thermore, the creation of gaps caused by the loss of soil and riparian vegetation as a result of gully erosion leads to fragmentation of the riparian zones. Riparian habitat degradation resulting from channelization typically decreases small mammal diversity and abundance [13-15] and alters species composition [16]. The high frequency of gully erosion adjacent to channelized agricultural streams and the severity of the resulting erosion have resulted in the installation of thousands of drop pipes in northwest Mississippi [9,17].

Installation of drop pipes is often supported on a cost-share basis by the USDA Natural Resources Conservation Service, US Army Corps of Engineers, and other action agencies in the United States [18]. The drop pipe structure consists of an earthen dam placed across eroding gullies with an embedded L-shaped metal pipe that transports runoff to the stream (**Figure 1**). Drop pipe installation halts gully erosion and allows for the incidental development of riparian habitat that reconnects riparian zones fragmented by gully erosion [19,20]. Previous assessments have documented differences in fish, amphibian, reptile, bird communities, and avian nest predators among riparian habitats established by drop pipes [17,21-24]. Previous findings also confirmed that forested riparian wetlands had greater vertebrate species richness (*i.e.*, combined species richness of all vertebrate classes) than riparian habitats impacted by gully erosion and terrestrial riparian habitats created by drop pipe installation [19,25]. Differences in small mammal communities and populations among riparian habitats established by drop pipe installation have not been examined.

Previous studies [15,26] have evaluated small mammal responses to streamside management zones and stream restoration within forested watersheds in the southeastern United States. However, information on small mammal which responses to riparian habitat restoration within

Figure 1. Aerial photograph of newly installed drop pipe structure.

agricultural watersheds in this region is lacking. We sampled small mammals over a one-year period from four riparian habitat types established as a result of drop pipe installation in northwestern Mississippi to examine small mammal habitat use within these restored riparian habitats. Specifically, our research questions were: 1) Does small mammal community structure differ among habitat type and season?; and 2) Does the abundance, weight, number of recaptures, and movement patterns of hispid cotton rats (*Sigmodon hispidus*) differ among habitat type and season?

2. Materials and Methods

2.1. Habitat Classification and Description

A pre-study survey of 180 drop pipe sites within the Yazoo River watershed indicated that restored riparian habitats fit one of four discrete types on the basis of habitat area, pool volume, and vegetative structure (**Table 1**). Subsequent plant censuses and total station surveys [17] supported our initial habitat classification. Type I habitats were the smallest riparian patches and were composed mostly of herbaceous vegetation (**Table 1**). The four dominant plant species within Type I habitats were bermuda grass (*Cynodon dactylon*), goldenrod (*Solidago* spp.), paspalum grass (*Paspalum* spp.), and panic grass (*Panicum* spp.). Type II habitats were larger riparian patches than Type I habitats (**Table 1**) and were composed of herbaceous vegetation mixed with shrubs and saplings. The four dominant plant species within Type II habitats were Japanese honeysuckle (*Lonicera japonica*), goldenrod, white ash (*Fraxinus americana*), and blackberry (*Rubus argutus*). Type III habitats were riparian patches larger than Type II habitats (**Table 1**) and characterized by the presence of an ephemeral pool surrounded by a ring of woody vegetation. The four most frequently occurring plant species were black willow (*Salix nigra*), bermuda grass, ragweed (*Ambrosia artemisiifolia*), and kudzu (*Pueraria lobata*). Type IV habitats were characterized by having the greatest habitat area, permanent pools, greatest plant species richness (**Table 1**), and an input channel extending into the adjacent field. Vegetation within Type IV habitats consisted of woody and herbaceous vegetation, and the four most frequently occurring plant species were blackberry, goldenrod, partridge pea (*Cassia fasiculata*), and bermuda grass. The amount of woody vegetation within a site varied more among Type IV habitats than Type III habitats. Type IV habitats ranged from sites composed of predominantly herbaceous vegetation with a few mature trees > 2 m tall to sites that contained pools and input channels surrounded by mature trees > 2 m tall. Type IV habitats contained the largest trees and common woody species included black willow, American elm (*Ulmus*

Table 1. Mean total habitat area (m^2), mean maximum pool volume (m^3), mean plant species richness (plant richness), and mean vertical structure of woody vegetation index within four restored riparian habitat types in northwestern Mississippi, June 1994 to July 1995.

Habitat Type	Habitat Area	Pool Volume	Plant Richness	Vertical Structure Index[b]
Type I (n = 4)	600	15	22	0.01
Type II (n = 4)	1000	41	22	0.21
Type III (n = 4)	1300	426	26	0.30
Type IV (n = 4)	3700	1343	46	0.20

[a]See [17] for description of sampling methods for all response variables; [b]Index of vertical structure indicates dominance of woody vegetation greater than 1.8 m tall and ranges in scores from 0 (site lacking woody vegetation > 1.8 m tall) to 1 (site dominated by woody vegetation > 1.8 m tall) [17].

americana), American sycamore (*Platanus occidentalis*), and sweet gum (*Liquidambar styraciflua*).

The pre-study survey of 180 drop pipe sites also found that Type I habitats occurred most frequently (61%), followed in abundance by Type III (21%), Type II (11%), and Type IV (7%) habitats [9]. We selected four sites of each habitat type within the Long and Hotophia Creek watersheds in Panola County, Mississippi (lat 34°9' to 34°33'N, long 89°43' to 90°11'W) as study sites (total 16 sites). Both watersheds were predominantly agricultural watersheds primarily devoted to cotton (*Gossypium hirsutum*) production. All study sites were adjacent to channelized agricultural streams and agricultural fields. We attempted to control for the potential influence of land use by choosing sites that were adjacent to cotton fields. Fifteen sites were adjacent to cotton fields, but logistical matters required us to select one site adjacent to a corn (*Zea mays*) field.

2.2. Small Mammal Trapping

Small mammals were sampled from 28 June 1994 to 20 July 1995 with sherman folding live traps (7.6 cm wide by 8.9 cm tall by 22.9 cm long) and pitfall traps (19 L buckets that are 34.9 cm tall and 28.6 cm in diameter) buried flush with the ground. The combined use of live traps and pitfall traps enabled us to sample the entire small mammal community more effectively. Pools within Type III and Type IV habitats prevented the use of the standard grid trapping array. Therefore, we established a trapping transect along the perimeter of each Type I, II, III, and IV site. The trapping transect encircled the entire site of Type I, II, and III sites. The large size of Type IV sites made surrounding the entire site impractical and the trapping transect was placed so it encircled the large pools within these sites. Placement of live traps and pitfall traps were alternated along the trapping transect and each trap was located 5 m apart. All traps were at least 1 m from the edge of all habitat types and the adjacent agricultural field. The total number of traps within a site varied, but the number of traps per transect length required to encircle the site (Type I, II, III) or pool (Type IV) was similar. Wood covers were placed above pitfall

traps and aluminum covers were placed over live traps to reduce trap mortality. Pitfall traps were unbaited and live traps were baited with a combination of peanut butter, rolled oats, and mixed seed the first trapping period of summer 1994. Rolled oats were used as bait for subsequent trapping periods because of problems with ants.

Trapping periods were summer 1994 (28 June to 1 July 1994; 18 July to 27 July 1994), fall 1994 (8 November to 16 November 1994), winter 1995 (21 February to 28 February 1995), spring 1995 (17 April to 26 April 1995), and summer 1995 (13 July to 20 July 1995). Trapping was conducted for 7 to 9 nights during each season within all sites. Traps were checked daily during each season and were closed between seasons. All animals captured were identified to species and released. Cotton rats were also weighed with spring scales and individually marked with ear tags (#1, National Band and Tag Co.).

2.3. Data Analyses

We calculated species richness (the number of species captured), evenness (the reciprocal of the Simpson's Index divided by species richness) [27], Shannon Diversity Index [28], and abundance (the number of captures) from every site during each season. Our use of two sampling methods with different capture efficiencies and the variability in sampling effort among habitat types and seasons (**Table 2**) required the use of an index to standardize our sampling effort. First, the sampling effort (*i.e.*, number of live trap nights and number of pit fall trap nights) within all sites and sampling periods for each sampling method was ranked. The sampling effort index is then calculated by summing the rank sampling effort of all sampling methods used in each site during each season [24]. We then standardized species richness and abundance values by dividing the values of these variables by the sampling effort index.

We calculated cotton rat abundance (# of individuals captured/sampling effort index), percent of juveniles (# of individuals < 100 gm in weight/total # of individuals) captured, mean cotton rat weight, and mean number of

Table 2. Mean number of trap nights for live traps (LT) and pitfall traps (PF) within four restored riparian habitat types in northwestern Mississippi during each season from June 1994 to July 1995.

	Type I		Type II		Type III		Type IV	
	LT	*PF*	*LT*	*PF*	*LT*	*PF*	*LT*	*PF*
Summer 1994	22	24	66	68	96	94	105	110
Fall 1994	22	24	82	84	102	100	120	126
Winter 1995	19	21	72	74	89	88	105	110
Spring 1995	25	27	92	95	115	113	135	142
Summer 1995	19	21	72	74	89	88	105	110

recaptures from each site during each season. We also calculated the average distance moved by an individual cotton rat (AVDI) with at least one recapture within each site. AVDI is the sum of all distances moved by an individual cotton rat divided by the number of days elapsed between the first and last recapture within a season [29]. We chose an index that averages movements over time rather than one that averages movements over number of recaptures because all individuals may not be captured during every trapping period [29]. We then calculated the mean AVDI from each site during each season.

We used a two factor analysis of variance (ANOVA) coupled with the Student-Neuman-Keuls (SNK) test to examine if differences in community (species richness, evenness, Shannon Diversity Index, abundance) and population response variables (cotton rat abundance, weight, percent juveniles, number of recaptures, movement) occurred among habitat types and seasons. The assumptions of normality and/or equal variance were not met for any community response variable and four of the population response variables (cotton rat abundance, weight, percent juveniles, number of recaptures). We rank transformed these eight response variables to conduct the two factor ANOVA. The use of rank transformation in conjunction with a parametric test is the equivalent of a nonparametric two factor ANOVA [30]. Additionally, some sites did not yield any cotton rat recaptures and resulted in missing values that prevented us from testing if an interaction effect occurred within the ANOVA analyses of mean number of recaptures and mean AVDI. ANOVA analyses were conducted using SigmaStat 3.1 for Windows [31] and a significance level of P < 0.05. Detrended correspondence analyses (DCA) were conducted on percentages of captures of small mammal species from each season to examine if species composition differed among habitat types in each season. To reduce the influence of rare species on the DCA results we omitted species that occurred in < 10% of all collections and selected the option in PC-ORD that results in downweighting of species that occurred < the frequency of the most common species/5 (*i.e.*, <15 to 20%) [32]. DCA analyses were conducted using PC-ORD for Windows [32].

3. Results

3.1. Community Responses

Ten small mammal species from 1743 captures occurred within a total of 12,446 trap nights. The three most frequently captured small mammals in all habitat types were hispid cotton rat, woodland vole (*Microtus pinetorum*), and the marsh rice rat (*Oryzomys palustris*) (**Table 3**). Species richness, evenness, Shannon Diversity Index, and abundance differed (P < 0.05) among habitat types (**Table 4**). Mean species richness was the greatest in Type II habitats and the least in Type IV habitats (**Table 5**). Mean evenness and abundance were the greatest in Type III habitats and the least in Type I habitats (**Table 5**). Mean Shannon Diversity Index within Type II, III, and IV habitats was greater than diversity within Type I habitats (**Table 5**). Species richness, Shannon Diversity Index, and abundance also differed (P < 0.05) among seasons (**Table 4**). Mean species richness, Shannon Diversity Index, and abundance were the least in summer 1994, increased in the fall 1994, and then either declined in the spring 1995 (species richness), declined in the summer 1995 (Shannon Diversity Index), or did not decline during the remainder of the study (abundance) (**Table 5**). None of the community response variables exhibited a significant interaction effect of habitat type and season (**Table 4**). Species composition did not differ among habitat types due to the variability in small mammal species composition that occurred among sites within habitat types (**Figure 2**). Type I habitats exhibited greater within habitat variability in species composition than the other habitat types from summer 1994 to spring 1995 (**Figure 2**). Variability in species composition was reduced within all habitat types in the summer 1995 (**Figure 2**).

3.2. Population Responses

We captured 386 individual cotton rats within all habitat types, and we marked and released 200 cotton rats between summer 1994 and spring 1995. One hundred and twenty cotton rats were recaptured within the same site at least once during a season. Only 32 from a possible 184

Table 3. Number of small mammal species captured within four restored riparian habitat types in northwestern Mississippi, June 1994 to July 1995.

Species	Type I	Type II	Type III	Type IV
Hispid cotton rat (*Sigmodon hispidus*)	77	273	209	320
Woodland vole (*Microtus pinetorum*)	1	37	115	64
Marsh rice rat (*Oryzomys palustris*)	2	38	96	52
Cotton mouse (*Peromyscus gossypinus*)	1	50	60	29
White-footed mouse (*Peromyscus leucopus*)	2	29	39	32
House mouse (*Mus musculus*)	23	23	19	18
Southern short-tailed shrew (*Blarina carolinensis*)	6	20	34	22
Southeastern shrew (*Sorex longirostris*)	0	10	7	12
Least shrew (*Cryptotis parva*)	2	8	4	5
Golden mouse (*Ochrotomys nuttalli*)	0	0	3	1

Table 4. P values from two factor analysis of variance tests conducted to determine if small mammal community and population response variables differed among habitat type, season, or the interaction of habitat type and season. P values < 0.05 are bolded.

	Habitat Type	Season	Interaction Effect
Community			
Species richness	**0.049**	**<0.001**	**0.848**
Evenness	**0.035**	0.212	**0.684**
Shannon diversity index	**<0.001**	**<0.001**	**0.108**
Abundance	**0.040**	**<0.001**	**0.209**
Population (Cotton Rat)			
Abundance	0.071	**<0.001**	**0.175**
Percent juveniles	0.296	**<0.001**	0.032
Weight	**<0.001**	**<0.001**	<0.001
Number of recaptures	0.143	0.081	-
Average distance moved	0.080	0.916	-

Table 5. Habitat and season factor means (SD) for small mammal species richness, evenness, Shannon Diversity Index (H'), and abundance in restored riparian habitats in northwestern Mississippi, June 1994 to July 1995.

Factor	Level	Richness	Evenness	H'	Abundance
Habitat Type	Type I	0.07 (0.09) AB	0.26 (0.38) B	0.26 (0.42) B	0.28 (0.37) B
	Type II	0.06 (0.04) A	0.49 (0.28) AB	0.99 (0.53) A	0.30 (0.20) AB
	Type III	0.05 (0.03) AB	0.59 (0.23) A	1.09 (0.51) A	0.32 (0.18) A
	Type IV	0.04 (0.02) B	0.44 (0.29) AB	0.89 (0.62) A	0.23 (0.17) AB
Season	Summer 1994	0.02 (0.02) C	0.31 (0.43) A	0.30 (0.45) B	0.06 (0.06) B
	Fall 1994	0.07 (0.04) A	0.43 (0.27) A	1.04 (0.48) A	0.47 (0.24) A
	Winter 1994	0.08 (0.05) A	0.57 (0.20) A	1.19 (0.50) A	0.35 (0.20) A
	Spring 1995	0.05 (0.03) B	0.51 (0.29) A	1.06 (0.63) A	0.23 (0.13) A
	Summer 1995	0.06 (0.08) B	0.40 (0.31) A	0.44 (0.40) B	0.30 (0.30) A

[a]Different letters indicate differences (P < 0.05) in ranked means among habitat type or season for each response variable.

individuals (accounting for known trap mortalities occurring within a season) were recaptured within the same site at least once between seasons. None of the seasonal recaptures occurred within Type I habitats, 12 seasonal recaptures occurred within Type II habitats, nine occurred within Type III habitats, and 10 occurred within Type IV habitats. Cotton rat abundance did not differ (P > 0.05) among habitat types, but differed (P < 0.05) among seasons (**Table 4**). Mean cotton rat abundance was the least in summer 1994, winter 1994, and spring 1995 and was the greatest in fall 1994 and summer 1995 (**Figure 3**). Percent juvenile cotton rats and weight exhibited a significant (P < 0.05) interaction effect (**Table 4**) as trends in these response variables among habitat types

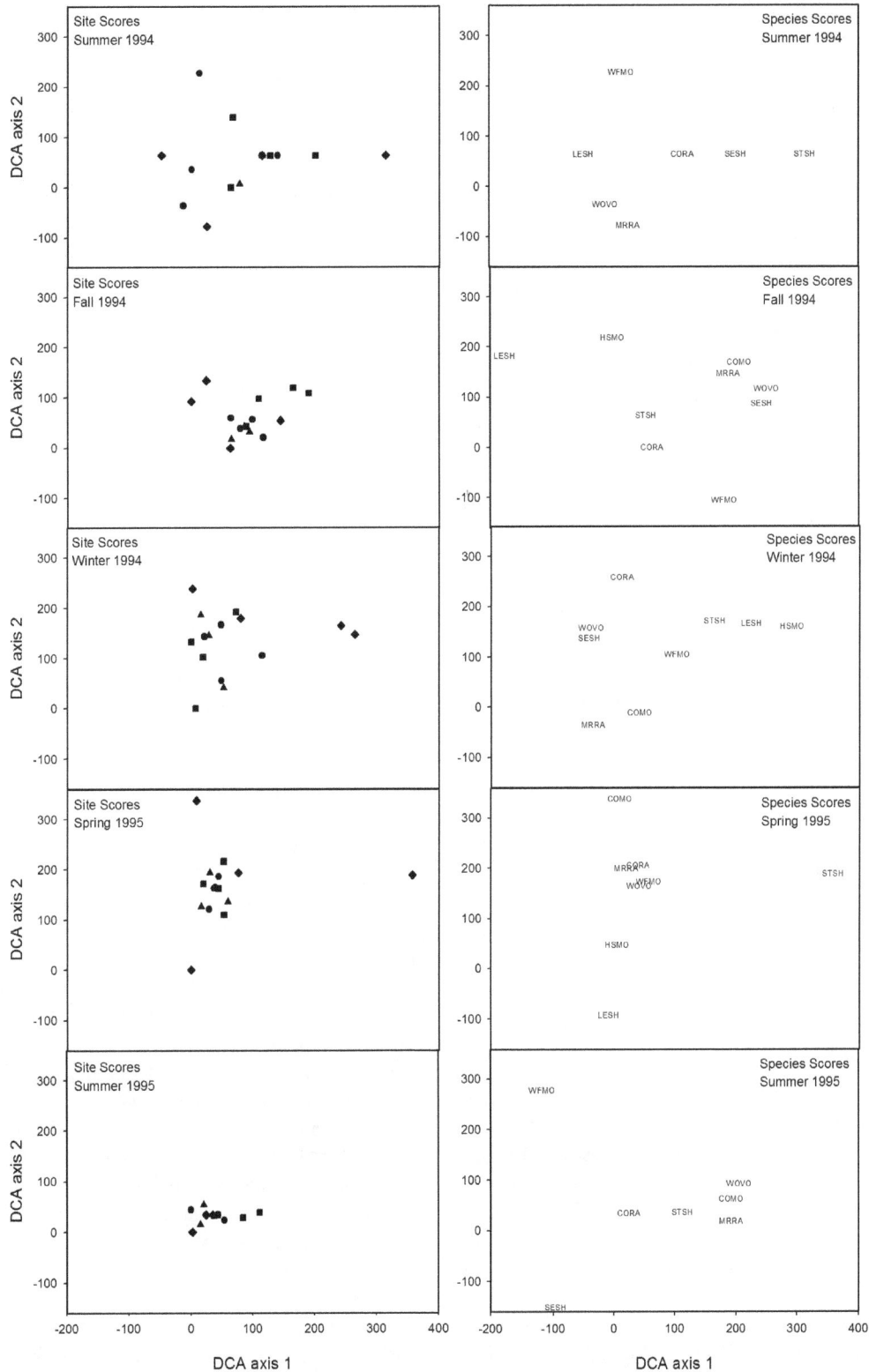

Figure 2. Site and species scores resulting from detrended correspondence analyses (DCA) of the percentage of small mammal species captured within four riparian habitat types in northwestern Mississippi during the summer 1994, fall 1994, winter 1994, spring 1995, and summer 1995. Habitat types are differentiated by different shapes within the figures: ♦: Type I sites; ●: Type II sites; ■: Type III sites; ▲: Type IV sites. Species codes are: COMO-cotton mouse; CORA-cotton rat; HSMO-house mouse; LESH-least shrew; MRRA-marsh rice rat; SESH-southeastern shrew; STSH-southern short-tailed shrew; WFMO-white-footed mouse; WOVO-woodland vole.

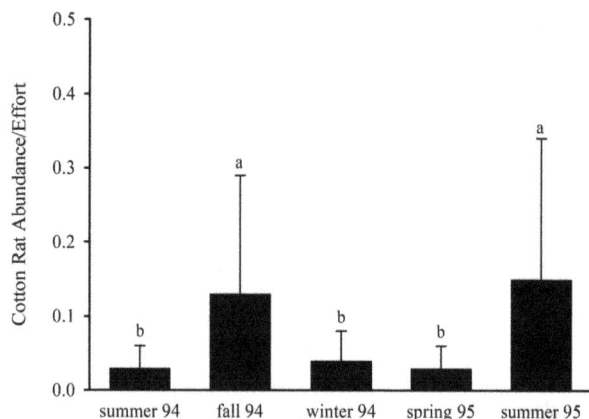

Figure 3. Mean cotton rat abundance within restored riparian habitats in northwestern Mississippi during summer 1994, fall 1994, winter 1994, spring 1995, and summer 1995.

differed among seasons. Mean percent juveniles did not differ among habitat types from summer 1994 to spring 1995 (**Figure 4**). Mean percent juveniles captured within Type I habitats was greater than the mean percent juveniles captured in the other three habitat types during summer 1995 (**Figure 4**). Mean cotton rat weight was greater in Type IV habitats than Type I, II, and III habitats during summer 1994 (**Figure 5**). No differences in mean cotton rat weight occurred among habitat types in fall 1994 (**Figure 5**). Mean weight was the least in Type I habitats, the greatest in Type II habitats, and intermediate in Type III and IV habitats in winter 1994 (**Figure 5**). Mean weight in Type II, III, and IV habitats was greater than mean weight in Type I habitats in spring and summer 1995 (**Figure 5**). Number of individuals recaptured and AVDI did not differ (P > 0.05) among habitats or seasons (**Table 3**).

4. Discussion

4.1. Community Responses

Riparian habitats established by drop pipe installation contained nine small mammal species of the possible 16 native species present within the watersheds of our study sites [33] and only one non-native species (house mouse) was captured. Small mammal diversity within the southeastern United States is depauperate compared to other regions of the United States [34]. The levels of small mammal diversity and species composition we documented within these restored riparian habitats was similar to those found in other created agricultural wetlands [35], restored riparian zones [34], and managed pine forests [26,35-39] in the southeastern United States. Native small mammal species not captured within our study sites included eastern mole (*Scalopus aquaticus*), eastern chipmunk (*Tamias striatus*), eastern gray squirrel (*Sciurus carolinensis*), fox squirrel (*Sciurus niger*), southern fly-

ing squirrel (*Glaucomys volans*), eastern harvest mouse (*Reithrodontomys humulis*), and the eastern woodrat (*Neotoma floridana*). Absence of these species from our study sites likely resulted from low capture vulnerability of these species to our trapping methods and/or their habitat preferences.

In general, Type II, III, and IV habitats exhibited greater species richness, evenness, diversity, and abundance than Type I habitats. We attribute these changes in community structure to the greater habitat area and vegetative structure within Type II, III, and IV habitats compared to Type I habitats. Small mammal communities within remnant habitats (*i.e.*, riparian zones, shelter belts, etc.) in agricultural landscapes are primarily influenced by habitat area and vegetative structure [40]. Small mammal species richness and abundance within riparian zones of agricultural streams in Quebec increased with increasing riparian widths and increasing vegetative structure [41]. Species richness of all small mammals and those preferring woodland habitat increased with increasing habitat area within fragmented forested riparian habitats and woodlots in Illinois [42]. Small mammal species richness and diversity increased with increasing habitat area and structural diversity within hedgerows in agricultural fields on Prince Edward Island, Canada [43] and within shelterbelts in Minnesota [40].

However, we did not observe a consecutive increase in species richness, evenness, diversity, and abundance from Type II to Type IV habitats despite increases in habitat area and vegetative structure. We were surprised that Type IV habitats with the greatest habitat area and vegetative structure did not exhibit the greatest species richness, diversity, and abundance values. We suspect these results stem from the influence of increasing pool development (*i.e.*, volume) that occurs from Type II to Type IV habitats. Little is known about the influence of increasing pool size on small mammal communities within riparian zones. Increased precipitation and flooding have been observed to decrease small mammal diversity and the abundance within riparian habitats in Mississippi and Missouri [44,45]. Increasing pool size within restored riparian habitats may also result in decreased small mammal diversity and abundance because it reduces the amount of terrestrial habitat within a site and the growth of herbaceous vegetation that is a critical habitat resource for many small mammals [45].

Seasonal changes in species richness, diversity, and species composition are likely a result of seasonal changes of cover and food resources within our sampling sites. We feel the seasonal changes in species richness and diversity were also influenced by the seasonal changes in cover and food resources within the adjacent agricultural fields. During the summer, crops present on the fields were approximately a meter tall, but little or no cover

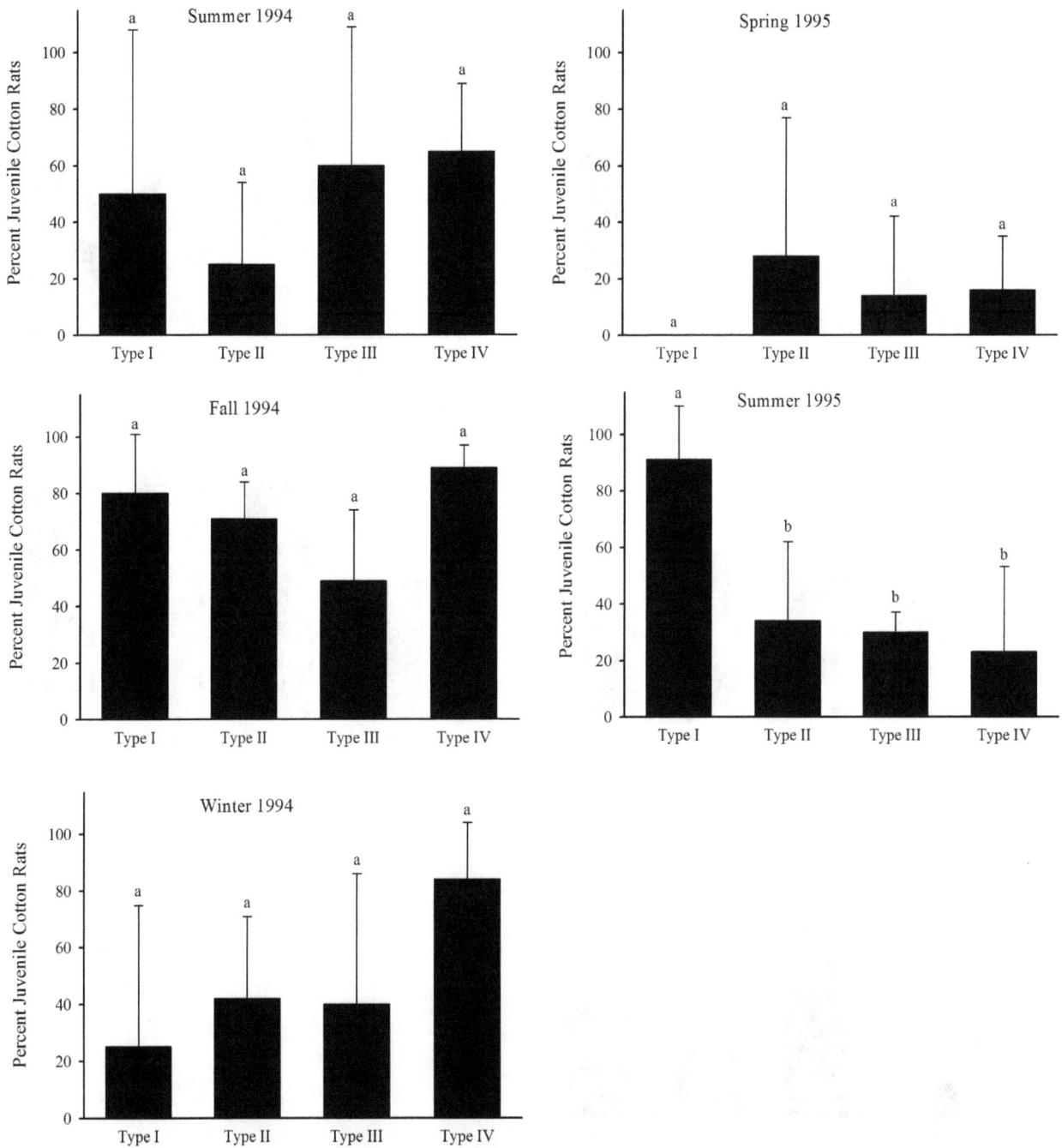

Figure 4. Mean percent juvenile cotton rats captured among four restored riparian habitat types in northwestern Mississippi during summer 1994, fall 1994, winter 1994, spring 1995, and summer 1995.

was present on the fields in the fall, winter, and spring. Therefore, observed declines in species richness and diversity during the summer may have occurred because small mammals were not limited to the riparian zone due to the increased cover and/or food resources available on agricultural fields.

4.2. Population Responses

Hispid cotton rats prefer grass dominated habitats, habi-

tats with thick cover, and have been captured in vegetation areas adjacent to ponds [46,47]. Specifically, cotton rats prefer habitats having stands of dense grass species greater than 0.25 m high associated with various scrub species [46]. All habitats created by drop pipes met these general habitat criteria, which may explain the numerical dominance of cotton rats in all habitat types (**Table 3**). Cotton rat abundance, weight, and movement patterns were less within unpreferred habitats (mowed, minimal

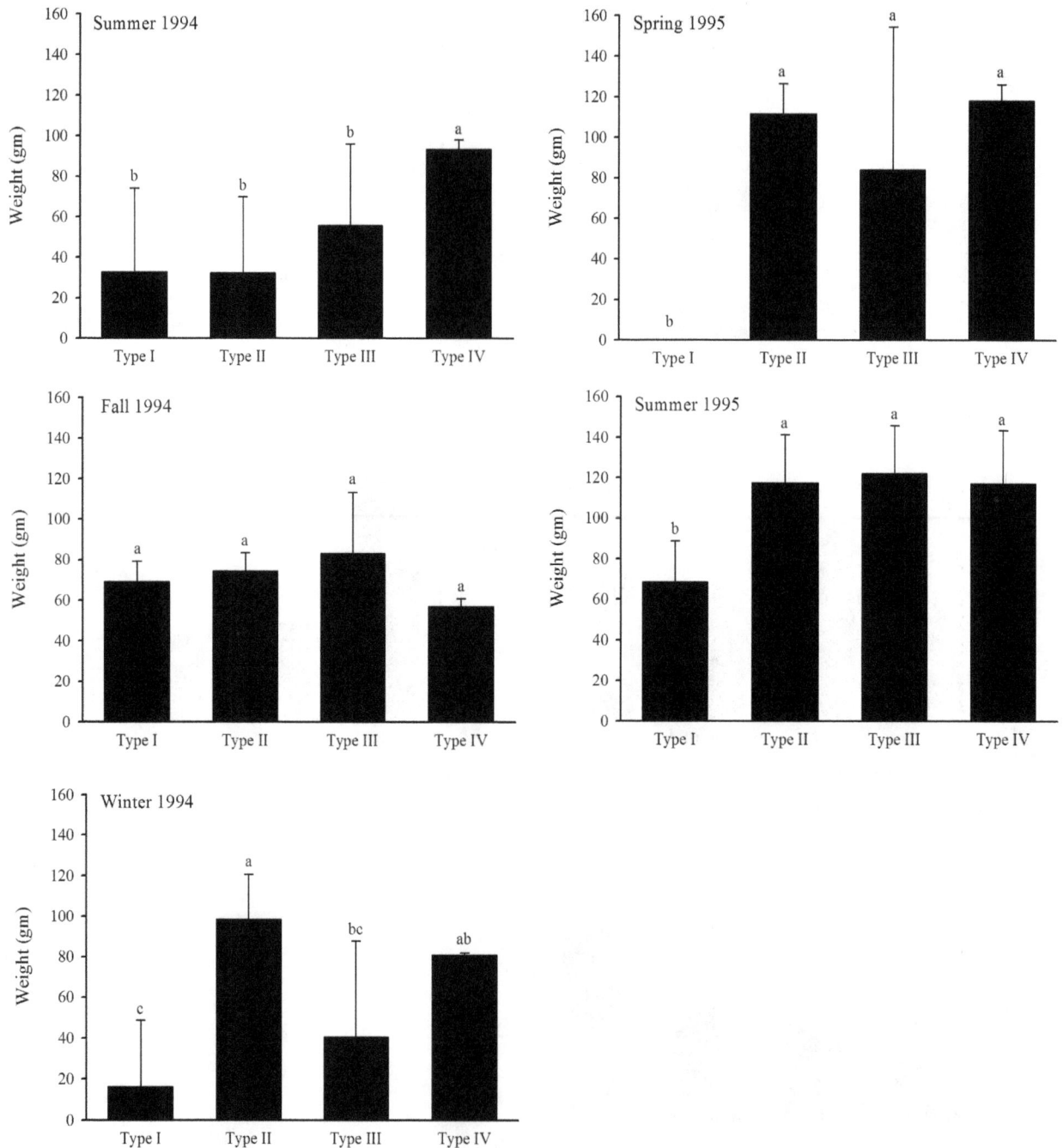

Figure 5. Mean cotton rat weight among four restored riparian habitat types in northwestern Mississippi during summer 1994, fall 1994, winter 1994, spring 1995, and summer 1995.

grass cover lacking shrub overstory) than preferred habitats (unmowed, dense grass cover with shrub overstory) in Texas coastal prairie [48]. Our results concur with these as we documented decreased cotton rat weights within small Type I habitats with reduced vegetative structure. We did not document differences in cotton rat abundance among restored riparian habitat types, but our results may have been obscured by gender differences as male and female cotton rats exhibit different habitat use patterns

[49,50]. We also documented greater percent of juvenile cotton rats in Type I habitats than the other three habitat types during summer 1995. Increased cotton rat abundance (adults and juveniles) also occurred during the summer 1995 and suggests the possibility that juvenile cotton rats were marginalized to the Type I habitats. Others in Oklahoma [46] and Georgia [49] have observed that smaller or juvenile cotton rats increase their usage of habitats with less cover during periods of increased

cotton rat abundances.

5. Conclusion

Environmental problems associated with stream channelization and subsequent channel incision and gully erosion occur within agricultural watersheds throughout the world [51,52]. Our results provide insights on how a gully erosion control structure and similar water control structures used nationally and internationally can be used to assist with the restoration of riparian zones of channelized agricultural streams. We observed that small mammal diversity, abundance, and hispid cotton rat weight were the greatest in the riparian habitat types (Type II, III, IV) having greater habitat size, vegetative structural diversity, and pool size than the smallest riparian habitat types (Type I) with the least vegetative diversity. Present drop pipe installation practices focus on erosion control without consideration of riparian restoration. Our results indicate small mammals will benefit the most from the establishment of Type II, III, and IV habitats that are the habitat types least frequently created by drop pipe installation. Modifying the drop pipe installation design to ensure that riparian habitats larger than 1000 m^2 with pool volumes greater than 41 m^3 and greater than 20% coverage of woody vegetation greater than 1.8 m tall are established within riparian zones adjacent to channelized agricultural streams in northwestern Mississippi will assist with the conservation of small mammals within these degraded riparian zones. Our findings also suggest that increasing the amount of terrestrial habitat associated with Type IV habitats containing large pools may promote increased diversity and abundance of small mammals. Future research on small mammals within riparian zones, agricultural fields, and other habitat types present within agricultural watersheds is necessary for developing watershed management plans that will benefit both wildlife and agriculture.

6. Acknowledgements

We thank the Vicksburg District US Army Corps of Engineers for providing partial cooperative funding and the USDA Natural Resources Conservation Service for landowner and site information. We also thank Eric Dibble, Norman Fausey, and Jeanne Jones for providing helpful comments on an earlier draft of this manuscript. Scott Knight and John Wigginton assisted with the start-up phase of this research. F. Douglas Shields Jr. provided data on the vegetative structure and assistance with analyzing the cotton rat movement data. Special appreciation goes to Chip Butts, Ezekiel Cooper, Kim Damon, Ken Kallies, Jonathan Maul, Todd Randall, Sam Testa, and John Wigginton for their assistance with field work.

REFERENCES

[1] R. J. Naiman, H. DeCamps and M. Pollock, "The Role of Riparian Corridors in Maintaining Regional Biodiversity," *Ecological Applications*, Vol. 3, No. 2, 1993, pp. 209-212.

[2] C. R. Hupp, "Riparian Vegetation, Channel Incision, and Ecogeomorphic Recovery," In: S. S. Y. Wang, E. J. Langendoen and F. D. Shields Jr., Eds., *Management of Landscapes Disturbed by Channel Incision*, Center for Computational Hydroscience and Engineering, University of Mississippi, University, Mississippi, 1997, pp. 3-11.

[3] S. C. Spackman and J. W. Hughes, "Assessment of Minimum Stream Corridor Width for Biological Conservation: Species Richness and Distribution Along Mid-order Streams in Vermont, USA," *Biological Conservation*, Vol. 71, 1995, pp. 325-332.

[4] E. D. Fleharty and K. W. Navo, "Irrigated Cornfields as Habitat for Small Mammals in the Sandsage Prairie Region of Western Kansas," *Journal of Mammalogy*, Vol. 64, No. 3, 1983, pp. 367-379.

[5] A. T. Doyle, "Use of Riparian and Upland Habitats by Small Mammals," *Journal of Mammalogy*, Vol. 71, No. 1, 1990, pp. 14-23.

[6] E. W. Chapman and C. A. Ribic., "The Impact of Buffer Strips and Stream-Side Grazing on Small Mammals in Southwestern Wisconsin," *Agriculture, Ecosystems, and Environment*, Vol. 88, No. 1, 2002, pp. 49-59.

[7] J. D. Osbourne, J. T. Anderson and A. B. Spurgeon, "Effects of Habitat on Small Mammal Diversity and Abundance in West Virginia," *Wildlife Society Bulletin*, Vol. 33, No. 3, 2005, pp. 814-822.

[8] S. C. Happ, G. Rittenhouse and G. C. Dobson, "Some Principles of Accelerated Stream and Valley Sedimentation," Technical Bulletin No. 695, United States Department of Agriculture, Washington DC, 1940.

[9] F. D. Shields Jr., S. S. Knight and C. M. Cooper, "Rehabilitation of Watersheds with Incising Channels," *Water Resources Bulletin*, Vol. 31, No. 6, 1995, pp. 971-982.

[10] R. Woodburn, "Science Studies a Gully," *Soil Conservation*, Vol. 15, No. 1, 1949, pp. 11-13, 22.

[11] E. H. Grissinger, J. B. Murphey and N. L. Coleman, "Planned Gully Research at the USDA National Sedimentation Laboratory," In: D. G. DeCoursey, Ed., *Proceedings of the Natural Resources Modeling Symposium*, USDA Agricultural Research Service, Publication ARS-30, 1985, pp. 475-478.

[12] J. Poesen, J. Nachtergaele, G. Verstraeten and C. Valentin, "Gully Erosion and Environmental Change: Importance and Research Needs," *Catena*, Vol. 50, No. 2-4, 2003, pp. 91-133.

[13] E. E. Possardt and W. E. Dodge, "Stream Channelization

Impacts on Songbirds and Small Mammals in Vermont," *Wildlife Society Bulletin*, Vol. 6, No. 1, 1978, pp. 18-24.

[14] J. S. Barclay, "Impact of Stream Alterations on Riparian Communities in South-Central Oklahoma," Report # FWS/OBS-80/17, Office of Biological Services, US Fish and Wildlife Service, Washington DC, 1980.

[15] T. T. Brown, T. L. Derting and K. Fairbanks, "The Effects of Stream Channelization and Restoration on Mammal Species and Habitat in Riparian Corridors," *Journal of the Kentucky Academy of Sciences*, Vol. 69, No. 1, 2008, pp. 37-49.

[16] A. R. Geier and L. B. Best, "Habitat Selection by Small Mammals of Riparian Communities: Evaluating Effects of Habitat Alterations," *Journal of Wildlife Management*, Vol. 44, No. 1, 1980, pp. 16-24.

[17] F. D. Shields Jr., P. C. Smiley Jr. and C. M. Cooper, "Design and Management of Edge-of-field water Control Structures for Ecological Benefits," *Journal of Soil and Water Conservation*, Vol. 57, No. 3, 2002, pp. 151-157.

[18] F. D. Shields Jr., P. C. Smiley Jr. and C. M. Cooper, "Modifying Erosion Control Structures for Ecological Benefits," *Journal of Soil and Water Conservation*, Vol. 62, No. 6, 2007, p. 157A.

[19] C. M. Cooper, P. C. Smiley Jr., J. D. Wigginton, S. S. Knight and K. W. Kallies, "Vertebrate Use of Habitats Created by Installation of Field-scale Erosion Control Structures," *Journal of Freshwater Ecology*, Vol. 12, No. 2, 1997, pp. 199-207.

[20] C. M. Cooper, F. D. Shields Jr., S. Testa III and S. S. Knight, "Sediment Retention and Water Quality Enhancement in Disturbed Watersheds," *International Journal of Sediment Research*, Vol. 15, No. 1, 2000, pp. 121-134.

[21] P. C. Smiley Jr., S. S. Knight, C. M. Cooper and K. W. Kallies, "Fish Richness and Abundance in Created Riparian Habitats of Channelized Northern Mississippi Streams," *Southeastern Fishes Council Proceedings*, Vol. 39, 1999, pp. 7-12.

[22] J. D. Maul, P. C. Smiley Jr. and C. M. Cooper, "Patterns of Avian Nest Predators and a Brood Parasite among Restored Riparian Habitats in Agricultural Watersheds," *Environmental Monitoring and Assessment*, Vol. 108, No. 1-3, 2005, pp. 133-150.

[23] P. C. Smiley Jr., J. D. Maul and C. M. Cooper, "Avian Community Structure among Restored Riparian Habitats in Northwestern Mississippi," *Agriculture, Ecosystems, and Environment*, Vol. 122, No. 2, 2007, pp. 149-156.

[24] P. C. Smiley Jr., S. S. Knight, F. D. Shields Jr. and C. M. Cooper, "Influence of Gully Erosion Control on Amphibian and Reptile Communities within Riparian Zones of Channelized Streams," *Ecohydrology*, Vol. 2, No. 3, 2009, pp. 303-312.

[25] P. C. Smiley Jr., C. M. Cooper, K. W. Kallies and S. S. Knight, "Assessing Habitats Created by Installation of

Drop Pipes," In: S. S. Y. Wang, E. J. Langendoen and F. D. Shields Jr., Eds., *Management of Landscapes Disturbed by Channel Incision*, Center for Computational Hydroscience and Engineering, University of Mississippi, University, Mississippi, 1997, pp. 887-892.

[26] D. A. Miller, R. E. Thill, M. A. Melchoirs, T. B. Wigley and P. A. Tappe, "Small Mammal Communities of Streamside Management Zones in Intensively Managed Pine Forests of Arkansas," *Forest Ecology and Management*, Vol. 203, No. 1-3, 2004, pp. 381-393.

[27] B. Smith and J. B. Wilson, "A Consumer's Guide to Evenness Indices," *Oikos*, Vol. 76, No. 1, 1996, pp. 70-82.

[28] A. E. Magurran, "Ecological Diversity and its Measurement," Croom helm, London, 1988.

[29] G. N. Cameron and W. B. Kincaid, "Species Removal Effects on Movements of *Sigmodon hispidus* and *Reithrodontomys fulvescens*," *American Midland Naturalist*, Vol. 108, No. 1, 1982, pp. 60-67.

[30] W. J. Conover, "Practical Nonparametric Statistics," John Wiley, New York, 1999.

[31] Systat Software, "SigmaStat 3.1 for Windows," Point Richmond, California, 2004.

[32] B. McCune and M. J. Mefford, "Multivariate Analysis of Ecological Data version 4.01. MjM Software," Gleneden Beach, Oregon, 1999.

[33] J. R. Choate, J. K. Jones Jr. and C. Jones, "Handbook of Mammals of the South-central States," Louisiana State University Press, Baton Rouge, Louisiana, 1994.

[34] L. D. Wike, F. D. Martin, H. G. Hanlin and L. S. Paddock, "Small Mammal Populations in a Restored Stream Corridor," *Ecological Engineering*, Vol. 15, No. S1, 2000, pp. S121-S129.

[35] T. A. Whitsitt and P. A. Tappe, "Temporal Variation of a Small-Mammal Community at a Wetland Restoration Site in Arkansas," *Southeastern Naturalist*, Vol. 8, 2009, pp. 381-386.

[36] M. S. Mitchell, K. S. Karriker, E. J. Jones and R. A. Lancia, "Small Mammal Communities Associated with Pine Plantation Management of Pocosins," *Journal of Wildlife Management*, Vol. 49, No. 4, 1995, pp. 875-881.

[37] R. E. Masters, R. L. Lochmiller, S. T. McMurry and G. A. Bukenhofer, "Small Mammal Response to Pine-Grassland Restoration for Red-Cockaded Woodpeckers," *Wildlife Society Bulletin*, Vol. 26, No. 1, 1998, pp. 148-158.

[38] N. L. Constantine, T. A. Campbell, W. M. Baughman, T. B. Harrington, B. R. Chapman and K. V. Miller, "Effects of Clearcutting with Corridor Retention on Abundance, Richness, and Diversity of Small Mammals in the Coastal Plain of South Carolina, USA," *Forest Ecology and Management*, Vol. 202, No. 1-3, 2004, pp. 293-300.

[39] R. W. Perry and R. E. Thill, "Small Mammal Responses

to Pine Regeneration Treatments in the Ouachita Mountains of Arkansas and Oklahoma, USA," *Forest Ecology and Management*, Vol. 219, No. 1, 2005, pp. 81-94.

[40] R. H. Yahner, "Small Mammals in Farmstead Shelterbelts: Habitat Correlates of Seasonal Abundance and Community Structure," *Journal of Wildlife Management*, Vol. 47, No. 1, 1983, pp. 74-84.

[41] C. Maisonneuve and S. Rioux, "Importance of Riparian Habitats for Small Mammal and Herpetofaunal Communities in Agricultural Landscapes of Southern Quebec," *Agriculture, Ecosystems and Environment*, Vol. 83, No. 1-2, 2001, pp. 165-175.

[42] D. L. Rosenblatt, E. J. Heske, S. L. Nelson, D. M. Barber, M. A. Miller and B. MacAllister, "Forest Fragments in East-central Illinois: Islands or Habitat Patches for Mammals?" *American Midland Naturalist*, Vol. 141, No. 1, 1999, pp. 115-123.

[43] M. Silva and M. E. Prince, "The Conservation Value of Hedgerows for Small Mammals in Prince Edward Island, Canada," *American Midland Naturalist*, Vol. 159, No. 1, 2008, pp. 110-124.

[44] M. J. Chamberlain and B. D. Leopold, "Effects of a Flood on Relative Abundance and Diversity of Small Mammals in a Regenerating Bottomland Hardwood Forest," *Southwestern Naturalist*, Vol. 48, No. 2, 2003, pp. 306-309.

[45] A. G. Elliott and B. G. Root, "Small Mammal Responses to Silvicultural and Precipitation-Related Disturbances in Northeastern Missouri Riparian Forests," *Wildlife Society Bulletin*, Vol. 34, No. 2, 2006, pp. 485-601.

[46] J. W. Goertz, "The Influence of Habitat Quality upon the Density of Cotton Rat Populations," *Ecological Monographs*, Vol. 34, No. 4, 1964, pp. 359-381.

[47] G. N. Cameron and S. R. Spencer, "*Sigmodon hispidus*," *Mammalian Species*, Vol. 158, 1981, pp. 1-9.

[48] S. R. Spencer and G. N. Cameron, "Behavioral Dominance and Its Relationship to Habitat Patch Utilization by the Hispid Cotton Rat (*Sigmodon hispidus*)," *Behavioral Ecology and Sociobiology*, Vol. 13, No. 1, 1983, pp. 27-36.

[49] W. Z. Lidicker, J. O. Wolff, L. N. Lidicker and M. H. Smith, "Utilization of a Habitat Mosaic by Cotton Rats During a Population Decline," *Landscape Ecology*, Vol. 6, No. 4, 1992, pp. 259-268.

[50] G. N. Cameron and W. B. Kincaid, "Mechanisms of Habitat Selection by the Hispid Cotton Rat (*Sigmodon hispidus*)," *Journal of Mammalogy*, Vol. 89, No. 1, 2008, pp. 126-131.

[51] A. Brookes, "Channelized Rivers: Perspectives for Environmental Management," Wiley, Chichester, 1988.

[52] S. Y. Wang, E. J. Langendoen and F. D. Shields, "Management of Landscapes Disturbed by Channel Incision," Center for Computational Hydroscience and Engineering, University of Mississippi, Mississippi, 1997.

13

Hydrological Modeling Using GIS for Mapping Flood Zones and Degree Flood Risk in Zeuss-Koutine Basin (South of Tunisia)

Khemiri Sami[1], Ben Alaya Mohsen[1], Khnissi Afef[2], Zargouni Fouad[1]

[1]Department of Geology, University of Sciences of Tunis, Manar, Tunisia,; [2]Water Research and Technology Center, Solimane, Tunisia.

ABSTRACT

This study lies within the scope of a strategy of prevention from inundations by the contribution of new technology in stage of the hydrological and geomorphological modeling for protection against the floods in a medium of weak at the average risk in South-eastern Tunisia, starting from the catchment area of Zeuss-Koutine. Considering the lack of studies we were brought to extract the area catchment in question, and to deduce its geomorphological and hydrometric characteristics, starting from the digital terrain model. We could obtain, by overlaying maps of slopes, indices and flows, the hydrological zonation of the catchment area of Zeuss-Koutine. The hydrological study of the basin's slopes of Zeuss-Koutine is not lying out that very little physical information rests primarily on cartographic processes. The use of the latter can be regarded as an allowing indicator, by the crossing of the explanatory factors of the surface flow (slopes and direction of flow), to define a set of homogeneous hydrological zones in the level of the hydrological characteristics (average slopes, altitudes, roughness, etc). It is mainly a question of better taking account of the physical properties of the basins slopes.

Keywords: Degree Flood Risk; GIS; Geodatabase; Flood Zones; Hydrological Modeling

1. Introduction

Prevention from natural risks represents a big challenge for humanity. Especially during last decade, there are brutal climatic changes around the world and any place could, at any moment, be flooded or stormed. In Tunisia, the climate is generally characterized by a high meteorological variability from north to south. It is qualified as an arid to semi-arid under Mediterranean influences toward the north and Saharan influences southward. This difference is due to the existence of a climatic barrier separating these two zones and known as Tunisian Ridge [1]. Such climate type with low rain laid, predominately, Tunisia at the exhibition of floods. The average annual rainfall, in southern Tunisia, doesn't exceed 80 mm however the evapotranspiration is about 2500 mm. The negative hydrologic balance of these two parameters and the rarity of runoff contribute very little to the deep groundwater recharge. Despite the previously mentioned characteristics and the position of Tunisia in global climatic zones, floods have affected and might affect the south of our country in the area of Gabes-Medenine and specifically Zeuss-Koutine basin purpose of this study [1].

The detailed preliminary study and the choice of the solution proposed and the technique employed for the fight and protection against these floods are extremely important. The prevention is a paramount phase in protection, and it requires several studies; geomorphologic, hydrological, topographic, hydrogeomorphologic, hydrogeological and modeling histories of the floods.

2. Methods and Workflow

The hydrological study of the Zeuss-Koutine basin was conducted on the basis of the modeling, by the use of

Geographical Information System (GIS) and, more precisely, the contribution of Digital Elevation Model of Ground (DEM) and its derivative. The DEM is used to define the slopes, the hydrographic network, the delimitation of the basin's slopes and the extraction of the physical characteristics and water streaming related to this basin.

The whole digital results got in the form of layers of numerical information (GRID) on the areas catchment of Zeuss-Koutine and the water ways of surface constitute basic and useful information for hydrological modeling.

The conjugation of the various classes of parameters obtained starting from these cards streaming of surface, of the slopes (relief), of the hydrographic network, the physical characteristics of the basin and the flow directions and accumulations [2], allowed us to establish maps which will be stacked to other layers such as those of pedology, geology, vegetation and infrastructures. All results and collected parameters will help constitute a data bank usable for a hydrological modeling with space discretization, to delimit homogeneous zones and to prepare a digital cartographic support of the flood plains and degree risks along the basin.

The flood mapping was developed as a result of the layering of several parameters covering geographical areas capable of being flooded according to three scenarios (flood low probability, flood medium probability and flood high probability). Through the application of the formula of Lee (2007), we were able to extract the degree of flood risk in the study area per pixel.

The choice of the technique employed and analyzes it physical and socio-economic operation area catchment to allow (**Figure 1**):

- To locate and to gradually choose the sectors on which it is desirable and possible to carry out preventive actions for protection against the floods.

- To quantify their effect on the risings and to appreciate their limits and costs.

The cartography of the flood plains and the realization of the maps of the degree risks require several types of studies.

These studies complementary and are connected.

The unit makes it possible to release an adapted and coherent vision of the operation of the catchment area and the problems of installation.

Principal the studies of preventions are the following

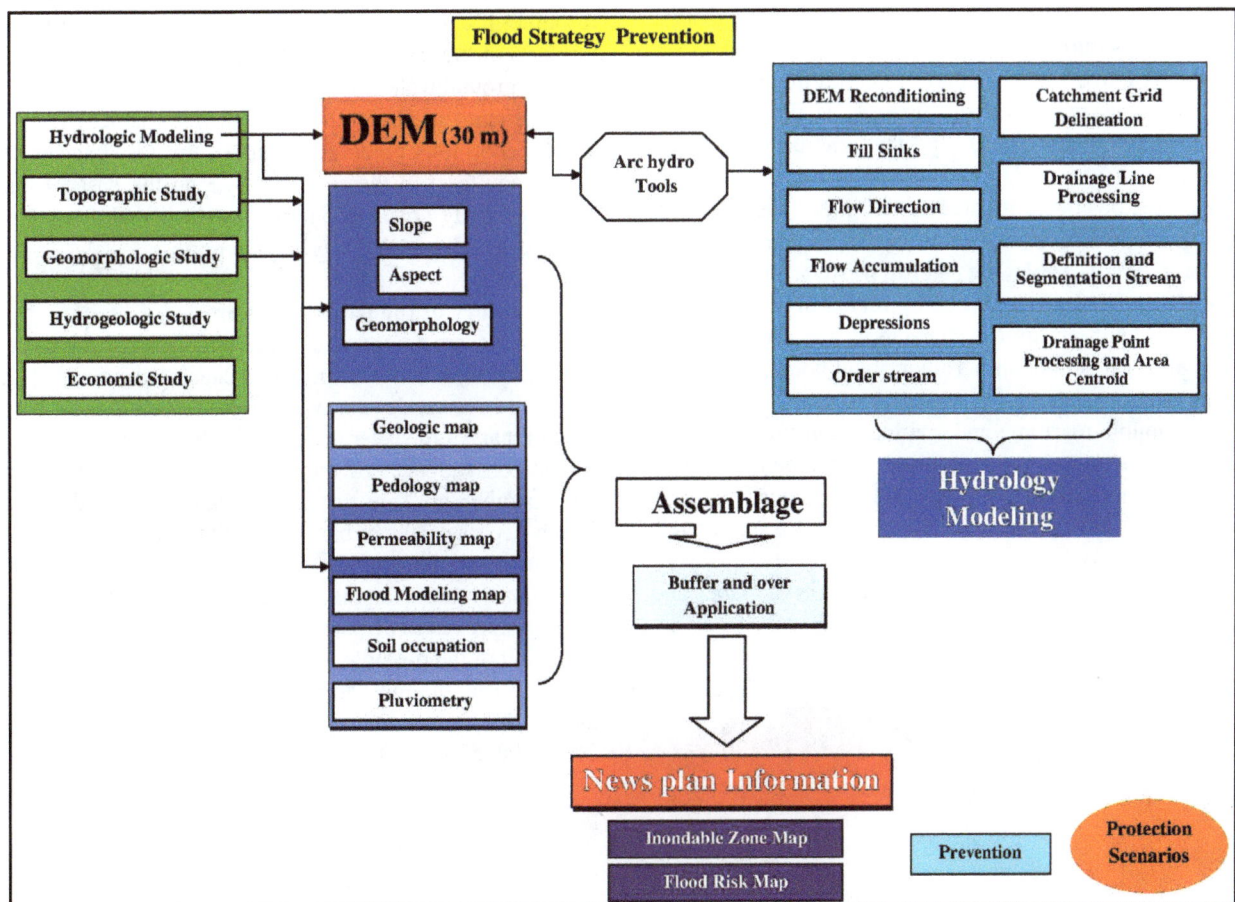

Figure 1. Chart showing the work flow of this study.

ones:

- The hydrological study
- The topographic study
- The geomorphologic and hydrogeologic study
- The hydraulic study
- The socio-economic study

The aim of this work is the geometrical, geomorphological and hydrological modeling of the Zeuss-Koutine basin. We will proceed to the interpretation of existent data, the exploitation of the derivatives of the digital terrain and the creation of new layers and overlay plans GRID information in digital form created by interpolation [3].

The results obtained will be used to carry out basic information tools of prevention for fixing priorities for flood management in the basin of Zeuss-Koutine.

On the basis of the results obtained, means and solutions to fight against the floods will be proposed. Thus several types of development and dams will be defined at the end of this study to protect the Zeuss-Koutine basin in the case of probable serious flooding.

2.1. Study Area

2.1.1. Geographical and Administrative Settings

The Zeuss-Koutine basin is situated in South-Eastern Tunisia (**Figure 2**). It is a part of coastal plain of Jeffara in the southeastern Golf of Gabes. It is limited by the latitudes 37°10' and 37°50', and the longitudes 8°50' and 9°20' [4].

This sector covers an area of 920 km^2. The Northern limit is formed by a line joining sebkhet Oum Ezassar to Henchir Fredj area. The North-Western, and South-Eastern limits are represented by reliefs of northern Dahar. The Southern limit is characterized by the reliefs of Tebaga of Medenine until Tadjeras and bordered by Mednine fault.

The double maritime and continental influence on our area generates a great variability of temperatures and precipitations space and time [5]. The annual average temperature in this area is of 20°C whereas annual average pluviometry is below 200 mm/an.

The annual evapotranspiration in stations of Gabes and Medenine exceeds 1300 mm. [6].

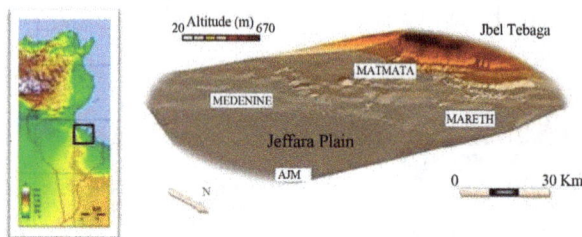

Figure 2. Study area.

2.1.2. Climatic Evolution in Tunisia

Climate of Tunisia is of the semi-arid type with very great variations in temperature and pluviometry in time that in space. The pluviometric mode is very variable in term of duration and intensity of the rains [7]. Rains of very strong intensity which can generate devastating risings of the natural environment have been witnessed [8].

Tunisian climate is mainly Mediterranean. The pupils can distinguish 3 great climatic units:

- At North, a "wet" zone (400 to 600 mm/year) which feeds the vast hydrographic network of Medjerda.
- In the center, on both sides of the "dorsale", precipitations spread out of 250 with 400 mm: it is the beginning of a semi-arid climate, with semi-steppe vegetation development.
- In the south, an arid region (less 250 mm/year) even less than 150 mm in southern of Douz) whose "chotts" and the dunes represent the landscape dominating.

With this distribution, it is necessary to add the strong irregularity inter and intra-annual (less than 30 days of rain/year in the south) and the risks posed by the risings (spring and autumn, and not only at north).

The country has approximately 4.6 billion m^3 water resources: 60% run on the surface, 40% are underground but 80% of these resources are located in the north while 70% of ground waters are in the south.

The country is also equipped with water tables in north and an enormous potential consisted the fossil sheets of water (aquifers) in the south such as in the case of our sector of study (Zeuss-Koutine). One will point out that salinity is higher (salt 1.5 g/liter) in north.

2.1.3. History of the Risings in Tunisia

The risings are temporal natural changes water's level caused by abundant precipitations causing the floods [9]; they can occur under various weather conditions and are integral part of the river mode.

The consequences of the floods vary from a medium with another and are primarily related to the distance which separates the urban areas, the agricultural lands and the installations of the floodplains [10], it is thus the preliminary study for the prevention against these risks of floods proves to be paramount for the protection or the reduction of the possible damage.

In Tunisia, the most important risings recorded during the last fifty years are those of 1962, 1969, 1973, 1986, 2003, 2004, 2006, 2007 and 2012.

The risings observed during the last two years caused phenomena ever known with very high damage [11].

The efforts made by the persons in charge (will control) make it possible to equip the country with a broad hydraulic infrastructure: 27 stopping's, 200 stopping's hill, 766 lakes hill and more than 3000 drillings and 151,000 wells of surface mobilizing 83% of the unit of the exploitable water resources. These efforts remain always insufficient in front of floods [12] which can affect South-east Tunisia and precisely the area of Djeffara. This later is marked by elevated chains of Tebbaga, Matmata, Mareth and relayed directly at the east and north east to the Zeuss-Koutine basin objective of this study [13]. The rain waters and streaming are at the origin of the floods.

The area of Zeuss-Koutine as all Tunisian South-east presents climatic conditions and geomorphologic which prevent the formation of the risings naturally. These conditions are supposed as follows [14]:

- Irregular annual pluviometry and weak 100 < P< 200 mm, rains are generally stormy and torrential [15];

- High Evapotranspiration EVT and important hydric deficit;

- Winds of various directions, often dessicants and with considerable dynamics [16];

- Specific vegetation with halophyte species;

- Almost complete absence of the high layers (forests) [17].

2.2. Bases Studies for the Definition of a Strategy of Protection (Prevention)

The prevention against the floods Zeuss-Koutine basin requires several studies; principal studies for the choice of a strategy of protection are as follow [18].

2.2.1. Hydrological Study at the Level of Catchment Area

Within the framework of the clarification of scenarios of protection of the basin Zeuss-Koutine, hydrological modeling aims at the definition of events of reference. This definition must be based on three types of information and simultaneous analyses:

- A first analysis consists in defining homogeneous hydrological zones, according to the climatologic (rain…) and morphometric characteristics of the catchment area (relief, slopes, aspect, form) and of its aptitude for the streaming (geology, pedology, occupation of the ground) according to the hydrometric data available [19];

- The second analysis relates to the detailed treatment of chronicles of measurement of flow;

- The last stage consists in describing the intensity of the risings in an unspecified point of the catchment area which will be by mesh in our study constituting the target site on which one seeks to determine the degree risk of rising.

2.2.2. Topographic Study

The technical studies, particularly the hydraulic study, require the knowledge of specific topographic data in an adapted way, the geometry of the ground and that of the works and installations present and projected.

2.2.3. Geomorphologic and Hydrogeological Studies
1) The hydro geomorphologic study

The hydrogeomorphologic method is a first phase of analysis [20] of the natural environments and valleys.

It is interested to describe and high light the whole of the bed of water way and boxing sound (terraces and slopes), as well as the various establishments likely to disturb the flows [21], by accelerating or by slowing down them [22].

It makes it possible to deduce the flood plains by the rare risings with exceptional as well as the values approximate from the physical parameters of the floods.

The DEM in 3D of the basin proposes a visualization of the space provision of the various beds of a waterway and their context. In this case, the modern alluvial plain is framed, on a side by a slope with steep slope, and other by a terrace [23].

- **Minor bed**

Including the bed at low water, it is the bed of the frequent risings. It corresponds to the principal channel of the waterway. It is generally borrowed by the annual rising, known as believed of full-edge, flooding only the low sectors and closest to the bed. This bed is easily locatable since it is delimited by more or less continuous abrupt banks.

- **Average bed**

It is the bed of the average risings. It ensures the transition between the major bed and the minor bed limited by slopes. It corresponds to the bed occupied by the frequent risings with averages (periods of return ranging between 2 and 10 years) which can have an important speed and a solid load.

- **Major bed**

It is the broadest bed, which corresponds to the alluvial plain and which is function of rare and exceptional risings. It recovers old average and minor beds and offers morphological characteristics much simpler. It presents one modeled flatter, and is located below boxing it.

This shutter attempts to study the fine evolution of the bed of the waterway (**Figure 3**), as well in plan as in profile. This shutter is particularly useful for the establishment and the design of brought closer protections (dams, protections of banks) and of works to be built in the bed (work of derivation, stopping).

2) The hydrogeologic study

It is interested under investigation of the relations between the aquifers and the river. This shutter is justified if the aquifers are likely to influence the mode of the risings or to produce floods by increase of aquifers in the

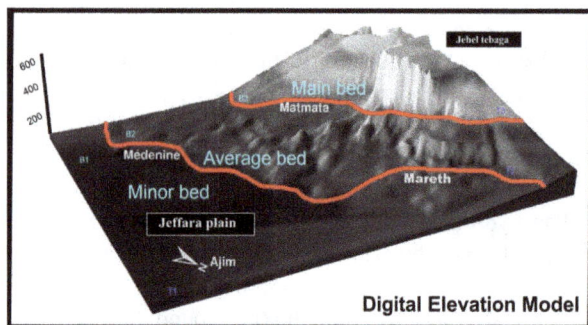

Figure 3. Hydrogeomorphologic classification.

protected sectors. This isn't applicable in the sector of Zeuss-Koutine. Deep aquifers are always over exploited and will not reach surface even in the years with a high pluviometry [24].

2.3. Choice of a Strategy of Prevention

After having determined the relative importance of the challenges and the economic appreciation of the damage which is associated for them, passage with a true analysis cost benefits, when it was tried, was done on the basis of assumption, whose principal one is that the damage which could be monetized reflect the whole of the impacts of the floods.

Among the costs, must appear the investment costs which reflects the cost of the techniques of preventions but as well they brought up to date costs of maintenance.

3. Results and Interpretation

3.1. DEM-Arc Hydro Tools

The terrestrial reproduction of the forms constitutes to translate, in a plan in 2D topographic surface 3D. The forms of representations of the relief are multiple: dimensioned points, contours, sights in prospect.

The development of the digital cartography and Geographical Information Systems SIG, allowed the creation of a new form of representation of altimetry information under digital format still called Digital Elevation Model of the ground DEM which is characterized by a considerable extensibility and flexibility on the level of its exploitation from which we can derive a multiplicity of products: contours, cards of slopes, exposure or of inter visibility, volumes, sights in prospect...

For hydrological studies of the areas catchment the DEM provides important information at the level of hydrological modeling by the extraction of the new plans of information allowing us having a detailed idea about the hydrographic networks, flows of water and physical characteristics of the basin. Several software SIG was interested in creation of the DEM; in our case we chose Arc-

Gis and its extension Arc Hydro Tools for the development of hydrological model.

Arc Hydro is a data model and toolset for integrating geospatial and temporal water resource information run within ArcGIS geographic information system. Although implemented in a commercial GIS environments.

All the calculations, carried out by the extension Hydro Arc, have as a base a digital elevation model (DEM).

The DEM chosen in this study is that elaborate following the interpolation by triangulation TIN of the vectorized level lines on scale 1/50,000 of the topographic maps of Matmata, Mareth, Ajim and Medenine; on the other hand, its precision doesn't make it possible to use it in a reliable way, it is for that one selected ace to use in more the DEM of the SRTM with resolution of 30.

This DEM is appeared as a grid whose elements are squares of 30 meters side thus an altitude is defined every 30 meters.

3.2. Hydrologic Modeling: DEM Derivatives

From the altimetry information of the DEM, derivatives maps and information's will be calculated, to realize a morphological analysis of the study area, by the construction of slope map, charts of orientations and sunning charts of intervisibility, profiles, geological cuts-off and charts of hydraulic installations. Other maps and information have a hydrological interest such as the delimitation of the basins slopes [25], the extraction of subbasin, the automatic generation of the hydrographic network like and we developed other charts of modeling like the direction and accumulation flow.

3.2.1. Slope Map
Measurements of dips are essential data for the geologist and in particular for the structural analyst and for the hydrogeologist. They allow quantifying the geological objects observed on the surface and modeling them.

As a result, the slope map confirms the predominance of platform structure which corresponds to the Djeffra plain, represented by yellow and green color on the map.

3.2.2. Aspect Map
This map shows slope aspect in Zeuss-Koutine basin. It uses different colors to indicate the direction of the steepest slope (**Figure 4**).

We notice the abundance of the blue sky color, blue dark, orange and red which indicates a NS (North-South) dip direction of the cells of the basin of Zeuss-Koutine. This information is very useful for the determination of the direction of major flow and after the direction of the installations of protection against the rising will be perpendicular to this Major flow direction.

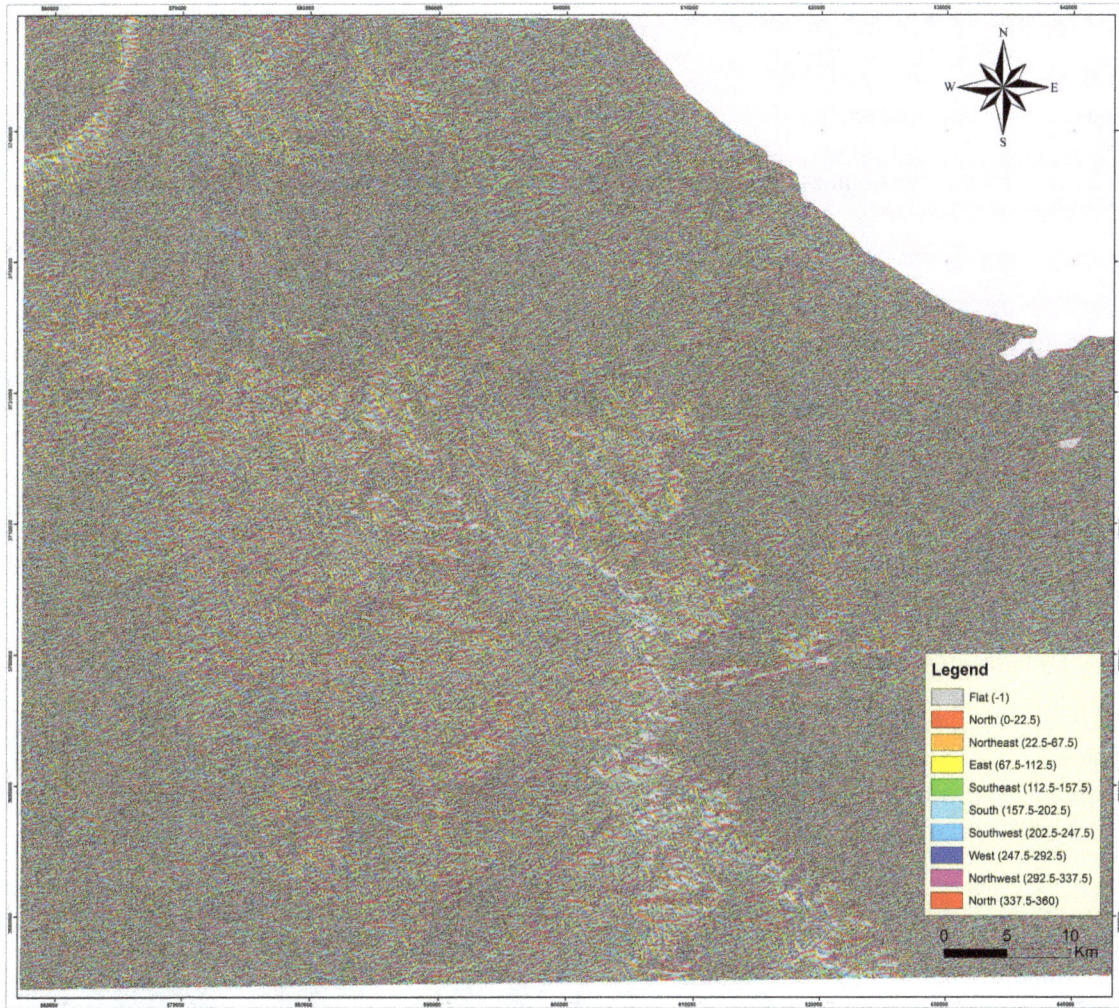

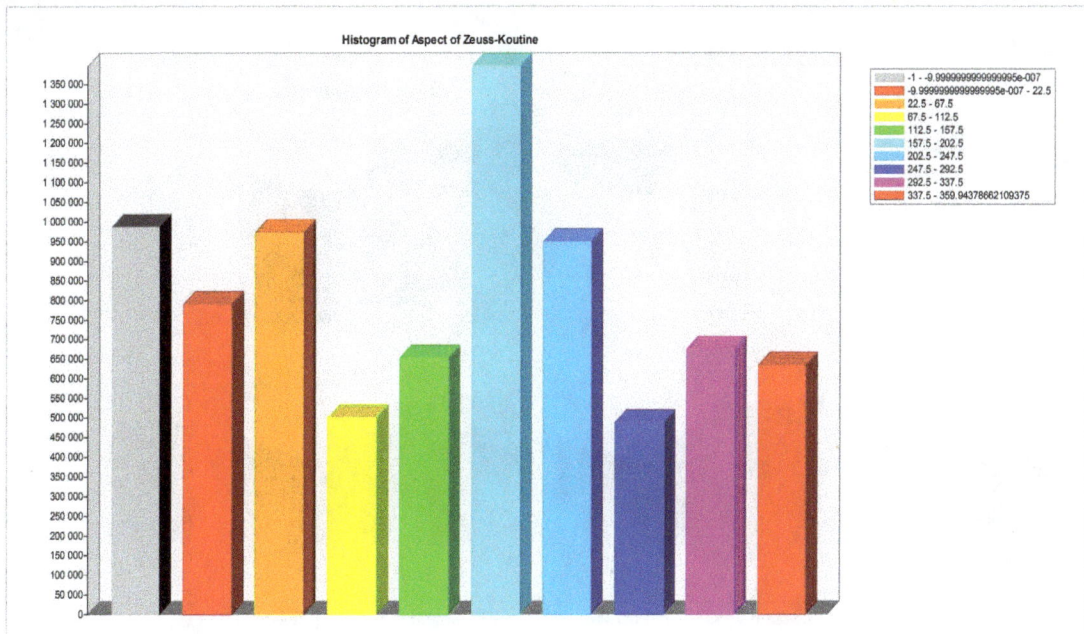

Figure 4. Aspect slope.

3.2.3. Flow Characteristics

The tools of hydrological modeling proposed by spatial analysis make available methods to describe the physical components of a surface or basin.

Adaptation of DEM is a necessary step before hydrologic modeling, the main possibilities are:

- Adaptation of DEM to the actual position of the stream;

- Fill the "holes".

In this study (**Figure 5**), we will perform drainage analysis on a Digital Elevation Model DEM (30 m) of Zeuss-Koutine basin and provide a digital representation of watershed characteristics used in hydrologic modeling. The DEM are used to derive several data sets that collectively explain the drainage of a catchment. Raster analysis is performed to create data on DEM Reconditioning, flow direction, flow accumulation, stream definition, stream segmentation, and watershed delineation.

- **DEM Reconditioning**

This function modifies a DEM by imposing linear features onto it (burning/fencing), we have chosen the hydrographic network of Zeuss-Koutine basin like a linear file for this application [26].

- **Fill Sinks**

Among the problems we met for the reproduction runoff Zeuss-Koutine maps, it's when the cell is surrounded by higher elevation cells, and the water is trapped in that cell and cannot flow. The Fill Sinks function modifies the elevation value to eliminate these problems filling these grids.

- **Flow Direction**

One of the principal parameters which can be extracted is the flow direction whose objective is the creation of a raster of integers to the direction of flow starting from each cell towards its neighbor with the steepest descent which values vary from 1 to 255 [27]. As a result, this chart is explained by the fact that each cell contains the direction (coding of Freeman) of water flow (Chiles J-p.2004).

Using the "Arc hydro" tool, we made the flow map of the study area. The cells flow to their nearest neighbor along 1 of 8 compass directions labeled as East = 1, SE = 2, S = 4, SW = 8, W = 16, NW = 32, N= 64, NE=128. Results were mostly more than 16, which mean a direction North West or North to South or South East... that is to say Jebel Tebbaga and anticlines around to the platform of Jeffara [28].

- **Flow Accumulation**

This function computes the flow accumulation grid that contains the accumulated number of cells upstream of a cell, for each cell in the input grid.

- **Stream Definition and Stream Segmentation**

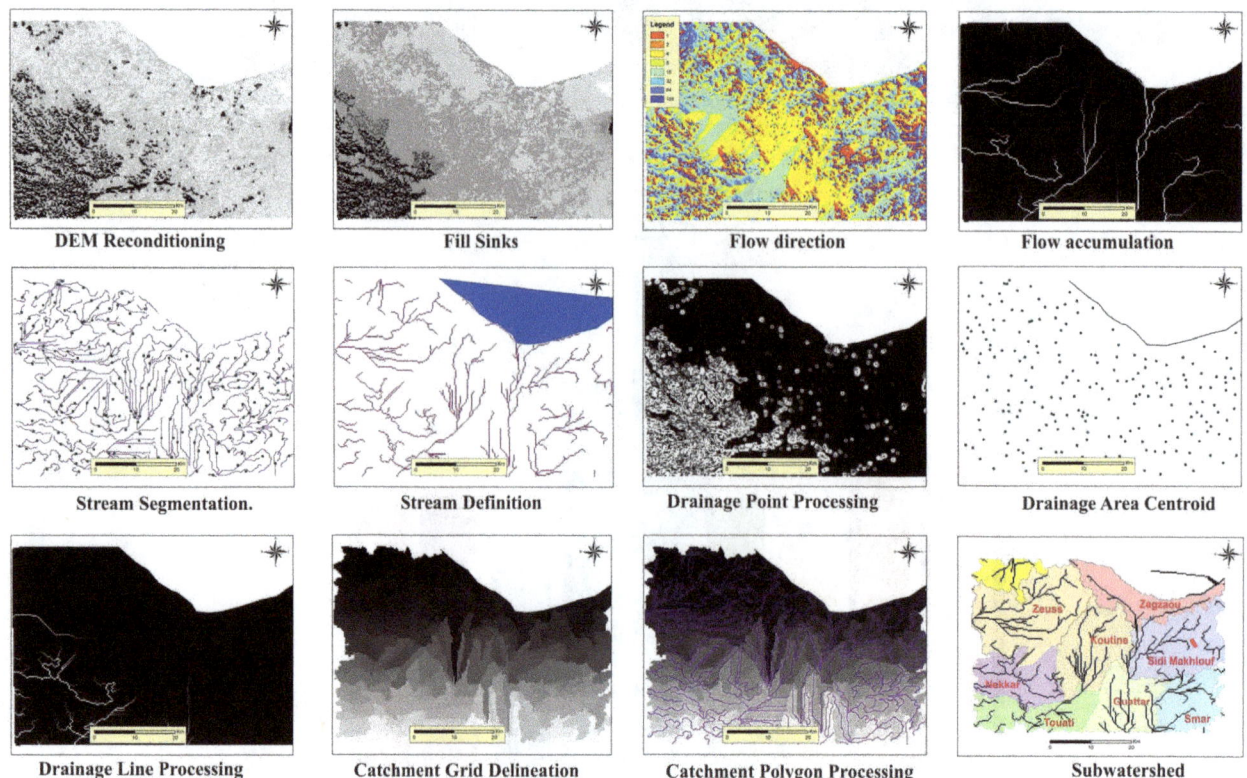

Figure 5. Different maps of flows extracted by Arc Hydro-tools.

With this function we creates a grid of stream segments that have a unique identification.

- **Drainage Point Processing and Area Centroid**

This function allows generating the drainage points associated to the catchments.

This function generates the Centroid of drainage areas as centers of gravity. It operates on a selected set of drainage areas in the input Drainage Area feature class. If no drainage area has been selected, the function operates on all the drainage areas.

- **Drainage Line Processing**

This function converts the input Stream Link grid into a Drainage Line feature class. Each line in the feature class carries the identifier of the catchment in which it resides.

- **Catchment Grid Delineation**

This function creates a grid in which each cell carries a value (grid code) indicating to which catchment the cell belongs. The value corresponds to the value carried by the stream segment that drains that area defined in the stream segment link grid.

- **Catchment Polygon Processing**

This function converts a catchment grid it into a catchment polygon feature.

- **Subwatershed Delineation**

This function allows delineating subwatersheds for all the points in a selected Point Feature Class. Input to the batch subwatershed delineation function is a point feature class with point locations of interest [29]. The Batch Point Generation function can be used to interactively create such a file.

- **Evacuation Point**

The cartography of the points of evacuation helps with determination of the troughs of low pressure constituting of the easily flooded zones (**Figure 6**).

- **Stream order (Strashler)**

The classification of the affluent helps with the determination of the degrees of influence and the contribution of each wadi (ramification) in the formation of the risings (**Figure 7**).

Generally the branches of greater order such as the wadi of order 4 are the main sources of the risings (Wadi Zeus, Om Zessar, Koutine…) this does not neglect the influence of the other derived wadis [30].

4. Discussion

4.1. Map of Flood Zones and Degree Flood Risk

The use of the SIG for the determination of the geomophometric and physical characteristics of the catchment area of Zeuss-Koutine, starting from the digital elevation model obtained following the digitalization of altitudes, allowed us the mapping of flood zone and degree flood risk. quantify certain physical and hydrological parameters sector of study, per units Nets or GRID or PIXEL, to have an explanation of the phenomenon of flood and its effects and to give scenarios of prevention for the fight against these catastrophes to short and long-term.

The hydrologic modeling within Arc Hydro tools is to represent the physical processes within each catchment of interest so that, when driven by atmospheric forcing (precipitation, temperature) and for known catchment physical characteristics (topography, land cover), the models generate stream flow hydrographs at the catchment outlet that reproduce the corresponding observed hydrographs.

Terrain Preprocessing uses DEM to identify the surface drainage pattern. Once preprocessed, the DEM and its derivatives can be used for efficient watershed delineation and stream network generation.

At the end of this part, the use of tools buffers with a distance of 1000 m and 1500 m and 2000 have allowed

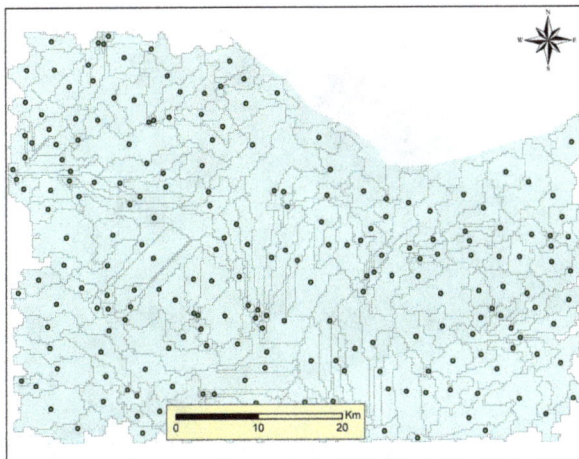

Figure 6. Evacuation point.

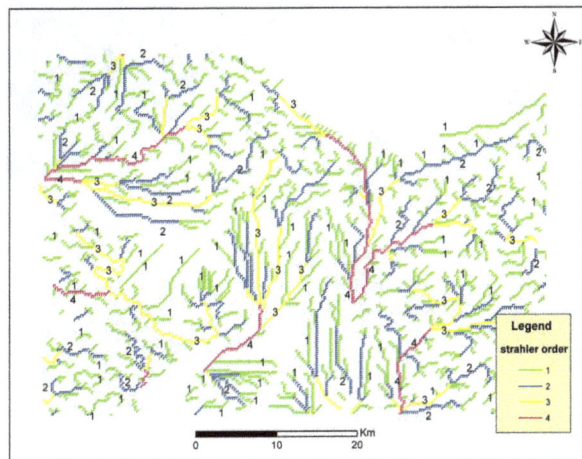

Figure 7. Strashler stream order.

us to classify the degree of intensity of each parameter (**Figure 8**).

The hydrological modeling of Zeuss-Koutine basin have provided a geomorphological idea of the basin by its subdivision into three bed the minor bed is the most flood zones and the more risky for the flood, then the main sources of flooding are the wadis of Zeuss, and Koutine Om Zessar and the waters become from the Jbel Tebaga in South West.

From the flow direction and accumulation map, we extract information about the mobility of surface water in Zeuss-Koutine.

This mobility is important due to the predominance of the sectors of low altitude (<250 m). All data are then used to build up a vector and GRID news plans of catchments, watershed and drainage lines. With this information, a geometric network is extracted and 9 Subwatershed was generated in the end: Zeuss basin, Koutine basin, Zigzaou basin, Sidi makhlouf basin, Guettar basin, Touati basin, Smar basin and Nekkar

basin.

The cartography of the flood zone (**Figure 9**) with plans of the degree risks of floods translating the intensity of the flood in each sector on the level of Zeuss-Koutine was deduced following the superposition from all extracted information.

The technique Buffer to different distance (500 m 1000 m - 1500 m - 2000 m - 3000 m); under ArcGIS on the various layers create to us aces allowed after space superposition and intercrossing of:

- To delimit the floodplains
- To chart a plan of degree risk of flood of Zeuss-Koutine

This work enabled us to produce tools of information of basic prevention to fix priorities as regards management of the floods: a chart of the floodplains and prevention plans of the risks.

The cards of the floodplains cover the geographical zones likely to be flooded according to 3 scenarios:

* Flood of weak probability or extreme scenarios of

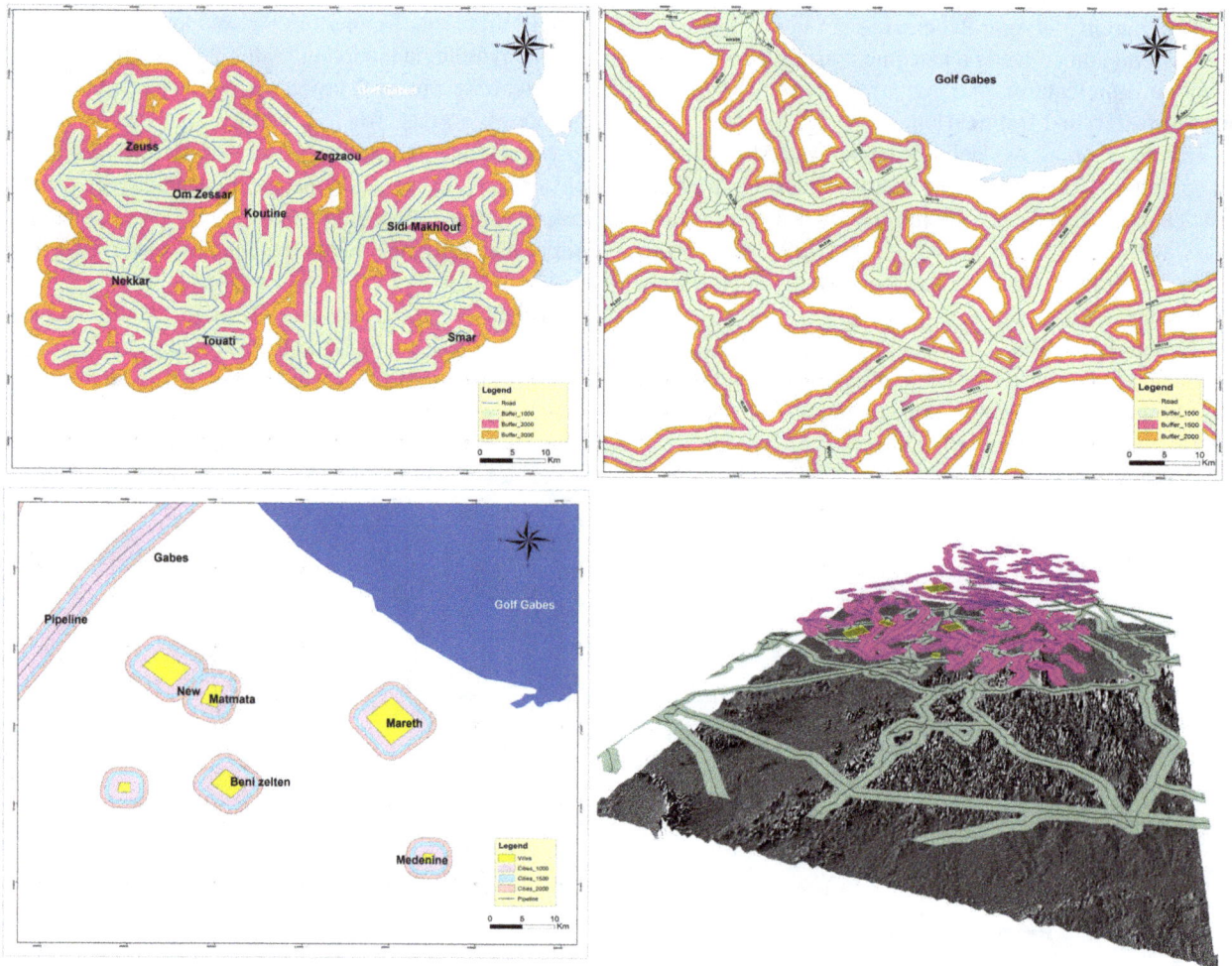

Figure 8. Buffer zone and overlays.

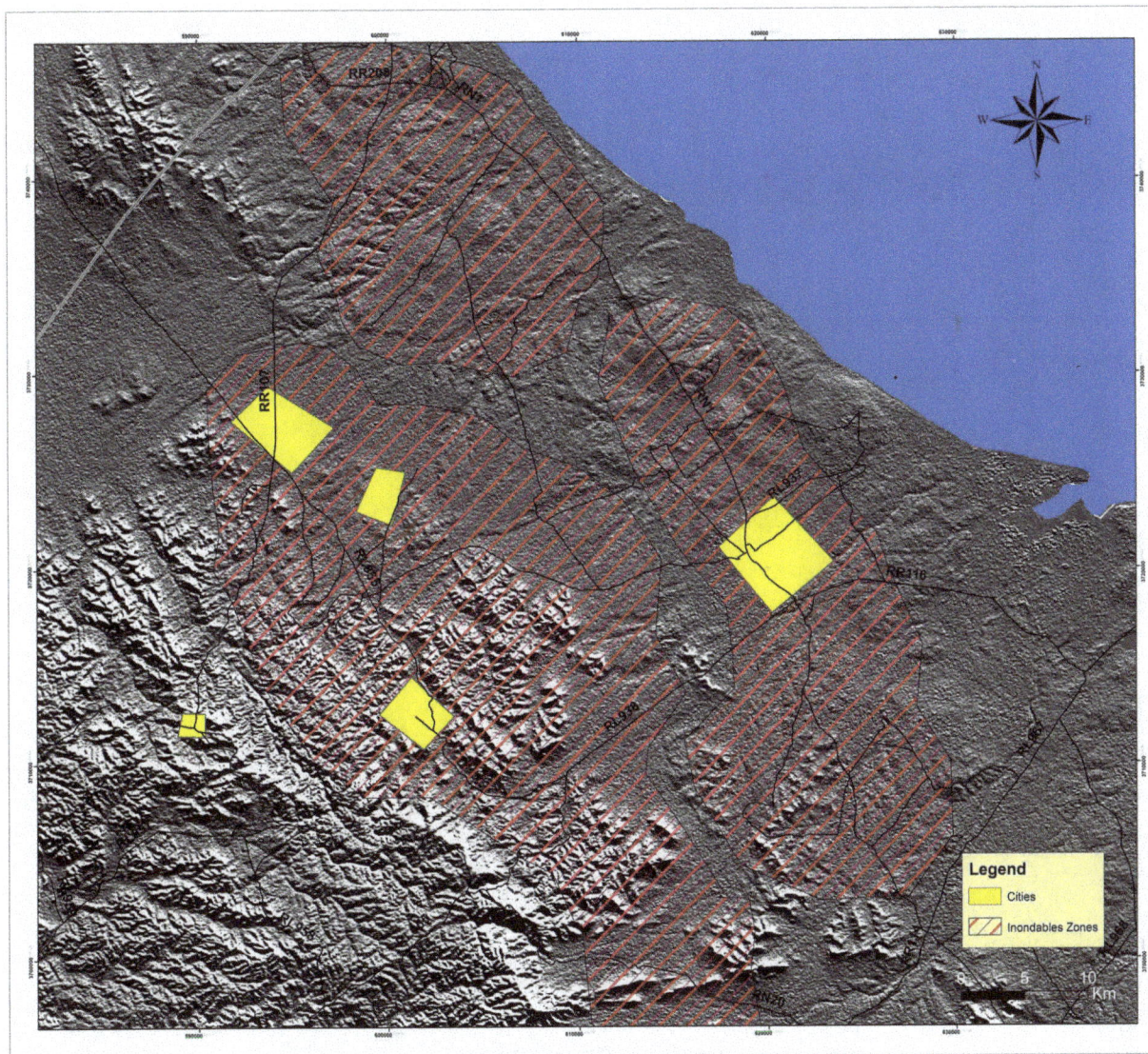

Figure 9. Flood zones.

events,

* Flood of average probability (correspondent at one period of higher or equal to return hundred years) and,

* If necessary, high flood probability.

For each scenario, the extent of the flood, the sectors easily flooded, the heights of water (or according to the case the water level) and possibly the current velocity or the flow of the rising, must be represented in carto-graphic form.

The maps of the degree floods risks show the intensity by sector of the potential negative consequences associated with the floods, *i.e.* saying the number of people and the extent of the goods likely to be affected by the risings, such as for example: agricultural economic activities, surfaces, transport networks. In this study, the calculation is based on the formula lee (2007).

RI = 2RH + SLOPE + RELIEF + 1/2 P 1/2 OS

RI = Risk of flooding

RH: Drainage

P: Permeability

OS: Land Use

All parameters in the formula must be in RASTER mode to obtain the map of Degree flood risk (**Figure 10**).

The results obtained show that the small bed or plain of Djeffara is the most easily flooded zone, and the sectors of Mareth and beni zeltene are exposed to the risks of floods. The upstream part of TouatiWadi presents executor probable flood which can be most catastrophic on the area of blessed Zeltene.

In the same way for Wadi Sidi Makhlouf and Oued Nekkar which present a danger to the areas successive of Mareth and Matmata.

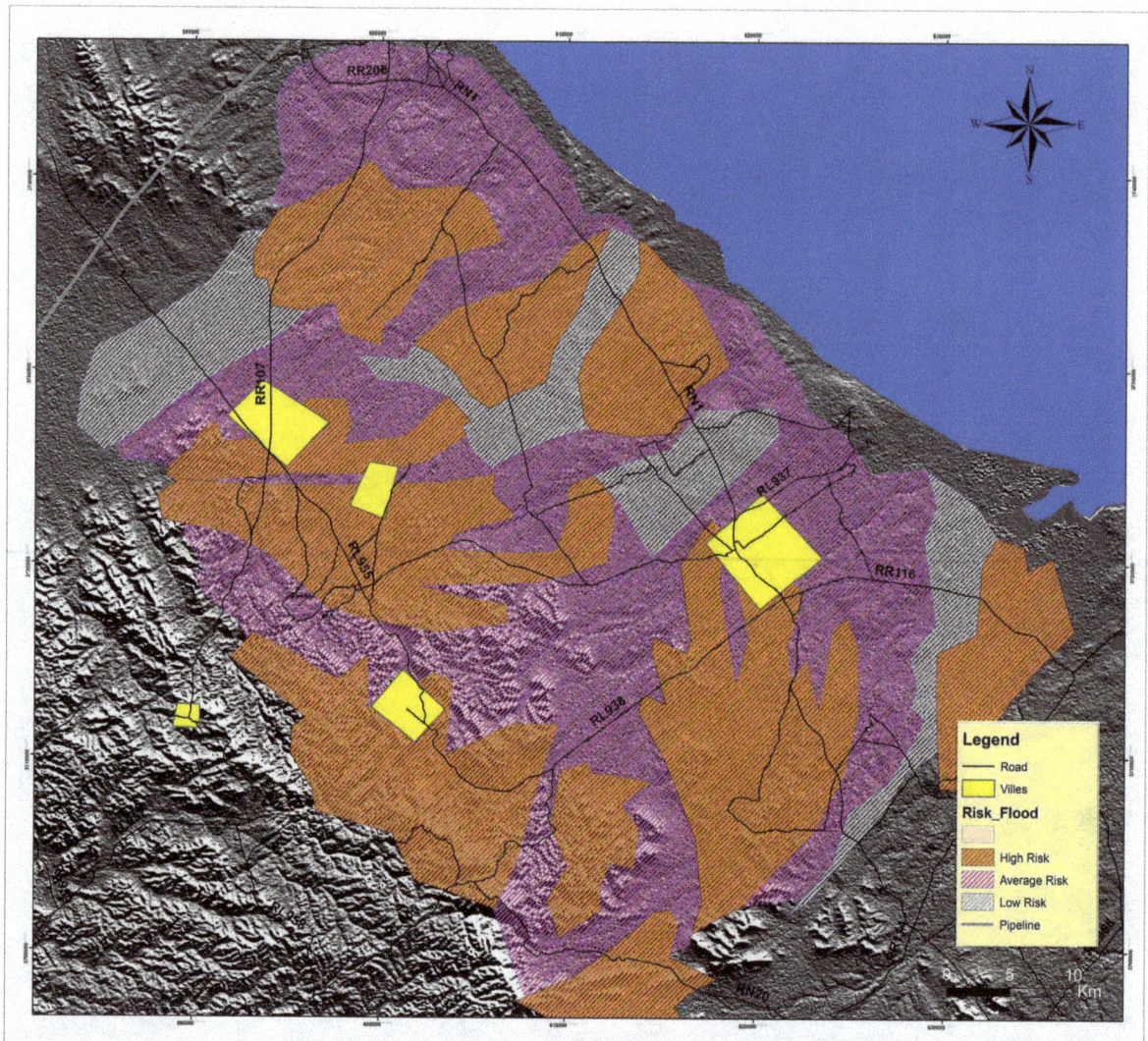

Figure 10. Degree flood risk.

4.2. Scenarios and Suggested Solution of Protection

Protection against the floods articulates about three shutters [31]: 1) the distant protection which consist in carrying out dam and lakes; 2) brought closer protection

against the flood which includes work of deviation of the wadi and waterways apart from the cities and the urban areas; 3) The cleansing of rain waters inside the urban areas and which consists in carrying out networks of cleansing.

The scenarios of continuation prevention have this job is multiple and alternatives but it is necessary to choose the scenario which is appropriate and which allows:

• To analyze the operation of installations as a preliminary, to guarantee that their operation especially on the geotechnical level [32], will not generate an on-

accident.

• The works can modify the perception of the risk on behalf of the residents. On the technical plan, the combination of many installations spatially distributed, and having a whole a local effect of rolling of the flows, must be analyzed on the whole of the catchment area. It is indeed advisable to make sure that the beneficial reduction in the risings obtained in a place will not involve a harmful synchronization elsewhere.

For the case of the basin of Zeuss-Koutine we see that best the solution of prevention against the flood and who aim at reducing the losses, in particular human and to protect the most exposed districts and the agricultural lands from the plain of Djeffara; is the installation of the dams of protection.

The dam of protection against the floods and channeled rivers is defined as a longitudinal work which does not have function to retain water but to prevent its flow. It creates a difference in water level between two parts of the same floodplain and that this difference creates "hydraulic head" which imposes the work on forces against which it must be correctly dimensioned to resist.

The dams are various types:

- Dams of protection against the river, longitudinal floods during water flow;
- Dams of belt in the inhabited places;
- Dams of estuaries and protection against the marine immersions;
- Channeled river shore embankments;
- Linear fill.

The ranking of the works is done according to the dangerousness with gradual obligations (H = plus great height between the top of the work and the original ground on the side of the protected area, P = population maximum residing in protected area plumb with the top):

- Classify A: H >= 1 m and P >= 50,000 inhabitants;
- Classify B: not classified of A and H >= 1 m and 1.000 <= P <= 50,000 inhabitants;
- Classify C: not classified of A and B and H >= 1m and 10 <= P <= 1.000 inhabitants;
- Classify D: either H < 1 m or P < 10 inhabitants.

The dams proposed for the basin of Zeuss-Koutine of study will be second class *i.e.* The Class B.

4.3. Various Types of Installations (Dams)

The installation of the dams for the prevention against the flood in the basin of Zeuss-Koutine or the Tunisian South generally; suitable technical solutions relating to the clean conditions of the site constitute on which will be realized: nature of the ground, slope of the original ground, surface quality of the ground and impluvium. In other words one can say that from a purely technical point of view the properties and the conditions of the site dictate the nature of the intervention.

The techniques of installation of the dams proposed on the level as of waterway can summarize as follows.

4.3.1. Dry Stone Cords

The dry stone cords consist into cubes of dry stone walls, two or three arranged, built in contours all the way along slope. They are carried out on the slopes upstream of the basins slopes, where the slope is very strong, vegetable cover is generally weak or absent, the streaming and the concentration of water are fast. Their crucial role is to break the speed of streaming of water and to limit its erosive capacity.

4.3.2. Dry Stone Thresholds

They are works carried out in masonries made in dry

stone staircase or sometimes using binder's vegetable in the thalwegs and Chaâbs. They make it possible to reduce the speed of streaming and the retention partial of water and certain sediments.

- **Manual tabias**

They are ground pads made manually by shovel and pick axe. The main aim of the realization of these works is to retain part of surface waters; the rest is evacuated by its outfalls towards the downstream.

- **Technical tabias**

They are bulkier works made downstream from the slopes and transversely gullies. The role is to partially retain runoff and fighting against erosion.

- **Jessours**

They are works made up of a ground pad compacted located in its upstream, it is equipped with pill way of surplus waters in the event of strong risings, this work can be side or central. Other techniques of installation of protection primarily applied to the system aquifers of Zeuss-Koutine can intervene at the time of the prevention such as; works of refills of the aquifers, works of spreading's of water of raw and wells filtering.

5. Conclusions

At the end of this work and following hydrological modeling carried out on the basin of Zeuss-Koutine, the got results enable us 1) to chart, in a homogeneous way on a whole catchment area, the easily flooded limits like all the natural or artificial elements, which can play a part on flow of the risings. It also appears the limits reached by the large known historical floods. 2) To determine the needs and the actions aiming at minimizing the possible risings and the most effective solution to reduce the damage due to the floods as well as the specific and dimensioned works protection to resist a probable rising. The prevention of the floods thus remains a pressing priority, for that the principles of the dynamic deceleration can be limited as follows:

- To moderate water running out on the slopes, which will delay of as much their arrival with the waterways; this to decrease the peak output and to delay the flows by it on a waterway, by slowing down water before its arrival in the bed, by mobilizing the natural capacities of damping in average bed and major bed, and by temporarily storing part of the volume of the rising in works;
- To reduce their acceleration in the beds of the waterways;
- To support, even restore, connection with the Appendices River and major bed in general; this allows the derivation of part of the flows in believed and the increase in the infiltrations towards the aquifers. This work is addressed particularly to these projects, for

which we can still simplify the concept and restrict it with the following technical objectives;

- To make sure that at the level of the area catchment, the implementation of the dynamic deceleration on a set of water ways does not generate one locally unexpected increase in the risks of flood, in particular by recombining of risings out of phase by the works.

As to the maintenance and monitoring of installations and done works, this methodology makes it possible to be freed from knowledge, never exhaustive, of the former risings, by drawing up all the zones likely to be flooded by over flow of waterway.

Thus, the floodplains of the area studied are defined in a homogeneous and perennial way. The corresponding cards make it possible for example to locate the zones of the territory which will be never flooded, therefore on which a sustainable development without risk of flood is possible. These various cards set of themes with alternatives of scales and formats digital GRID present a useful support for the cartography of the zones likely to be flooded and to thereafter meet the various needs for the managers of the territory, and it can be used like a prototype for other sectors of studies. The knowledge of these floodplains is not exhaustive; those which are charted it are as an indication and their limits remain approximate and variable with time.

REFERENCES

[1] B. Kartic, P. Moumita and B. Dr. Jatisankar, "Application of RS & GIS in Flood Management a Case Study of Mongalkote Blocks, Burdwan, West Bengal," *International Journal of Scientific and Research Publications*, Vol. 2, No. 11, 2012, pp. 1-9.

[2] B. Ben Baccar, "Contribution to the Hydrogeologic Study of the Multi-Layer Aquifer of Southern Gabes Thesis of Third Cycle Doctorate," University of Southern Paris, 1982, pp. 132-154.

[3] B. Ayo and B. Ibrahim, "Selection of Landfill Sites for Solid Waste Treatment in Damaturu Town-Using GIS Techniques," *Journal of Environmental Protection*, Vol. 2, No. 1, 2011, pp. 1-10.

[4] M. Ben Youssef and B. Peybernes, "New Biostratigraphy of the Lower Cretaceous of Southern Tunisia," *Newspaper of African Earth Sciences*, Vol. 5, No. 3, 1986, pp. 217-231.

[5] C. Bill and R. Simon, "Automated Rig Removal with Bayesian Motion Interpolation," The Foundry, London, 1998.

[6] S. Bouaziz, "Study of Breakable Tectonics in Punt Forms and the Saharian Atlas (Meridional Tunisi) Evolution of the Paleo Fields of Constraints and Geodynamic Implications," Thesis of State FAC Sc Tunis, Tunis, 1995.

[7] S. Wade, J. P. Rudant, A. M. Dia and J. Kouamé, "Application of Earth Observation Data and GIS to Urban Flood Magement—Case-Study of Saint Louis, Senegal (West Africa)," *Abstract United Nations Regional Workshop on the Use of Space Technology for Disaster Management for Africa*, Munich, 18-22 October 2004, pp. 221-250.

[8] R. Derouiche and H. Chaib, "Study by Digital Model of the Impact of Installations on the Refill of the Groundwater of Zeuss Koutine," General Direction of Resources Water (Direction Générale des Ressources en Eau DGRE), 1997.

[9] M. Fersi, "Hydrogeological Study of the River of Oum Ezassar in Koutine," General Direction of Resources Water (Direction Générale des Ressources en Eau DGRE), 1985.

[10] G. Dawod, M. Mirza and K. Al-Ghamdi, "GIS-Based Spatial Mapping of Flash Flood Hazard in Makkah City, Saudi Arabia," *Journal of Geographic Information System*, Vol. 3, No. 3, 2011, pp. 225-231.

[11] E. Gaubi, "Study of the Piezometry and the Geochemistry of the Groundwater of Zeuss Koutine (Area of Medenine)," Faculty of Sciences of Tunis FST, 1988.

[12] T. Hatzichristos and M. Giaoutzi, "Landfill Siting Using GIS, Fuzzy Logic and Delphi Method," *International Journal of Environmental Technology and Management*, Vol. 6, No. 1-2, 2006, pp. 218-231.

[13] F. Kamoun, "The Jurassic Level of the Pilot Tunisian South of the African margin of the Tethys Stratigraphy, Sedimentology and Micropaleontology," Specialty Thesis, University Paul Sabatier Toulouse III Review of Sciences of the Ground, Vol. 11, 1989, p. 338.

[14] W. Mamou, "Characteristics and Evolution of the Water Resources of the Southern Tunisia," State Doctor Thesis, University of Southern Paris, Paris, 1990.

[15] F. Monat, "Installation of a Geographical Information System (GIS) on the Basin of Merguellil (Tunisia) Training Course of Second Year Report," Dr. Bernard Jatiskander, Montpellier, 2000.

[16] Reports (2005-2011) of Drillings of the Groundwater of Zeuss Koutine. DGRE and SONEDE. General Direction of Resources Water (Direction Générale des Ressources en Eau DGRE), pp. 5-7.

[17] M. Ross and P. Tara, "Integrated Hydrologic Modeling with Geographic Information Systems," *Journal of Water Resources Planning and Management*, Vol. 119, No. 2, 1993, pp. 129-140.

[18] S. Saidi, S. Bouri and H. Ben Dhia, "Groundwater Vulnerability and Risk Mapping of the Hajeb-Jelma Aquifer (Central Tunisia) Using a GIS-Based DRASTIC Model," *Environmental Earth Sciences*, Vol. 59, No. 7, 2010, pp. 1579-1588.

[19] S. Saidi, S. Bouri, H. Ben Dhia and B. Anselme, "A GIS-Based Susceptibility Indexing Method for Irrigation

and Drinking Water Management Planning: Application to Chebba-Mellouleche Aquifer, Tunisia," *Agricultural Water Management*, Vol. 96, No. 12, 2009, pp. 1683-1690.

[20] P. Santra, U. K. Chopra and D. Chakraborty, "Spatial Variability of Soil Properties and Its Application in Predicting Surface Map of Hydraulic Parameters in an Agricultural Farm," *Current Science*, Vol. 95, No. 7, 2008, pp. 937-945.

[21] A. Sarangi, C. A. Madramootoo and P. Enright, "Comparison of Spatial Variability Techniques for Runoff Estimation from a Canadian Watershed," *Biosystems Engineering*, Vol. 95, No. 2, 2006, pp. 295-308.

[22] H. Yahyaoui, H. Chaieb and Mr. Ouessar, "Impact of Work of Conservation of Water and Grounds on the Refill of the Groundwater of Zeuss-Koutine (Medenine Tunisian South-East)," In: J. Graaff and Mr. Ouessar, Eds., *Toilets Harvesting in Mediterranean Zones Year Impact Assessment and Economic Evaluation*, Wageningen University, Wageningen, 2002, pp. 71-86.

[23] H. Yahyaoui and Mr. Ouessar, "Abstraction and Recharge Impacts on the Ground Water in the Arid Regions of Tunisia Case of Zeuss-Koutine Water Table," UNU Desertification Series, 2000, pp. 72-78.

[24] H. Yahyaoui, "Hydrogeologic Study of the Ground Water of the Catchment Area of the Rivers of Smar," General Direction of Resources Water (Direction Générale des Ressources en Eau DGRE), 1999.

[25] H. Yahyaoui, "Updating Balances of the Deep Groundwaters in the Governorate of Medenine," General Direction of Resources Water (Direction Générale des Ressources en Eau DGRE), 2000.

[26] H. Zammit, "Modeling of the Hydrogeology and the Salinity of the Groundwater of Zeuss Koutine," Ecole des Ingénieurs Tunis ENIT, 2002.

[27] P. P.Adhikary, H. Chandrasekharan, D. Chakraborty and K. Kamble, "Assessment of Groundwater Pollution in West Delhi, India Using Geostatistical Approach," *Environmental Monitoring and Assessment*, Vol. 167, No. 1-4, 2010, pp. 599-615.

[28] H. Ben Ouezdou, "Morphological and Stratigraphic Study of the Quaternary Formations of the Gulf of Gabes," *Reviews of Groundwater Science*, Vol. 5, No. 4, 1987, pp. 165-166.

[29] J. P. Dash, A. Sarangi and D. K. Singh, "Spatial Variability of Groundwater Depth and Quality Parameters in the National Capital Territory of Delhi," *Environmental Management*, Vol. 45, No. 3, 2010, pp. 640-650.

[30] E. Gaubi, "Hydrological Synthesis of the Triassic Sandstone Groundwater (Regions of Medenine and of Tataouine)," *Africain Journal Science*, Vol. 10, No. 5, 2011. pp. 13-14.

[31] B. W. Hermant, D. Dibyendu, V. R. Desai, B. Klaus and A. Rafig, "Morphometric Analysis of the Upper Catchment of Kosi River, Using GIS Techniques," *Saudi Society for Geosciences*, Vol. 7, No. 6, 2011, pp. 13-14.

[32] S. Mohan and R. Gandhimathi, "Solid Waste Charactrization and Assessment of the Effect of Dumping Site Leachate on Groundwater Quality: A Case Study," *International Journal of Environment and Waste Management*, Vol. 3, No. 1-2, 2009, pp. 65-77.

14

Response of Epilithic Diatom Communities to Downstream Nutrient Increases in Castelhano Stream, Venâncio Aires City, RS, Brazil

Juliara Stahl Böhm, Marilia Schuch, Adriana Düpont, Eduardo A. Lobo[*]

Biology and Pharmacy Department, Laboratory of Limnology, University of Santa Cruz do Sul, Santa Cruz do Sul, Brazil.

ABSTRACT

The Castelhano Stream Hydrographic Basin, located in the city of Venâncio Aires, RS, Brazil, shows an area of 675.3 km^2, highlighting the Castelhano Stream as their main water course. The stream is the main responsiblity for the local water supply; however, there are no published studies in the literature regarding their water quality. In this context, the present research aimed to assess the water quality of Castelhano Stream in terms of organic pollution and eutrophication, applying the Biological Water Quality Index (BWQI), which uses epilithic diatoms communities as bioindicators. Biological samples were collected at three sampling stations along the stream in the months of September, November and December 2012. The results showed 81 identified species, distributed in 30 genera. The water pollution levels detected ranged from "strong" (66.7%) and "very strong" (33.3%), with differences in species composition between sampling stations. The sampling station S1 in the upper reaches was characterized by the presence of indicative species of acidophilus and lentic environments with large amounts of organic matter. The sampling stations S2 and S3, in the intermediate and lower reaches, respectively, showed a substitution of species in the community, with the presence of highly tolerant taxa to organic pollution and eutrophication. The high pollution levels detected along the basin are related to the nutrients and high organic load originating from livestock, domestic and industrial waste, as well as excess fertilizers and agricultural inputs used in farming. The results demonstrate the necessity to implement mitigation measures to contain the processes of organic pollution and eutrophication detected due to the dangers offered to public health and the environment.

Keywords: Castelhano Stream Hydrographic Basin; Biological Water Quality Index (BWQI); Epilithic Diatoms; Organic Pollution; Eutrophication

1. Introduction

At highly industrialized regions in which water demand has increased, most part of domestic and industrial effluents, as well as chemical fertilizers and pesticides used in agriculture, are dumped directly into water bodies, reducing further the possibility of use water resources and drastically modifying the characteristics of aquatic ecosystems. This damage has been demonstrated in two ways: through the introduction of toxic substances into groundwater and through the phenomenon of artificial eutrophication that, besides reducing the water quality, produces significant changes in the metabolism of whole ecosystem [1].

Eutrophication is the increased concentration of nutrients in the aquatic ecosystems, especially phosphorus and nitrogen, which causes an increase in their productivity, characterized as a complex phenomenon because of their ecological basis. This increase in nutrients required for primary producers results in a massive increase of algae growth, which prevents the light penetration to the submerged vegetation, resulting in a massive amount of dead biomass. As a consequence, bacteria need large amounts of oxygen to decompose this material, reducing the oxygen concentration of the water [2,3].

Environmental monitoring studies in freshwater bodies at central region of Rio Grande do Sul have been demonstrated that these systems already show fairly advanced conditions of eutrophication [4-14]. Furthermore, accord-

[*]Corresponding author.

Response of Epilithic Diatom Communities to Downstream Nutrient Increases in Castelhano Stream, Venâncio Aires City, RS, Brazil

157

ing to Tundisi (2006), this condition characterizes the watercourses throughout southern Brazil [15]. Such problems highlight the importance of adoption criteria aimed to assess the pollution levels in aquatic ecosystems, through the use of bioindicators. In this context, epilithic diatoms have been recommended by researchers to assess the water quality due to the sensitivity of this group in relation to a variety of environmental conditions [16].

The city of Venâncio Aires, located at Taquari-Antas Hydrographic Basin, RS, Brazil, has a main source of supply for human consumption the Castelhano Stream; however, there are no published studies in the literature regarding their water quality. In this context, the present research aimed to assess the water quality of Castelhano Stream in terms of organic pollution and eutrophication, applying the Biological Water Quality Index (BWQI), which uses epilithic diatoms communities as bioindicators.

2. Materials and Methods

Venâncio Aires city is located in the central depression of Rio Grande do Sul state (**Figure 1**), in the northeast mountain range (29°39'30"—South latitude and 52°8'41" —North latitude). Inserted at Taquari-Antas hydrographic basin (98%), has an area of 773,239 km^2 and a population density of 65,964 inhabitants [17]. Their main water course is the Castelhano Stream, which has a watershed of 675.3 km² and a length of over 100 km [18]. This stream is the main responsible for the local water supply, the most part it's surrounded by small and medium rural farms, with subsistence crops (rice, tobacco, corn), livestock activities, as well as areas of forest remnants.

Three scientific expeditions were performed along the stream, in September, November and December of 2012, where samples of epilithic diatoms were collected. Three sampling stations were selected along the stream (**Figure 1**), station 1 located upper reaches, station 2 in an intermediate reaches, and station 3 in the low reaches.

For qualitative and semi-quantitative diatom analyses, samples were scrubbed off the upper surface of submerged stones about 10 - 20 cm in diameter using a toothbrush and fixed with formalin [19]. Diatom samples were cleaned with sulphuric and hydrochloric acids and mounted on microscopic slides with Naphrax. All individuals found in random transects under light microscopy across each permanent slide were identified and counted, up to a minimum of 600 valves, using an Olympus BX-40 microscope. The taxonomic references Metzeltin & Lange-Bertalot (1998, 2007), Metzeltin & García-Rodríguez (2003), Metzeltin et al. (2005) and Rumrich et al. (2000) were used for species identification [20-24]. Following the criterion of Lobo & Leighton (1986) the quantitatively important species (abundant species), were

indicated [25]. Voucher samples were stored in the DIAT-UNISC Herbarium at the University of Santa Cruz do Sul, RS, Brazil. Based in the classification of diatoms for southern Brazilian rivers proposed by Lobo et al. (2004a) the Biological Water Quality Index (BWQI) was calculated for all sampling sites and dates [6].

Descriptive statistics was used to tabulate the data and its graphical illustration [26]. In order to assess quantitatively the similarity between the sampling stations, from abundant species, the hierarchical clustering method of Ward was used (minimum variance method) to identify homogeneous groups [27]. The data were processed using the statistical program PAST [28,29].

3. Results and Discussion

In relation to epilithic diatom composition, 81 species were identified belonging to 30 genera, 27 taxa, distributed in 17 genera were considered abundant. The results obtained from the Biological Water Quality Index (BWQI) indicated that the water pollution levels ranged between "strong" (66.7%) and "very strong" (33.3%), in the three sampling stations.

As illustrated in **Figure 2**, the sampling stations S1 and S3, corresponding to the upper and lower reaches of the stream, respectively, showed a "strong" pollution level in 100% of the samples collected. The sampling station S2, in the intermediate section, was characterized by having the highest contamination levels, since 100% of the samples showed a "very strong" pollution.

These high pollution levels observed are due to the presence in high percentages of tolerant species to organic pollution [16] and eutrophication [6], for example *Achnanthidium minutissimum* sensu lato, *Achnanthidium exiguum* var. *constrictum* (Grunow) Anderson, Stoermer e Kreis, *Cocconeis placentula* var. *lineata* (Ehrenberg) Van Heurck, *Diadesmis contenta* (Grunow ex V. Heurck) Mann, *Eolimna minima* (Grunow) Lange-Bertalot, *Fallacia monoculata* (Hustedt) Mann, *Geissleria aikenensis* (Patrick) Torgan & Oliveira, *Gomphonema gracile* Ehrenberg, *Gomphonema parvulum* (Kützing) Kützing, *Luticola goeppertiana* (Bleisch) Mann, *Mayamaea atomus* (Kützing) Lange-Bertalot, *Navicula cryptotenella* Lange-Bertalot, *Navicula rostellata* Kützing, *Navicula symmetrica* Patrick, *Nitzschia palea* (Kützing) Smith, *Planothidium lanceolatum* (Brébisson ex Kützing) Lange-Bertalot, *Sellaphora pupula* sensu lato e *Surirella angusta* Kützing.

Among these, *G. parvulum* was abundant in 6 of the 9 samples, equivalent to 66.7% (**Figure 3**). This species showed the higher abundance at the sampling station S1, corresponding to the upper section, reaching the maximum percent of relative abundance of 71.4% in September. This specie is tolerant to organic pollution, being

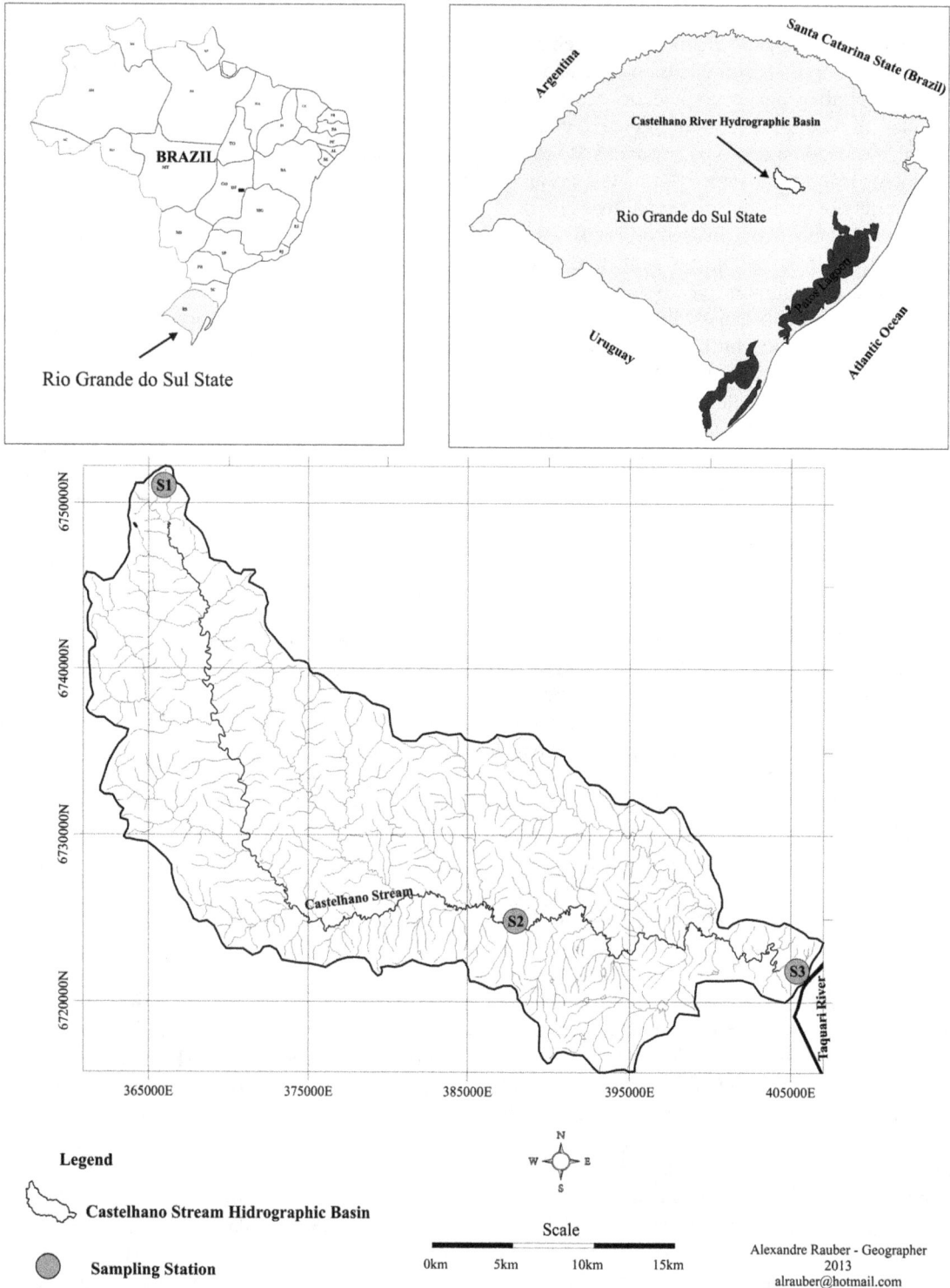

Figure 1. Map of the study area and localization of the Castelhano Stream Hydrographic Basin, in the state of Rio Grande do Sul, Brazil, and the three sampling stations (S1 - S3) along the stream.

indicative of α-mesosaprobic conditions [16]. At streams located in Mato Leitão, RS, Lobo *et al.* (1999) classified this species as belonging to both α-mesosaprobic and polysaprobic conditions [30]. In the same study area,

Rodrigues & Lobo (2000) reported the occurrence of this species in β-mesosaprobic environments [31]. Kobayasi & Mayama (1989) and Lobo *et al.* (1995) classified this taxa as highly tolerant to organic pollution in rivers stud-

Response of Epilithic Diatom Communities to Downstream Nutrient Increases in Castelhano Stream,
Venâncio Aires City, RS, Brazil

159

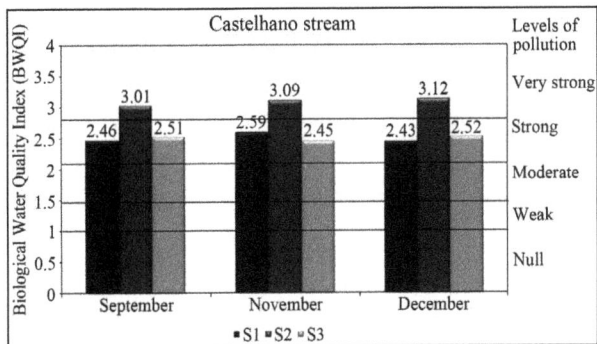

Figure 2. Water quality assessmentin three sampling stations (S1, S2, S3) along the Castelhano Stream, using theBiological Water Quality Index (BWQI).

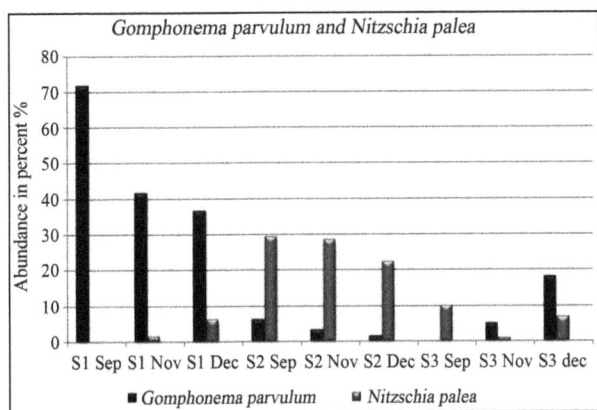

Figure 3. Relative abundance (%) of G. *parvulum*and N. *palea* species insamples collectedin Castelhano Stream, Venâncio Aires, RS.

ied in Japan [32,33].

N. palea was equally abundant in 6 of the 9 samples, equivalent to 66.7% (**Figure 3**). This species showed the higher abundance at sampling stations S2 and S3, corresponding to the intermediate and lower sections, respectively, reaching the maximum percent of relative abundance of 29.4%, at the sampling station S2, in September. This species is considered tolerant to organic pollution, being indicative of α-polisaprobic conditions [16]. Moreover, Van Dam *et al.* (1994) stated that *N. palea* corresponds to polysaprobic taxa, indicating hypereutrophic conditions [34]. At Gravataí River, this species was found in all samples collected from the upper reaches to the lower section, however, the highest densities were recorded in the lower section highly polluted [12].

Differently, Souza (2002), in a study realized at Monjolinho River, São Carlos, Brazil, *G. parvulum* was found in places where the physical and chemical characteristics of the river were considered oligosaprobic (negligible pollution), with a highly abundance of 95.1%, while in polysaprobic conditions the abundance was low, about 10.6% [35].

Salomoni *et al.* (2011) analyzing epilithic diatoms at Gravataí River, RS, highlighted *G. parvulum* as the most abundant species in the sampling stations 1, 2 and 3, corresponding to the upper reaches of the river, with a relative abundance of 37%, 78% and 48%, respectively, condition which led to the classification of this taxa in Group C, characterizing oligotrophic/mesotrophic environments [36]. The authors argue that the species morphology may change as a result of genetic variability, as well as the ecological variation may result in different ecotypes, which would explain the many responses assigned to the same species [13]. Clearly, a more detailed study of ecology, physiology and morphotypes of *G. parvulum* in Southern Brazilian Rivers is required.

Another important factor is the relationship between diatom communities, pH and conductivity of the basin. The pH causes a physiological stress directly on the diatoms, and also strongly influences other chemical variables of the water. The high conductivity, in turn, is related to the emergence of species known to be resistant to heavy metal pollution and have been frequently recorded in eutrophized waters with high organic pollution and low dissolved oxygen levels [37].

These authors, studying urban streams at São Paulo, SP, Brazil, observed diatom communities continuously distributed along the gradient of conductivity and pH. Moderately polluted regions, characterized by low conductivity, showed species like *Eunotia bilunaris* (Ehrenberg) Mills, *Fragilaria capucina* Desmazieres, *Gomphonema angustatum* (Kützing) Rabenhorst, *Pinnularia gibba* (Ehrenb.) Grunow and *Ulnaria ulna* (Nitzsch) Compere, while heavily impacted sites, characterized by a high conductivity and pH slightly acidic, showed species such as *G. parvulum*, *S. pupula* and *N. palea*. These species that have a great development in polluted areas can also occur in relatively clean waters, once the species show the upper limits of tolerance to pollution and not the lower limits [38].

The dendrogram of **Figure 4** clearly shows the biological condition of the sampling stations studied, separating the upper reaches, of the intermediate and lower reaches. At upper reaches it was observed that the algal community was predominantly composed by species from genera *Eunotia*, highlighting *E. pseudosudetica*, Metzeltin, Lange-Bertalot & Garcia-Rodriguez, *E. subarcuatoides* Alles, Nörpel & Lange-Bertalot and *E. cf. veneris*, besides taxa such as *Achnanthes microcephala* (Kützing) Grunow, *L. goeppertiana* (Bleisch) Mann, *N. cryptotenella* Lange-Bertalot and *G. parvulum* (Kützing) Kützing. The genera *Eunotia* has been referred in literature as acidophilus, and characteristic of lentic waters [37,39].

In relation to the species composition of the middle and lower reaches, it was observed the replacing of algal

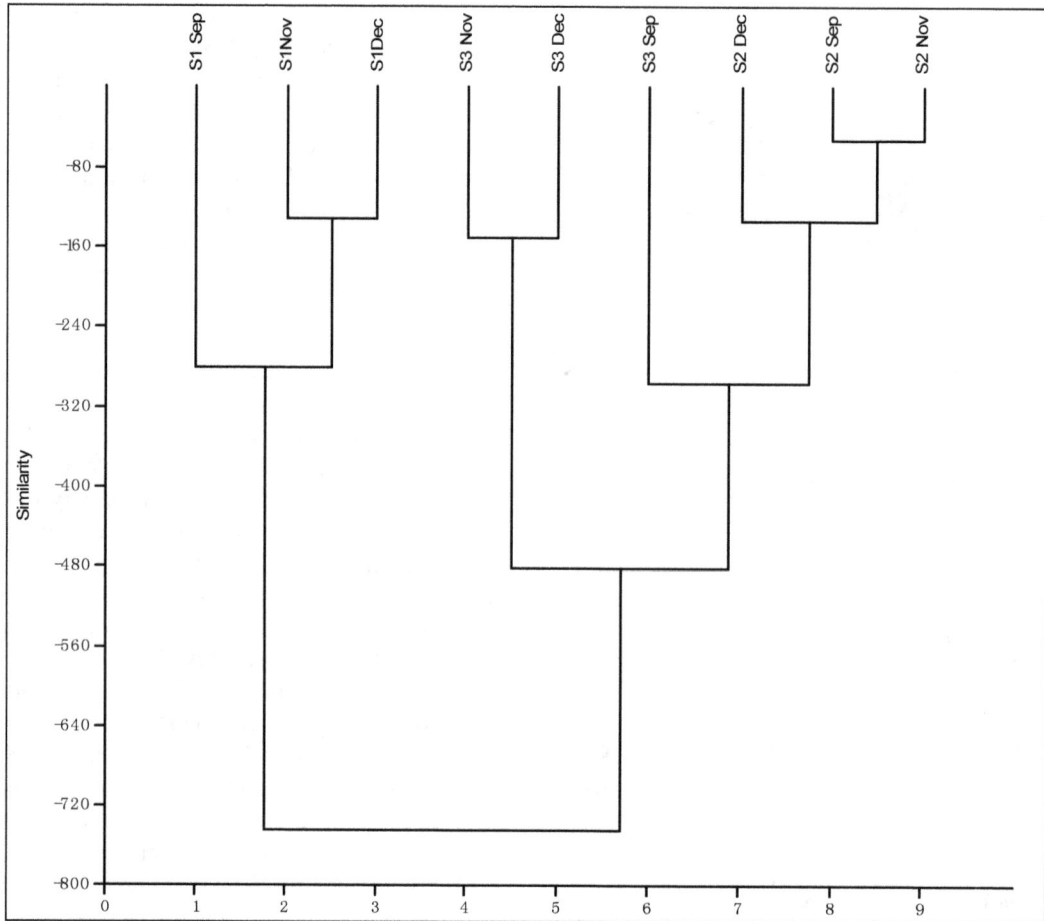

Figure 4. Cluster analysis of the three sampling stations (S1, S2, S3) along the Castelhano Stream, Venâncio Aires, RS.

community by species of bigger tolerance to organic pollution and eutrophication, such as *Nitzschia palea* (Kützing) Smith, *Gomphonema parvulum* (Kützing) Kützing, *Sellaphora pupula* sensulato, *Surirella angusta* Kützing, *Mayamaea atomus* (Kützing) Lange-Bertalot, *Planothidium lanceolatum* (Brébisson ex Kützing) Lange-Bertalot, *Achnanthidium exiguum* var. *constrictum* (Grunow) Anderson, Stoermer e Kreis.

As referred, *G. parvulum* is a characteristic species of environments with a high degree of eutrophication [6], and is also typical from lentic waters [39]. So, their high abundance can be attributed to the large amount of organic matter in the environment from domestic animals present in the site. This condition justified the grouping of station 1, with the presence of species that reveal acidic environments, devoid of riparian vegetation, with lentic waters and excess of organic matter.

The pollution levels "very strong" and "strong" in the sampling stations S2 and S3, respectively, can be justified by the geographical location of them, since they receive many tributaries coming from the urban area, that carry domestic and industrial effluents. The pollution

sources observed come from brooks tributaries of Castelhano Stream, these go through the city and are responsible for carry the diluted wastewater. The city still doesn't have absolute sewerage collection system, *i.e.*, a specific plumbing to domestic sewage. Therefore, the destination of wastewater treated or not, is mostly plumbing and galleries of rainwater, discharging in small streams and brooks causing severe environmental impacts such as eutrophication [40].

4. Conclusions

The response of epilithic diatom communities to downstream nutrient increases in Castelhano Stream Hydrographic Basin, RS, can be characterized by the presence of indicative species from acidophilus and lentic environments, with a lot of organic matter in the upper reaches, followed by the species substitution in the community in the intermediate and lower reaches, with the presence of highly tolerant species to organic pollution and eutrophication. These high pollution levels detected along the basin may be related to the nutrients and organic load originating from the livestock, domestic and

Response of Epilithic Diatom Communities to Downstream Nutrient Increases in Castelhano Stream, Venâncio Aires City, RS, Brazil

161

industrial sewage, excess of fertilizers and agricultural inputs used in farming.

It is important to note that the city of Venâncio Aires is in progress to the establishment of a Sewage Treatment Plant, which will treat wastewater from a part of the urban area, according to the goals proposed in the Municipal Sanitation Plan of the city [40]. In this way, it's expected an improvement in the water quality at the hydrographic basin of the city, since the results obtained in this study clearly demonstrate the need to implement mitigation measures to contain the process of organic pollution and eutrophication detected, considering the potential risks in terms of public health and environment.

REFERENCES

[1] F. A. Esteves, "Fundamentals of Limnology," Interciência LTDA, Rio de Janeiro, 2011.

[2] C. S. Galli and P. S. Abe, "Availability, Pollution and Eutrophication of Water," Associação Instituto Internacional de Ecologia e Gerenciamento Ambiental, São Carlos, 2012, pp. 165-174. http://www.abc.org.br/IMG/pdf/doc-816.pdf

[3] S. M. Branco, "Water and Preservation," Editora Moderna, São Paulo, 2001.

[4] M. A. Oliveira, L. Torgan, E. A. Lobo and A. Schwarzbold, "Association of Periphitic Diatom Specieson Artificial Substrate in Lotic Environments in the Arroio Sampaio Basin, RS, Brazil: Relationships with Abiotic Variables," Revista Brasileira de Biologia, Vol. 61, No. 4, 2001, pp. 523-540.

[5] C. E. Wetzel, E. A. Lobo, M. A. Oliveira, D. Bes and G. Hermany, "Epilithic Diatoms Related to Environmental Factors in Different Reaches of the Rivers Pardo and Pardinho, Pardo River Hydrographic Basin, RS, Brazil: Preliminary Results," Caderno de Pesquisa Série Biologia, Vol. 14, No. 2, 2002, pp. 17-38.

[6] E. A. Lobo, V. L. M. Callegaro, G. Hermany, D. Bes, C. E. Wetzel and M. A. Oliveira, "Use of Epilithic Diatoms as Bioindicators from Lotic Systems in Southern Brazil, with Special Emphasis on Eutrophication," Acta Limnologica Brasiliensia, Vol. 16, No. 1, 2004, pp. 25-40.

[7] E. A. Lobo, D. Bes, L. Tudesque and L. Ector, "Water Quality Assessment of the Pardinho River, RS, Brazil, Using Epilithic Diatom Assemblages and Faecal Coliforms as Biological Indicators," Vie & Milieu, Vol. 54, No. 2/3, 2004, pp. 115-126.

[8] E. A. Lobo, V. L. Callegaro, C. E. Wetzel, G. Hermany and D. Bes, "Water Quality Study of Condor and Capivara Streams, Porto Alegre Municipal District, RS, Brazil, Using Epilithic Diatoms Biocenoses as Bioindicators," Oceanological and Hydrobiological Studies, Vol. 33, No. 2, 2004, pp. 77-93.

[9] E. A. Lobo, V. L. Callegaro, G. Hermany, N. Gomez and L. Ector, "Review of the Use of Microalgae in South America for Monitoring Rivers, with Special Reference to Diatoms," Vie & Milieu, Vol. 54, No. 2/3, 2004, pp. 105-

114.

[10] E. A. Lobo, C. E. Wetzel, L. Ector, K. Katoh, S. Blanco and S. Mayama, "Response of Epilithic Diatom Community to Environmental Gradients in Subtropical Temperate Brazilian Rivers," Limnetica, Vol. 29, No. 2, 2010, pp. 323-340.

[11] G. Hermany, A. Schwarzbold, E. A. Lobo and M. A. Oliveira, "Ecology of the Epilithic Diatom Community in a Low-Order Stream System of the Guaíba Hydrographical Region: Subsidies to the Environmental Monitoring of Southern Brazilian Aquatic Systems," Acta Limnologica Brasiliensia, Vol. 18, No. 1, 2006, pp. 25-40.

[12] S. E. Salomoni, O. Rocha, V. L Callegaro and E. A. Lobo, "Epilithic Diatoms as Indicators of Water Quality in the Gravataí River, Rio Grande do Sul, Brazil," Hydrobiologia, Vol. 559, No. 1, 2006, pp. 233-246.

[13] S. E. Salomoni, O. Rocha, G. Hermany and E. A. Lobo, "Application of Water Quality Biological Indices Using Diatoms as Bioindicators in the Gravataí River, RS, Brazil," Brazilian Journal of Biology, Vol. 71, No. 4, 2011, pp. 949-959.

[14] M. Schuch, E. F. Abreu-Júnior and E. A. Lobo, "Water Quality of Urban Streams, Santa Cruz do Sul, Rio Grande do Sul, Based on Physical, Chemical and Biological Analyses," Bioikos, Campinas, Vol. 26, No. 1, 2012, pp. 3-12.

[15] J. G. Tundisi, "Limnology and Water Resources Management in Brazil," Projeto Brasil das Águas, 2006. http://www.brasildasaguas.com.br/

[16] E. A. Lobo, V. L. M. Callegaro and E. P. Bender, "Use of Epilithic Diatoms as Water Quality Indicators in Rivers and Streams of the Guaiba Hydrographic Region, RS, Brazil," EDUNISC, Santa Cruz do Sul, 2002.

[17] IBGE, "Population Census IBGE—Brazilian Institute of Geography and Statistics," 2010. http://censo2010.ibge.gov.br

[18] E. Collischonn, "Climatology and Urban Space Management: The Case of a Small City," Mercator, Vol. 9, No. 1, 2010, pp. 53-70.

[19] H. Kobayasi and S. Mayama, "Most Pollution Tolerant Diatoms of Severely Polluted Rivers in the Vicinity of Tokyo," Japanese Journal of Phycology, Vol. 30, 1982, pp. 188-196.

[20] D. Metzeltin and H. Lange-Bertalot, "Tropical Diatoms of South America I," Iconographia Diatomologica, Vol. 5, No. 1, 1998.

[21] D. Metzeltin and H. Lange-Bertalot, "Tropical Diatoms of South America II," Iconographia Diatomologica, Vol. 18, No. 1, 2007.

[22] D. Metzeltin and F. García-Rodríguez, "Las Diatomeas Uruguayas," Montevideo, DI.R.A.C, 2003.

[23] D. Metzeltin, H. Lange-Bertalot and F. García-Rodríguez, "Diatoms of Uruguay," Iconographia Diatomologica, Vol. 15, No. 1, 2005.

[24] U. Rumrich, H. Lange-Bertalot and M. Rumrich, "Diatomeen der Anden. Von Venezuela bis Patagonien

(Feurland)," *Iconographia Diatomologica*, Vol. 9, No. 1, 2000.

[25] E. A. Lobo and G. Leighton, "Community Structures of Planktonic Phytocenosis in the Lower Reaches of Rivers and Streams of Central Zone of Chile," *Revista Biología Marina, Valparaíso*, Vol. 22, 1986, pp. 1-29.

[26] S. D. Callegari-Jacques, "Biostatistics. Principles and Applications," Porto Alegre, Artmed, 2006.

[27] J. F. Hair, R. E. Anderson, R. L. Tatham and W. C. Black, "Multivariate Data Analysis," Porto Alegre, Bookman, 2005.

[28] O. Hammer, D. A. T. Harper and P. D. Ryan, "PAST: Paleontological Statistics Software Package for Education and Data Analysis," *Palaeontologia Electronica*, Vol. 4, No. 1, 2001.

[29] J. L. Valentim, "Numerical Ecology: An Introduction to Multivariate Analysis of Ecological Data," Interciência, Rio de Janeiro, 2000.

[30] E. A. Lobo, A. B. Costa and A. Kirst, "Water Quality Assessment of Sampaio, Bonito and Grande Streams, City of Mato Leitão, RS, Brazil, According to the Resolution CONAMA 20/86," *Revista Redes, Santa Cruz do Sul*, Vol. 4, No. 2, 1999, pp. 129-146.

[31] L. M. Rodrigues and E. A. Lobo, "Analysisof Epilithic Diatoms Community Structure in the Sampaio Stream, City of Mato Leitão, RS, Brazil," *Caderno de Pesquisa, Série Botânica, Santa Cruz do sul*, Vol. 12, No. 2, 2000, pp. 5-27.

[32] H. Kobayasi and S. Mayama, "Evaluation of River Water Quality by Diatoms," *The Korean Journal of Phicology*, Vol. 4, 1989, pp. 121-133.

[33] E. A. Lobo, K. Katoh and Y. Aruga, "Response of Epilithic Diatom Assemblages to Water Pollution in Rivers Located in the Tokyo Metropolitan Area, Japan," *Freshwater Biology*, Vol. 34, No. 1, 1995, pp. 191-204.

[34] H. Van Dam, A. Mertens and J. Sinkeldam, "A Coded Checklist and Ecological Indicator Values of Freshwater Diatoms from the Netherlands," *Netherlands Journal of Aquatic Ecology*, Vol. 28, No. 1, 1994, pp. 117-133.

[35] M. G. M. Souza, "Analysis of Epilithic Diatoms Community Structure in the Monjolinho River (Impacted by Organic Pollution), City of São Carlos, São Paulo, Southeast Brazil," Ph.D. Dissertation, Universidade Federal de São Carlos, São Carlos, 2002.

[36] S. E. Salomoni, O, Rocha, G. Hermany and E. A. Lobo, "Application of Water Quality Biological Indices Using Diatoms as Bioindicators in the Gravataí River, RS, Brazil," *Brazilian Journal of Biology*, Vol. 71, No. 4, 2011, pp. 949-959.

[37] T. Bere and J. G. Tundisi, "Weighted Average Regression and Calibration of Conductivity and pH of Benthic Diatom Assemblages in Streams Influenced by Urban Pollution, São Carlos, SP, Brazil," *Acta Limnologica Brasiliensia*, Vol. 21, No. 3, 2009, pp. 317-325.

[38] T. Bere, "Benthic Diatom Community Structure and Habitat Preferences along an Urban Pollution Gradient in the Monjolinho River, São Carlos, SP, Brazil," *Acta Limnologica Brasiliensia*, Vol. 22, No. 1, 2010, pp. 80-92.

[39] R. S. Moro and C. B. Fürstenberger, "Catalog of the Main Ecological Parameters of Non-Marine Diatoms," UEPG, Paraná, 1997.

[40] Venâncio Aires, "Municipal SanitationPlan," Law No. 5023 of November 30, 2011, Venâncio Aires, 2011.

15

Seasonal Succession of the Plankton and Microbenthos in a Hypertrophic Shallow Water Reservoir at Modra (W Slovakia)

Marta Illyová[1], František Hindák[2], Alica Hindáková[2], Eva Tirjaková[3], Ján Machava[4]

[1]Institute of Zoology, Slovak Academy of Sciences, Dúbravská Cesta 9, Bratislava, Slovakia; [2]Institute of Botany, Slovak Academy of Sciences, Dúbravská Cesta 9, Bratislava, Slovakia; [3]Department of Zoology, Faculty of Natural Sciences, Comenius University, Bratislava, Slovakia; [4]Catholic University in Ružomberok, PF, Hrabovská Cesta 1, Ružomberok, Slovakia.

ABSTRACT

The seasonal development of the phytoplankton, phytobenthos, zooplankton, and microbenthos in a high eutrophised intravilan water reservoir was studied. Finally, 25 genera with 44 species of Cyanobacteria/Cyanophytes and 67 genera with 102 species as well as infraspecific taxa of different groups of microscopic algae were identified. The phytoplankton in most parts of the water basin was strongly dominated by green colonial alga *Golenkiniopsis longispina*. From October until December a cyanophyte species *Aphanocapsa delicatissima* with typical cell dimensions of picoplankton/ was found in large amounts/predominated. As early as spring, a plankton bloom in all its components was observed. At that time, also a high concentration of total phosphorus was recorded, which in the second half of April dropped rapidly. The concentration of chlorophyll-*a* increased from 162.7 µg/L in March to 2322 µg/L in September. Massive occurrence of benthic protozoa in the plankton, as a consequence of anoxia, has been observed. Further, the detritivore and omnivore ciliate species *Coleps hirtus* dominated in the microbenthos. Altogether 74 of ciliate taxa were detected. Their abundance and biomass reached peak in April, but these steadily decreased from May until the end of the year. Extreme values of zooplankton density (54,016 ind/L) were recorded in spring followed by a sudden fall in summer and autumn. The contribution of rotifers (*Brachionus* spp., *Filinia longiseta*) in the total zooplankton density and biomass was 98%. Relatively a low species richness of crustaceans (4 Cladocera and 3 Copepoda) was observed.

Keywords: Cyanobacterial Water Blooms; Eutrophication; Ciliates; Zooplankton; Shallow Ponds

1. Introduction

Cultural eutrophication is the Earth's most widespread water quality problem. It causes harmful algal blooms, fish kills and many related problems in fresh waters that are adjacent to areas with large human populations [1].

The small size and shallow water bodies are less stable than larger lakes, and thus very sensitive to any human intervention. Pollution from agriculture and sewage is recognized as having a significant negative impact on water quality and aquatic biota [2], besides the fish stock [3]. The biological reaction of aquatic system to nutrient enrichment is the eutrophication, the eventual conesquence of which is the development of primary production to nuisance proportions [4]. Free dissolved phospho-rus and nitrogen are important nutrients for photosynthetic organisms [5], mainly for cyanobacterial blooms [6, 7]. Eutrophication causes considerable changes in biochemical cycles and biological communities [8]. Community interaction in pelagic food webs is affected by large scale of physical, chemical and biological processes and are govern by nutrient limitation, competition, predation and other ecological forces [9,10]. In shallow waters, trophic level interactions are complicated by detritus pathways and influences from the sediments [11].

In this paper we describe the seasonal development of phytoplankton, algal picoplankton, cyanobacterial bloom, ciliates and metazooplankton in small hypertrophic urban reservoir in 2009 and try to elucidate some interactions between food web components.

2. Material and Methods

The water basin which is situated within the area of Modra (town in W Slovakia, 48°18'55.28"N, 17°19'2.4"E) was originally built and created as a flood control reservoir for the town. The access path and the drive-way for the fire truck machinery to the basin as well as the concrete edges along the whole circumference of its area are clear evidence of this reality "**Figure 1**". However, for the past decades, it has served as a fishpond. The water basin lies at an altitude of 144 meters and its surface covers 0.55 hectares. The maximum depth of this reservoir, at high level water conditions, is 2 meters and it has neither regular direct inflow nor outflow. Due to this fact, the only way how the water basin obtains water is from snow-melt during winter and spring seasons and from rainfall throughout the year. Furthermore, a certain amount of nutrition gets into the basin from local people by feeding the fish.

Our research was carried out from February to December 2009. During this investigation period samples were taken at monthly intervals. The physical and chemical water parameters are as follows: water temperature (°C), pH, oxygen content (mg·L) and oxygen saturation (%). All the parameters were measured near the surface in situ by a multi-functional instrument WTW 80 1i according to relevant working methods and processes. The content of dissolved phosphorus (TRP) (µg/L) was evaluated according to standard analytical techniques (STN EN ISO 6878); total phosphorus (TP) and total nitrogen (TN) were determined by a spectrophotometer DR 2800; and the chlorophyll-a concentration was evaluated by means of the standard method (ISO 10260:1992). Phytoplankton samples were taken by plankton net in mesh size of 10 µm. Cyanobacteria and algae were determined merely from fresh samples, diatoms were defined from permanent slides as well. For the determination of phototrophic microorganisms several monographs were used [12-17].

Ciliophora were studied in plankton and benthos. Plankton samples were collected from a single spot in the studied basin, using a take-off apparatus placed on a 3-meter long telescopic pole, while benthic samples were taken from four sampling sites. The analysis of samples was conducted in vivo by means of a light microscope within 8 hours after sample collection. Taxonomically difficult species were examined also in protargol-impregnated slides whose preparation followed the protocol as described by Foissner [18]. Quantitative evaluation of ciliate abundances included enumerating of active ciliates in 10 subsamples, each with a volume of 10 µL. The obtained data were consequently recalculated to 1 mL. The estimation of the total biomass was based on the mean biomass values for particular species as given in

Figure 1. Water reservoir in Modra; right bottom (Photo F. Hindák).

the determination atlases of Foissner *et al*. [18-22].

The metazooplankton samples for the qualitative analysis were taken by vertical tows from the bottom by a plankton net (70 µm mesh size). The qualitative samples were taken with a 2 L Patalas sampler from various depths of the basin (surface, middle and bottom). The entire water volume (10 L) was filtered through a net of 70 µm mesh size and preserved with 4% formalin. The zooplankton density (ind/L) was assessed in a 1 mL Sedgwick-Rafter chamber. The biomass (g/m^3) was established as wet weight and it was calculated from the recorded average body lengths and the body length/biomass ratio using tables assembled from several bibliographic sources [23-27].

3. Results

3.1. Physical and Chemical Variables

In the season of 2009, high annual mean values of nitrogen, phosphorus and chlorophyll-a were recorded "**Table 1**". It is notable that during the first months of our investigation, from end of March until May, low values of oxygen were recorded due to respiration having been in process during the night before "**Figure 2**". For instance, in March 27 the value for oxygen was 6.07 mg/L and in April 03 we even witnessed oxygen of 4.25 mg/L. Over the entire period of our research, high pH values were measured in the water basin peaking in April with pH 10.81 and in May with pH 10.42. The concentration of total reactive phosphorus (TRP) in the water reached 42.7 µg/L after the ice melted (February 25). The maximum value of TRP was recorded in early March "**Figure 3**". Although the highest inflow concentration of TRP, 1240.8 µg/L was supposed on April 08, it occurred earlier, in April 03 with a high value of TRP concentration 652 µg/L. Meanwhile, in the course of April this value was gradually decreasing and at the end of the month (April 29) it reached 180 µg/L. In the second half of the

Table 1. Annual average values and ranges for selected variables in 2009.

Variable (unit)	Mean	Minimum	Maximum
Water temperature (°C)	16.6	1.3	26
pH value	9.34	8.06	10.81
Oxygen (mg/L)	12.88	4.24	25.97
Oxygen (%)	131	7.5	315
SRP (μg/L)	310.77	14.38	1244.80
Total phosphorus (μg/L)	614	75	1550
Total nitrogen (mg/L)	9.51	0.56	22.85
BOD_5 (mg/L)	7.49	5.15	9.67
CHSK (mg/L)	147.8	23.7	534.3
SO_4 (mg/L)	93.5	60.4	141.7
NO_3 (mg/L)	5.8	0.32	16.2
Cl (mg/L)	45.14	0.62	227
Na (mg/L)	21.2	18.3	25.4
K (mg/L)	31.5	22.9	37.1
Mg (mg/L)	20.8	14.5	31.1
Ca (mg/L)	31.4	10.1	82
Chlorophyll-a (μg/L)	947.3	208.3	2322.4
Conductivity (mS/cm)	553	460	802

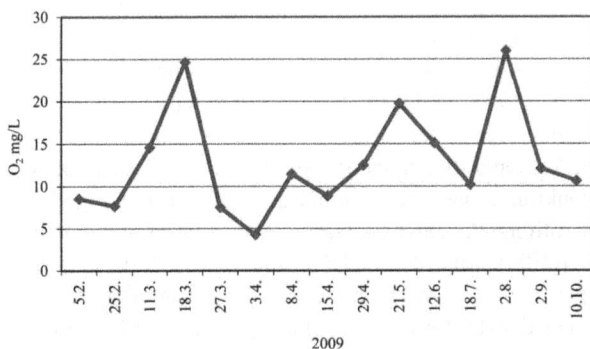

Figure 2. Seasonal development of oxygen (mg/L) in 2009.

vegetation season, from August and November, the TRP values tended to be low "**Figure 3**". During the season the concentration of chlorophyll-a was increasing strongly ranging from the lowest values in March 162.7 μg/L up to maximum values in September 2322 μg/L; high values over 1000 μg/L also persisted in October as well as in November.

3.2. Phytoplankton

Centric diatoms *Stephanodiscus hantschii* and *Cyc-*

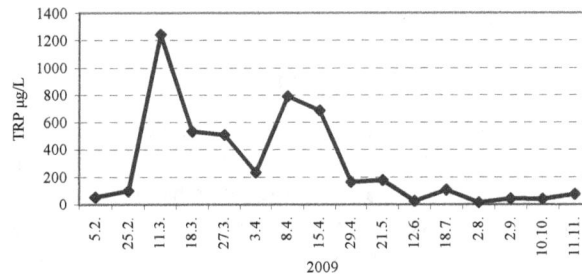

Figure 3. Seasonal development of soluble reactive phosphorus (TRP) (μg/L) in 2009.

lostephanos invisitatus together with cyanobacteria *Microcystis ichtyoblabe* and *Aphanizomenon gracile* dominated the phytoplankton from early spring (March-April) and this fact had a major effect on the colour of the water which appeared to be brown greenish. At the end of April and in May the mass development of the green colonial alga *Golenkiniopsis longispina* was manifested and the colour of water changed into green or dark green. Sporadically and usually in small numbers also some other chloroccalean algae were observed from the genera Scenedesmus, Oocystella, Monoraphidium, and species *Coenococcus planctonicus, Dictyosphaerium tetrachotomum, Kirchneriella obesa, Lagerheimia longiseta, L. wratislaviensis, Micractinium pusillum, Pediastrum boryanum, Pseudodictyosphaerium minutum, Scenedesmus pectinatus, Siderocelis ornata, Siderocystopsis fusca* and *Tetraedron caudatum* [28].

It was interesting to see and it should certainly be pointed out that some groups of algae, e.g. diatoms, desmids or algal flagellates were almost missing in summer and autumn. Only occasionally a green neustonic film, caused by *Chlamydomonas debaryana*, was formed at the edge of the reservoir or *Euglena viridis* was concentrated at the bottom of the reservoir. Periphyton was composed of filamentous microorganisms from cyanobacteria *Homoeothrix janthina, Oscillatoria janus, O. princeps, Phormidium tenue, Calothrix fusca*, from green algae *Oedogonium* sp., *Aphanochaete repens* and zygnematophycean algae from the genera Mougeotia, Spirogyra and Zygnema. The cyanobacterial water bloom started to be obvious and evident from May. At the beginning it was dominated by the colonies of *Microcystis ichtyoblabe*, later on (from July to early August) by a rare nostocalean species, *Cylindrospermopsis raciborskii*. Terminal heterocytes were formed very rarely; thus the majority of filaments resembled similar species—*Raphidiopsis mediterranea*. From August until the end of the season in December colonies of chroococcal cyanobacterium *Aphanocapsa delicatissima* with cells of picoplanktic size (1 - 2 μm in diameter) dominated very strongly. Taxa *Microcystis botrys, M. aeruginosa* and *Anabaenopsis milleri* ranked among the group of accompanying species of cyanobacterial water blooms.

3.3. Ciliates and Microbenthos

Totaly 74 taxa of ciliates were identified during the year 2009 [29]. The highest species richness of ciliates was recorded in March (12). On the other hand, the lowest number of species (5 - 8 species) was observed during the second half of the year. The seasonal changes in biomass and abundance of ciliates are shown in "**Figure 4**". Cell abundance ranged from 305 to 10,570 ind/mL. The mean ciliate number was 2595 ind/mL. Biomass ranged from 2870 to 296,690 mg/mL, averaging 71,491 mg/mL. At the beginning of the vegetation season, *i.e.* in March, the community structure was stabilized with an average density of 300 - 400 ind/mL and a moderate domination of members of the genus *Vorticella*. In April, a boom of ciliate growth was noticed (over 10,000 ind/mL), which was primarily the result of a mass occurrence of the prostomatid *Coleps hirtus*. Also *Pseudovorticella natans* was recorded during early spring at a relatively high abundance (400 ind/mL). Moreover, a few large ciliate species, such as *Linostomella vorticella* and *Paramecium caudatum* were observed. In May, the ciliate density in the plankton dropped approximately to 2500 ind/mL. *Coleps hirtus* fell back, but the bacterivorous *Cinetochilum margaritaceum*, which feeds also on small aglae, became dominant. Some benthic species, *Spirostomum teres* and *Holophrya teres*, were noted in the plankton. This is a consequence of anoxia in the benthos, and this phenomenon became very conspicuous in June when the ciliate density in the plankton (≈6000 ind/mL) was several times higher than that in the benthos (1000 ind/mL). During this period there was a comparatively poor species spectrum (5 - 8 species) with *Coleps hirtus* and *Cinetochilum margaritaceum* being dominant both in the plankton and in the benthos. In the July plankton, there was an overall decrease of ciliate densities (≈1200 ind/mL), but the number of species increased moderately to 12. The community was equable and no species was dominanting over other ones. Small bacterivorous species, such as *Ctedoctema acanthocryptum*, *Cyclidium glaucoma* and *Cinetochilum margaritaceum* are characteristic for this period.

Species of algal and cyanobacterial diet, especially, *Frontonia leucas* and *Halteria grandinella* began to occur in the summer communities. Further, algivorous and bacteriovorous species (*Pseudocohnilembus pusillus*, *Cinetochilum margaritaceum*) replaced the predatory species in benthos. Frontonia leucas was the most dominant algivorous species in the summer months, but disappeared in September when the bacteriovorous species *Dexiotricha tranquilla* became dominant both in the plankton and in the benthos. As concerns predatory

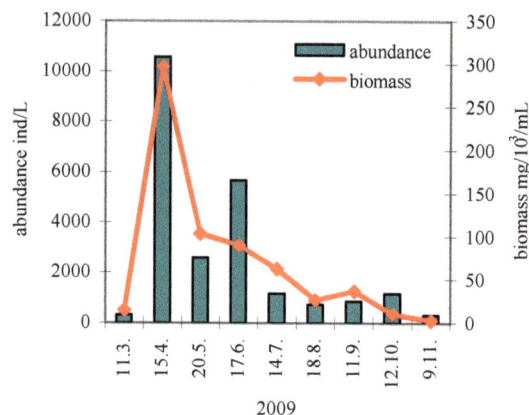

Figure 4. Seasonal development of abundance (ind/mL) and total biomass (mg·10⁻³ m/L) of ciliates in 2009.

gymnostomate ciliates, members of the genus Actinobolina occurred mainly in spring and summer. At the same time, a high abundance of colonial peritrichous ciliates of the genus *Epistylis* (*E. chrysemydis* and *E. enzii*) was recorded mostly in the benthos. By contrast, typical planktobionts such as *Rimostrombidium*, *Limnostrombidium*, *Tintinidium* and *Hastatella* were noted only rarely and at low abundance.

3.4. Metazooplankton and Rotifers

The season of changes in biomass and abundance of zooplankton are showen in "**Figure 5**", the changes in biomass of rotifers, cladocerans and copepods in "**Figure 6**". The abundance of the zooplankton ranged during the sampling period from 102 ind/L up to 54,016 ind/L. The metazooplankton was dominated by rotifers, which contributed 98% of its total density and 91.6% of the total zooplankton biomass. Accordingly, it can be stated that the seasonal dynamics of the abundance of total net zooplankton copies the seasonal dynamics of the abundance of rotifers. As early as March, a very unusual abundance of rotifers was recorded (15,680 ind/L). Moreover, this high density production kept increasing and in April reached extremely high values (53,712 ind/L) with the dominance of the genus *Polyarthra* (92%). Although in April the biomass of rotifers was high, it reached its maximum level in July. We started to witness a species dominance of *Brachionus budapestinesis* (57%) and *Filinia longiseta* (36%). In the second half of the vegetation season a considerable drop of rotifers was observed "**Figure 5**". While at the beginning of autumn the dominance of *B. budapestinensis* (85%) was still prevailing, in November only two species (*Keratella quadrata* and *Cephalodella gibba*) of a very low abundance were found.

The proportion of cladocerans to total quantity was considered low and their ratio to the total biomass indi-

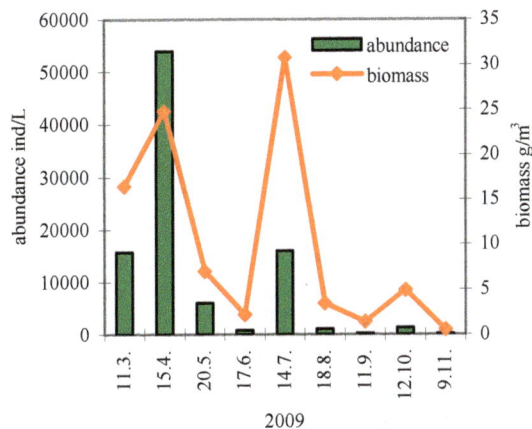

Figure 5. Seasonal development of abundance (ind/L) and total biomass (g/m) of net zooplankton in 2009.

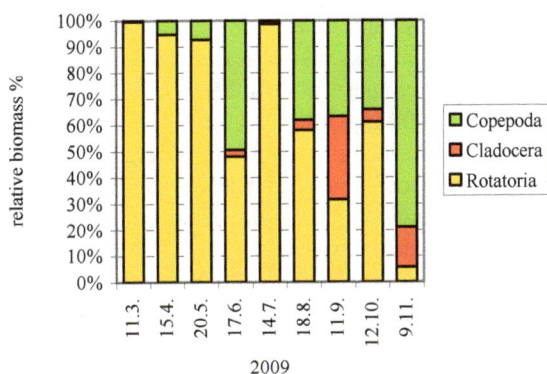

Figure 6. Seasonality of biomass (%) proportions of all groups of net zooplankton in 2009.

cated 1.4%; however, during the autumn months their relative biomass increased due to a regress of rotifers "**Figure 6**". The most important cladoceran species was *Alona rectangula* with abundance ratio from 2 ind/L to 44 ind/L. In addition to Alona, also a rare occurrence of *Chydorus sphaericus* was observed. The contribution of copepods in a total zooplankton biomass came to 7%. Throughout the whole season the copepods were present mostly as nauplii and small copepodites, adult being rare (*Acanthocyclops trajani*).

4. Discussion

There are some evidence about the high eutrophication of water body in the reservoir: 1) cyanobacterial bloom and high chlorophyll-a content as a consequence of a high phosphorus content; 2) changes in species composition of planktonic communities, decrease of quantity and species biodiversity at all plankton communities and microbenthos; 3) anoxia at the bottom of reservoir indicated by protozoan community.

1) The bloom of filamentous cyanobacteria in the half of the summer is an indication of the eutrophic to the

hypertrophic state of the Modra reservoir. For example, *Cylindrospermopsis raciborskii* which is adapted to low light conditions is able to fix atmospheric nitrogen [30]. The small reservoir at Modra can be described as a typical pond with massive technical modification and with no original macrovegetation [31]. It can be assumed that its water is rich in nutrients and tied to sediments what is very typical for all small water basins and fishponds [3], or urban lakes [32]. The high concentration of chlorophyll-a and the dominance of cyanobacteria as well as of green algae (namely Chlorococcales) over the vegetation period indicated a high trophic level in the water basin and the high rate of primary production. It is generally agreed that if there is a sufficient source of nutrients in water, phosphorus particularly, an excessive phytoplankton production occurs cf. [33]. Phosphorus is often found to be the limiting nutrient in inland fresh waters [10,34]. Values in excess of 30 µg/L ortho-phosphate phosphorus (PO_4-P) in river waters and in excess of 20 µg/L total P (Ptot) in lakes are considered by the Irish EPA to lead to eutrophication [35].

However, as far as spring we measured and recorded far higher values for the reactive as well as total phosphorus concentration which largely determined the excessive production of phytoplankton. This corresponds well with the findings published by [36] that found out a linear correlation between TP and the maximum phytoplankton biomass at the beginning of vegetation season. In spring, in addition to diatoms, a proliferation of cyanobacteria from the genus *Microcyctis* was observed at Modra. For the period of winter the green colour of water showed itself through the ice which is not an unusual phenomenon. The production of cyanophytes below the ice at the temperature of 3°C was also observed by Maršálek [37], or by Kiss and Genkal [38]. The high values of phosphorus remained until the middle of April which was followed by a sudden drop presumably due to consumed phosphorus at the massive phytoplankton development. This manifested itself also with high values of alkaloids, as they increase the values of pH in their surroundings even to 9 - 11 by their metabolic activities [37].

2) During the period of our investigation 25 genera with 44 species of cyanophytes and 67 genera with 102 species as well as infraspecific taxa of different groups of microscopic algae were identified [28]. Five are the first records from the territory in the Slovak Republic cf. [39], *i.e.* two species of cyanobacteria: *Synechococcus mucicola* Joosten, *Synechocystis endophytica* (G. M. Smith) Joosten, and three species of green algae: *Nautococcus mamillatus* Korshikov, *Bicuspidella incus* Pascher, *Desmatractum indutum* Pascher [28]. However, the phytoplankton diversity was generally lower in comparison

with other eutrophic waters in Western Slovakia cf. [40].

Eutrophication causes considerable changes in biochemical cycles and biological communities [8]. In the Modra reservoir, the response of planktonic and benthic ciliates on the phytoplankton development was manifested as a change in the ciliate community structure [29]. The omnivorous species were replaced by algivorous and bacterivorous ones. Ciliates seem to be very flexible in reaction to environmental changes since many of them are capable to alter their food sources upon the offer within a relatively short time. Dias and D' Agosto [41] found out that *Frontonia leucas* feeds especially on diatoms in oligotrophic waters and only rarely consumes bacteria and cyanobacteria. However, the diet of this species changes when saprobiological level increases. In Modra, abundance of *F. leucas* raised in summer after a massive development of cyanobacteria and green algae especially *Golenkiniopsis*. Due to the high feeding activity, this ciliate species could play a key role in reduction of water bloom [42]. These authors also showed that *Halteria grandinella* is another good candidate for reduction of water bloom, as it consumed over 70% of all cyanobacteria which were engulfed by ciliates. These findings are supported also by our observations in that *H. grandinella* and also *F. leucas* co-occurred in the period with a maximum developed cyanobacterial water-bloom. Also Vörösváry [43] recorded an increased occurrence of algivorous and bacterivorous species, especially from the genera *Chilodonella*, *Colpidium*, *Stylonychia*, *Coleps*, *Paramecium* and *Frontonia* as well as of species living on organic remnants (*Spirostomum*) in a stream with water bloom due to the pollution by sewage waters. Thus, these taxa can be potentially used in suppression of water bloom. Peritrichous filter feeders, which often occur at high abundances in benthos, could be also highly effective in elimination of water bloom [44].

The hypertrophic conditions of the water body in Modra were documented also by the net zooplankton community: the extremely high abundance of the rotifers in spring and extremely low planktonic crustaceans. The extreme abundance of rotifera at the beginning of the vegetation season draw the parallel between them and the maximum values in spring which is the typical standard for pond ecosystems [3]. The maximum density of rotifers in April reproduces a massive increase of diatoms and colonial green algae as the majority of species present in water are herbivorous filter feeders. Sládeček [45] in his research make a reference to a very eutrophic fishpond with abundance as high as 23,900 ind/L. High abundance values of rotifers in spring (over 30,000 ind/L) were also observed in naturally eutrophic two arms in the Morava River floodplain [46]. The spring phase of rotifer

development was followed by a rapid drop in its density which is quite typical for the summer season [47]. From August on, no rotifer development was spotted and the quantity of these species dropped significantly most likely due to the existence of cyanobacteria. One of the most undesirable aspects of cyanobacteria is an eventual production of toxins [33]. As Dumont [48] states, the extra cellular substances of cyanobacteria are toxic for rotifers so they repress their development. At that time the species typical for self-purification reservoirs of sewage (*Brachionus budapestinensis* and *Filinia longiseta*) ruled in the rotifer community, so did the ones which are the indicators of deteriorating saprobic conditions in waters [45].

After a spring development of algae we suppose the increase of density of filter feeder planktonic crustaceans [49], e.g. large cladocerans grazers and herbivorous copepods (*Eudiaptomus*). Large grazers are able to control phytoplankton biomass even under hypereutrophic condition (up to 1600 µg/L) [50]. But despite the reach food source in the water basin in Modra, there were neither herbivore copepods observed at all, nor large filter feeders from the genus *Daphnia*; moreover, we did not even record the species *Bosmina longirostris*. It can be explained by the lack of edible food here. The typical net zooplankton species for eutrophic water and fishponds are small cladocerans *Bosmina longirostris* and *B. coregoni*, and a high abundance of cyclopoid copepods *Thermocyclops* and *Eudiaptomus* [3,32,51]. Our findings documented the occurrence of only two small cladocerans *Alona rectangula* and *Chydorus sphaericus*, which were observed in plankton in a small abundance. Both species were spotted in plankton of extremely hypertrophic water bodies by Sládeček and Sládečková [52]. The poor abundance of planktonic crustaceans was most likely due to cyanobacteria which strongly proliferated as far back as the beginning of the vegetation season and were less edible for filter feeders than small algae. Particularly, the larger colonial and filamentous cyanobacteria cannot serve as food source for zooplankton because of its parameters [49] as well as its eventual toxicity [33]. Mayer *et al.* [32] also recorded a significant decrease in zooplankton biomass over the summer development of cyanobacteria which, as they claim, was determined by changes in food source and the increase of water temperature.

3) We assume that the direct consequence of the high production of phytoplankton was the excessive consumption of oxygen near the bottom of the basin which led to anoxia. Presumably, the adverse conditions (anoxia) at the bottom of the basin contributed to the occurrence of some benthic species, e.g. *Spirostomum teres* and *Holophrya teres*, in the plankton. Moreover, ciliate densities

in the June plankton (≈6000 ind/mL) were several times higher than those in the benthos, which is a rare phenomenon in the protozoan communities. Also Finlay [53] noted that ciliates can migrate from the benthic zone into the water column depending on the oxygen content. Moreover, he argued that species bigger than 150 μm, such as *Loxodes* and *Spirostomum*, do not migrate. However, we observed high abundances *of Spirostomum teres* not only in the benthos but also in the plankton during May. This essentially supports Finlay's migratory theory also for this species.

5. Conclusion

The seasonal development of the phytoplankton, phytobenthos, zooplankton and microbenthos of a highly eutrophised intravilan water reservoir was investigated. The cyanobacterial bloom and high chlorophyll-a content, as a consequence of a high phosphorus concentration, influenced the seasonal dynamic of the plankton and microbenthos communities. The highest concentration of TRP was recorded from March to early April. Subsequently the boom of the planktonic communities occurred in that ciliates and rotifers reached the highest abundance and biomass, and the green colonial alga *Golenkiniopsis longispina* manifested a mass development. The cyanobacterial water bloom, composed mainly from the colonies of *Microcystis ichtyoblabe*, started in May without any significant influence on the planktonic communities of the reservoir. However, the colonies of chroococcal picoplanktic cyanobacterium, with cells 1 - 2 μm in diameter, dominated very strongly during the summer. Consequently the densities of ciliates and other zooplankton dropped significantly. Paced plankton sank to the bottom and started to decompose, which resulted in oxygen depletion. As a consequence of anoxia, the massive and multiple occurrences of benthic protozoa in the plankton were observed in June. The ciliate densities in the plankton were several times higher than those in the benthos. From August on, in the second half of the vegetation season, a considerable drop in abundance of planktonic invertebrates was observed, most likely due to the presence of cyanobacteria.

6. Acknowledgements

This study was supported by APVV, project No. 0566-07, VEGA projects No. 1/0600/11 and 2/0113/13. This publication is the result of the project implementation: Comenius University in Bratislava Science Park supported by the Research and Development Operational Programme funded by the ERDF Grant number: ITMS 26240220086. The study was also supported by the project ITMS: 26240220049.

REFERENCES

[1]   D. W. Schindler and J. R. Vallentyne, "The Algal Bowl: Overfertilization of the World's Freshwaters and Estuaries," University of Alberta Press, Edmonton, 2008.

[2]   J. D. Allan, "Landscapes and Riverscapes: The Influence of Land Use on Stream Ecosystems," *Annual Review of Ecology, Evolution, and Systematics*, Vol. 35, 2004, pp. 257-284.

[3]   V. Kořínek, J. Fott, J. Fuksa, J. Lellák and M. Pražáková, "Carp Ponds of Central Europe," In: R. G. Michael, Ed., *Managed Aquatic Ecosystems*, Elsvier Sci. Pub. B.V., Amsterdam, 1987, pp. 26-61.

[4]   M. Salvia-Castellvi, A. Dohet, P. Vander Borght and L. Hoffman, "Control of the Eutrophication of the Reservoir of Esch-sur-Sûre (Luxembourg): Evaluation of the Phosphorus Removal by Predams," *Hydrobiologia*, Vol. 459, No. 1-3, 2001, pp. 61-71.

[5]   D. M. Anderson, P. M. Glibert and J. M. Burkholder, "Harmful Algal Blooms and Eutrophication: Nutrient Sources, Composition, and Consequences," *Estuaries*, Vol. 25, No. 4, 2002, pp. 704-726.

[6]   G. A. Codd, "Cyanobacterial Toxins, the Perception of Water Quality and the Priorisation of Eutrophication Control," *Ecological Engineering*, Vol. 16, No. 1, 2000, pp. 51-60.

[7]   Y. Yamamoto, H. Tsukada and Y. Matsuzawa, "Relationship between Environmental Factors and the Formation of Cyanobacterial Blooms in a Eutrophic Pond in Central Japan," *Algological Studies*, Vol. 134, No. 1, 2010, pp. 25-39.

[8]   E. Lelková, M. Rulík, P. Hekera, P. Dobiáš, P. Dolejš, M. Borovičková and A. Poulíčková, "The Influence of the Coagulant PAX-18 on *Planktothrix agardhii* Bloom in a Shallow Eutrophic Fishpond," *Fottea*, Vol. 8, No. 2, 2008, pp. 147-154.

[9]   S. R. Carpenter, "Complex Interactions in Lake Communities," Springer Verlag, 1988.

[10]   S. R. Carpenter, N. F. Caraco, D. L. Correll, R. W. Howarth, A. N. Sharpley and V. H. Smith, "Nonpoint Pollution of Surface Waters with Phosphorus and Nitrogen," *Ecological Applications*, Vol. 8, No. 3, 1998, pp. 559-568.

[11]   A. Herzig, "The Zooplankton of the Open Lake," In: H. Löffler, Ed., *Neusiedlersee. Limnology of a Shallow Lake in Central Europe*, Dr. W. Junk Publishers, Hague, 1979, pp. 281-336.

[12]   J. Komárek and B. Fott, "Chlorophyceae (Grünalgen), Ordnung: Chlorococcales," *Die Binnengewässer*, Vol. 16, No. 7, 1983.

[13] F. Hindák, "Colour Atlas of Cyanophytes," Veda, Bratislava, 2008.

[14] K. Krammer and H. Lange-Bertalot, "Bacillariophyceae, 1. Teil: Naviculaceae, " *Süßwasserflora von Mitteleuropa*, Vol. 2, No. 1, 1986.

[15] K. Krammer and H. Lange-Bertalot, "Bacillariophyceae, 2. Teil: Bacillariaceae, Epithemiaceae, Surirellaceae," *Süßwasserflora von Mitteleuropa*, Vol. 2, No. 2, 1988.

[16] K. Krammer and H. Lange-Bertalot, "Bacillariophyceae, 3. Teil: Centrales, Fragilariaceae, Eunotiaceae," *Süßwasserflora von Mitteleuropa*, Vol. 2/3, 1991a.

[17] K. Krammer and H. Lange-Bertalot,"Bacillariophyceae, 4. Teil: Achnanthaceae, Kritische Ergänzungen zu *Navicula* (Lineolatae) und *Gomphonema*," *Süßwasserflora von Mitteleuropa*, Vol. 12/4, 1991b.

[18] W. Foissner, H. Blatterer, H. Berger and F. Kohmann, "Taxonomische und Ökologische Revision der Ciliaten des Saprobiensystems Band I.," Informationsberichte des Bayer. *Landesamtes für Wasserwirtschaft*, Vol. 1, 1991.

[19] W. Foissner, H. Blatterer, H. Berger and F. Kohmann, "Taxonomische und Ökologische Revision der Ciliaten des Saprobiensystems Band II.," Informationsberichte des Bayer. *Landesamtes für Wasserwirtschaft*, Vol. 5, 1992.

[20] W. Foissner, H. Berger and F. Kohmann, "Taxonomische und Ökologische Revision der Ciliaten des Saprobiensystems Band III, " Informationsberichte des Bayer. *Landesamtes für Wasserwirtschaft*, Vol. 1, 1994.

[21] W. Foissner, H. Berger, H. Blatterer and F. Kohmann, "Taxonomische und ökologische Revision der Ciliaten des Saprobiensystems Band IV, Gymnostomatea, Loxodes, Suctoria," Informationsberichte des Bayer. *Landesamtes für Wasserwirtschaft*, 1995.

[22] W. Foissner, H. Berger and J. Schaumburg, "Identification and Ecology of Limnetic Plankton Ciliates," Informationsberichte des Bayer, Landesamtes für Wasserwirtschaft, Műnchen, Vol. 3/99, 1999.

[23] F. D. Morduchaj-Boltovskoj, "Materialy po Srednemu vesu Bespozvonochnykh Basseĭna Dona," Trudi Problemnih i Tematičeskih Sovešcanija, *Zool. Inst. 2, Problemy Gidrobiologii Vnutrennych Vod*, Vol. 2, 1954, pp. 223-214.

[24] S. N. Ulomskij, "Rol' Rakoobraznych v Obshchej Biomasse Planktona ozer (K Voprosu o Metodike Opredeleniya Vidovoj Biomassy Zooplanktona)," *Trudi Problemnih i Tematičeskih Sovešcanija, Zool. Inst.*, 1, *Problemy Gidrobiologii Vnutrennych Vod*, Vol. 1, 1951.

[25] S. N. Ulomskij, "Syroĭ ves Massovykh Form Nizshikh Rakoobraznykh Kamskogo Vodokhranilischa i Nekotorykh ozer Urala i Zaural'ya," *Trudi Uraljskogo otdeleniya Gosudarstvennogo Nauchno-Issledovateljskogo Institute Ozernogo i Rechnogo, Ribnogo Khozyaystva*, Vol. 5, 1961, pp. 200-210.

[26] A. Nauwerck, "Die Beziehungen Zwischen Zooplankton und Phytoplankton im See Erken," *Acta Universitatis Upsaliensis: Symbolae Botanicae Upsalienses*, Vol. 17, No. 5, 1963, pp. 1-163.

[27] B. Düssart, "Limnologie. L'étude des eaux Continentales, " Guathier-Villars, Paris, 1966.

[28] F. Hindák and A. Hindáková," Cyanobacteria and Algae of a Small Eutrophic Water Reservoir at Modra (W Slovakia)," *Bulletin Slovenskej Botanickej Spoočnosti*, Vol. 32, No. 2, 2010, pp. 129-135.

[29] E. Tirjaková, "Ciliated Protozoa Communities (Protozoa: Ciliophora) in Hypertrophic Shallow Water Reservoir in Modra (W Slovakia)," *Folia faunistica Slovaca*, Vol. 15, No. 1, 2010, pp. 33-41.

[30] W. Lampert and U. Sommer, "Limnoökologie," Thieme Veralg, Stuttgart, New York, 1993.

[31] M. Fulín, I. Hudec, E. Sitášová, V. Slobodník and P. Sabo, "Environmentálno-Ekonomické Vyhodnotenie Funkcií a Hospodárenia v Rybníkoch na Slovensku," Nadácia IUCN, Svetová únia Ochrany Prírody, Slovensko, IUCN Gland, Švajčiarsko a Cambrige, Veľká Británia, 1994.

[32] J. Mayer, M.T. Dokulil, M. Salbrachter, M. Berger, T. Posch, G. Pfister, A. K. T. Kirschner, B. Velimirov, A. Steitz and T. Ulbricht, "Seasonal sUccession and Trophic Relations between Phytoplankton, Zooplankton, Ciliate and Bacteria in a Hypertophic Shallow Lake in Vienna, Austria," *Hydrobiologia*, Vol. 342-343, 1997, pp. 165-174.

[33] B. Maršálek, V. Keršner and P. Marvan, "Vodní Květy Sinic," Nadatio flos-aquae, Brno, 1996.

[34] A. J. Wade, G. M. Hornberger, P. G. Whitehead, H. P. Jarvie and N. Flynn, "On modelling the Mechanisms That Control In-Stream Phosphorus, Macrophyte, and Epiphyte Dynamics: An Assessment of a New Model Using General Sensitivity Analysis," *Water Resource Research*, Vol. 37, No. 11, 2001; pp. 2777-2792.

[35] J. Lucey, J. J. Bowman, K. J. Clabby, P. Cunningham, M. Lehane, M. MacCarthaigh, M. L. McGarrigle and P. F. Toner, "Water Quality in Ireland (1995-1997)," Environmental Protection Agency, Wexford, 1999.

[36] P. J. Dillon and F. H. Rigler, "A Test of the Simple Nutrient Budget Model Predicting the Phosphorus Concentration in Lake Water," *Journal of the Fisheries Research Board of Canada*, Vol. 31, No. 11, 1974, pp. 1771-1778.

[37] B. Maršálek, "Hledání Achilovy Paty u Cyanobaktérií," In: M. Rulík, Ed., *Limnologie na Přelomu Tisíciletí*, Conference proceedings, Kouty nad Desnou, 2000, pp. 88-93.

[38] K. T. Kiss and S. I. Genkal, "Winter Blooms of Centric Diatoms in the River Danube and in Its Side-Arms near Budapest (Hungary)," *Hydrobiologia*, Vol. 269-270, No. 1, 1993, pp. 317-325.

[39] F. Hindák and A. Hindáková, "Checklist of Cyanobacteria and Algae of Slovakia," In: K. Marhold and F. Hindák, Eds., *Checklist of Non-Vascular and Vascular Plants of Slovakia*, Veda, Bratislava, 1998, pp. 12-100.

[40] F. Hindák and A. Hindáková, "Cyanophytes and Algae of Gravel Pit Lakes in Bratislava, Slovakia: A Survey," *Hydrobiologia*, Vol. 506-509, No. 1-3, 2003, pp. 155-162.

[41] R. J. P. Dias and M. D'Agosto, "Feeding Behavior of

Frontonia leucas (Ehrenberg) (Protozoa, Ciliophora, Hymenostomatida) under Different Environmental Conditions in a Lotic System," *Revista Brasileira de Zoologia*, Vol. 23, No. 3, 2006, pp. 758-763.

[42] K. Šimek, J. Bobková, M. Macek and J. Nedoma, "Ciliate Grazing on Picoplankton in a Eutrophic Reservoir during the Summer Phytoplankton Maximum: A Study at the Species and Community Level," *Limnology and Oceanography*, Vol. 40, No. 6, 1995, pp. 1077-1090.

[43] B. Vörösváry, "A Kalános Patak Csillós Véglényei," *Annales Biologicae Universitatis Szegediensis Szeged*, Vol. 1, 1950, pp. 343-387.

[44] J. A. Laybourn-Parry, "Protozoan Plankton Ecology," Chapman & Hall, London, 1992.

[45] V. Sládeček, "Rotifers as Indicators of Water Duality," *Hydrobiologia*, Vol. 100, No. 1, 1983, pp. 169-201.

[46] M. Illyová, "Zooplankton of Two Arms in the Morava River Floodplain in Slovakia," *Biologia*, Vol. 61, No. 5, 2006, pp. 531-539.

[47] M. Devetter, "Influence of Environmental Factors on the Rotifer Assemblage in an Artificial Lake," *Hydrobiologia*, Vol. 387/388, 1998, pp. 171-178.

[48] H. J. Dumont, "Biotic Factors in the Population Dynamics of Rotifers," *Archiv für Hydrobiologie, Beiheft Ergebnisse der Limnologie*, Vol. 8, 1977, pp. 98-122.

[49] U. Sommer, M. Gliwitcz, W. Lampert and A. Duncan, "The PEG-Model of Seasonal Succession of Planktonic Events in Fresh Waters," *Archiv fuer Hydrobiologie*, Vol. 106, No. 4, 1986, pp. 433-471.

[50] K. Vakkilainen, T. Kairesalo, J. Hietala, D. M. Balayla, E. Bécares, W. J. Van de Bund, E. Van Donk, M. Fernández-Aláez, M. Gyllström, L-A. Hansson, M. R. Miracle, B. Moss, S. Romo, J. Rueda and D. Stephen, "Response of Zooplankton to Nutrient Enrichment and fish in Shallow Lakes: A Pan-European Mesocosm Experiment," *Freshwater Biology*, Vol. 49, No. 12, 2004, pp. 1619-1632.

[51] Z. M. Gliwicz, "Studies on the Feeding of Pelagic Zooplankton in Lake with Varying Trophy," *Ekologia Polska*, Vol. 17A, 1969, pp. 663-708.

[52] V. Sládeček and A. Sládečková, "Biologie Stabilizačních nádrží," *X. Limnologická Konference* (*Conference Proceedings*), Stará Turá, 1994, pp. 178-190.

[53] B. J. Finlay, "Oxygen Availability and Seasonal Migrations of Ciliated Protozoa in a Freshwater Lake," *Journal of General Microbiology*, Vol. 123, No. 1, 1981, pp. 173-178.

16

Health Risk Assessment of Mobility-Related Air Pollution in Ha Noi, Vietnam

Vu Van Hieu[1,2], Le Xuan Quynh[3], Pham Ngoc Ho[1], Luc Hens[4]

[1]Research Centre for Environmental Monitoring and Modelling, Vietnam National University, Hanoi, Vietnam; [2]Department of Human Ecology, Vrije Universiteit Brussel, Brussels, Belgium; [3]Department of Geography, Vrije Universiteit Brussel, Brussels, Belgium; [4]Vlaamse Instelling voor Technologisch Onderzoek, Mol, Belgium.

ABSTRACT

Hanoi is the capital city of Vietnam and the second largest city of the country, just behind Ho Chi Minh City. During the last two decades, Hanoi developed fast and expanded steadily. Since the city acquired large parts of the surrounding provinces in 2008, Hanoi tripled its size and doubled its population. The new development aims to spread the concentrated population and economic activities to alleviate the stress caused by pollution and the decreasing quality of life of the residents. Hanoi has a very fast growing fleet of motor vehicles, at the rate of 12% - 15% annually. The fast transition from bikes to motorcycles and to cars results in a most serious environmental burden in particular on the air quality and human health. This paper overviews the air quality and pollution caused by road traffic in central Hanoi (5 old districts) and the related health outcomes due to particulate matters (PM_{10} and $PM_{2.5}$). It uses dose-response functions to quantify the number of extra deaths resulting from traffic-related particulate matters. The results are compared with those of other studies to assess the impacts of air pollution on human health in large, crowded and fast developing cities in Southeast Asia. Assessment of the health risk caused by traffic shows that mobility in Hanoi causes a high health burden. In 2009, mobility caused 3200 extra deaths by traffic related PM_{10}. The result shows that health impacts due to air pollution are by far larger than the number of fatalities due to traffic accidents.

Keywords: Hanoi; HIA; Risk Assessment; Traffic-Related Health Effects

1. Introduction

Hanoi is the capital of Vietnam and the second largest city of the country, just behind Ho Chi Minh City. During the last two decades, Hanoi developed fast and expanded steadily. Hanoi changed its boundaries 4 times in 1961, 1978, 1991 and 2008. After the last boundary modification, Hanoi covers the total area of 3348.5 km^2; has a population of 6.45 million people with an average density of 1926 people/km^2, distributed over 27 districts (9 urban and 18 rural) and 408 communes (as of 31 December 2008) [1].

Hanoi experienced a 11.7‰ population growth a year in 2000, which increased to 11.8‰ in 2005, 12.5‰ in 2008 and 12. 7‰ in 2009. The city is characterised by a fast urbanisation rate, achieving 5.6%/year during the period 2001-2005 but reduced to 3% during the period 2006-2009. By 2009, the urban area was 40.8%, an increase from 33.2% in 2000. Urban population accounts for around 41.3% of the total [1-3]. In 2006, Hanoi had an unemployment rate of 6.1%, which reduced to 5.4% in 2008. The city has about 3974 km of roads, of which 643 km within the 9 old districts (that account for 6.8% of the urban area).

Since the city acquired a large part of its surroundings in 2008, Hanoi tripled its size and doubled its population. The new development aims to spread the concentrated population and economic activities to the newly acquired areas to alleviate the stress currently put on the environment due to air, noise and water pollution and the decreasing quality of life of the residents. Hanoi is a highly polluted city as a result of the dense traffic and the industry that is still localised in the inner city.

Hanoi has a very fast growing fleet of motor vehicles, at the rate of 12% - 15% annually. Both cars and motorcycles grow rapidly. In 2000, around 46,200 cars/trucks circulated in the city, accounted for 9.5% of the country's total vehicle fleet. Its 865,232 motorcycles comprise 12.38% of the country's motorcycle fleet [3]. By the end of 2009, there are more than 2.76 million motorcycles in circulation. The number of cars doubled during the period 2005-2009, raising from nearly 150,000 cars in 2005 to more than 304,000 in 2009 [4]. The fast transition from bikes to motorcycles (since the middle of the 1990s) and to cars (since the late 1990s) results in a most serious environmental burden in particular on the air quality and the associated human health impacts.

Motorised mobility is a major source of air pollution. Studies in large cities in developing countries such as New Dehli, Bangkok, Beijing, Manila and Jakarta show a 40% - 80% contribution of vehicles to the total concentration of particulate matters [5-12]. Particulate matter (especially PM_{10}) and their effects on human health have been the subject of many epidemiological studies and reviews. The results consistently show that 24-hr average concentrations of particulate matters are related to daily mortality and daily hospital admissions [13-18]. Chronic exposure to PM_{10} is also linked to mortality [18-23].

In Vietnam, studies on health impacts of air pollution have been done recently in selected major cities, such as Ho Chi Minh City, Ha Noi and Hai Phong. Nguyen [24] studied the impacts of air pollution on human health in Hanoi using published dose-response functions to calculate long-term health impacts of Total Suspended Particles (TSPs) using air quality data for the period 1994-1998. Most of the other recent studies use health surveys as the main method to assess the health effects on the population. These health surveys are based on the presence/absence of chronic obstructive pulmonary diseases (COPDs) and respiratory diseases.

This paper overviews the air quality and pollution caused by road traffic in central Hanoi (5 old districts) and the related health outcomes due to particulate matters. It uses dose-response functions to quantify the number of extra deaths resulting from traffic-related air pollution. The results are compared with those of other studies to assess the impacts of air pollution on human health in large, crowded and fast developing cities like Hanoi.

2. Materials and Methods

2.1. Air Monitoring

Air quality monitoring data were obtained from the Environmental Monitoring Centre of the National Environmental Agency. They include information of air pollutants (CO, SO_2, NO_2 and TSP) and noise, measured quarterly during the period 2005-2009 at 5 monitoring locations. **Figure 1** shows the geographical distribution of the monitoring stations. They are located in a way that the data provide a fair idea of the average pollutant concentrations 5 districts downtown Hanoi, which are Ba Dinh, Dong Da, Hai Ba Trung, Hoan Kiem and Thanh Xuan.

Figure 1. Map of monitoring locations.

2.2. Health Risk Assessment

Health risk assessment using the classic 4-step paradigm:

- Hazard Identification has been done based on literature review and ground-checked through a health survey. Health impacts of various air pollutants, especially PM, benzene, and ground level ozone have been extensively described [5,6,17,19-23,25-34]. Most of the epidemiological evidence points to the link between exposure to air pollutants and respiratory and cardio-vascular diseases, such acute upper respiratory infections (acute pharyngitis, acute tonsillitis, acute laryngitis and tracheitis, etc.), influenza and pneumonia, acute bronchitis and bronchiolitis, and primary hypertension. In this study, the mortality effect of PM_{10} and the morbidity effects of $PM_{2.5}$ will be considered.

- Exposure assessment: due to insufficient data (such as detailed population density data, air quality values for small unit, and population activities indoor-outdoor, etc.), it is supposed that the whole of the population in the 5 assessed districts of Hanoi exposed to the same level of outdoor air pollution. The annual concentration of the monitoring station for the period 2005-2009 was used.

- Dose-Response assessment follows the dose-response function for PM_{10} established by prior research that has been published [18,20,22,23]. These data shown an increase in mortality of 4.3% per 10 $\mu g/m^3$ increase in PM_{10} (**Table 1**).

• Risk characterisation was assessed based on the number of extra cases of health outcome linked to traffic-attributable air pollution.

For $PM_{2.5}$, the number of restricted-activity days were calculated as proposed by Fisher et al. [23] at 9.1 cases per 100 persons per 1 $\mu g/m^3$ annual $PM_{2.5}$. The formula for calculation is:

$$N_{RAD} = E_{PM2.5} * \left(9.1 * P_{exp} / 100\right) \quad (1)$$

where N_{RAD} is the number of restricted-activity days; $E_{PM2.5}$ is the annual concentration of $PM_{2.5}$ and P_{exp} is the exposed population. The number of restricted-activity days experienced by an individual in the course of a year is an important measure of functional well-being. The definition of "restricted-activity days" is the average annual number of days a person experienced at least one of the following: 1) a bed day, during which a person stayed in bed more than half a day because of illness or injury related to traffic; 2) a work-loss day, on which a currently employed person missed more than half a day from a job or business; 3) a school loss day, on which a student 5 - 17 years of age missed more than half a day from the school in which he or she was currently enrolled; or 4) a cut-down day, on which a person cuts down for more than half a day on things he usually does.

For PM_{10}, the number of extra deaths was calculated using the formula [20] that allows to calculate the long-term impact of PM_{10} on mortality within the group aged over 30 years old:

$$Po = Pe / \left(1 + \left((RR-1)(EPM-BPM)/10\right)\right) \quad (2)$$

where:

- Po = baseline mortality per 1000 in the age group 30+, after deducting the air pollution effect (this will depend on the other variables).
- Pe = crude mortality rate per 1000 in the age group 30+. Due to the lack of specific data, the crude mortality rate for whole population of each assessed district will be used. Data comes from Vietnam General Statistical Office [2].
- E_{PM} = PM_{10} exposure level in the area of interest ($\mu g/m^3$).
- B_{PM} = threshold PM_{10} exposure level for mortality

Table 1. Dose-response relationship used in [20].

Health outcome	Relative risk (RR)[a]	95% CL[b]
Total mortality (adults > 30 years, excluding violent deaths)	1.043	1.026 - 1.061

[a]Relative risk associated with a 10 $\mu g/m^3$ increase in PM^{10}; [b]95% confidence level.

effect. In this study, with an assumed threshold for PM_{10} at 7.5 [23].

- RR = epidemiologically derived relative risk for a 10 $\mu g/m^3$ increment of PM_{10}, assuming a linear dose-response relationship above the threshold (B) for the age group 30+.

The increased mortality is calculated using the following formula:

$$D_{PM} = Po * \left(RR - 1\right) \quad (3)$$

where: D_{PM} = number of additional deaths per 1,000 people in the age group 30+ for a 10 $\mu g/m^3$.

The number of deaths due to PM_{10} is calculated as follows:

$$N_{PM} = D_{PM} * P_{30+} * \left(E_{PM} - B_{PM}\right)/10 \quad (4)$$

where: P_{30+} is the population over 30-years old (47.1% of the total population of Hanoi according to the 2009 Nationwide Population Survey and Population Projection until 2050)

3. Results

3.1. Air Quality

According to the Vietnamese National Environmental Agency, in Hanoi, the fleet of motor vehicles increases at the rate of 12% - 15% per year. Both the number of cars and of motorcycles increases rapidly. By the end of 2009, it is estimated that there are more than 2.76 million motorcycles in circulation.

The vehicle inventory for Hanoi done by the authors in 2011 shows that, by the end of April 2010, Hanoi traffic is dominated by motorcycles with more than 3.6 million units in operation. There are nearly 160 thousand cars and nearly 68 thousand trucks. 94% of the vehicle pool in Hanoi are motorcycles.

Motorcycles produce the most VOC for each kilometer travelled. They emit ten times more VOCs than a car and 1.5 times more than a bus [33]. The motorcycle fleet also emits more CO. As its capacity is lower than this of a car and much lower than this of a bus, consequently, the pollution level per passenger kilometre is the highest of all motorised traffic. Therefore, as the preferred mode of transport, accounts for 94% of the fleet, and around 85% of the total road length travelled (**Table 2**), motorcycles are the main source of air pollutant emissions in Hanoi city.

Cars become gradually more popular in Hanoi. Their number has nearly doubled between 2005 and 2009 (from 149,333 units in 2005 to 304,143 units in 2009) [4].

Based on the number of vehicles in circulation, pollution emissions can be assessed. **Table 2** shows the estimation of air pollution emissions from motorised traffic sources in Hanoi.

Table 2 also shows that motorcycles contribute the most to the total traffic emissions. They account for over 70% of the total TSP emissions and more than 95% of the total VOC emissions.

Hanoi has 5 air quality monitoring stations that were in operation during the period 2007-2009. Average annual air quality data at the stations is presented in **Table 3**.

In 2003, daily monitoring data for 365 days show that there were 359 days with PM_{10} concentration over 250 g/m^3 [35]. Monitoring results show that the daily concentrations of PM_{10} fluctuate with a large difference between day-time and night-time concentrations. During peak hours (around 8 am and 6 pm), the concentrations of PM_{10}, CO and NO_2 are highest, showing the impacts of the traffic on air quality. While during night-time (9 pm to 5 am next day), PM_{10} concentration is mostly below the standards (QCVN 05:2009), concentrations during

the day are 2 to 3 times higher than the standard [36].

The annual average concentrations of PM_{10} are higher than the standards for 3 consecutive years in 2007, 2008 and 2009. For NO_2 and SO_2, hourly concentrations are within the permissible levels but daily concentrations are higher than the standards.

3.2. Health Effects

In this study, the mortality attributed to PM_{10} and the morbidity of $PM_{2.5}$ are assessed. As for the mortality, the results provide a figure on additional (or extra) deaths caused by the increased of the concentrations of PM_{10}. The increased mortality due to PM_{10} exposure for the group over 30-years is presented in **Table 4**. Only the urban sections of Hanoi are taken into account in the calculation for area and population.

Table 2. Emission by vehicle types in Hanoi.

Vehicle type	Travelled distance (km)	Emissions (tonne/year) for 2009				
		TSP	SO_2	NO_x	CO	VOC
Motorcycle	5858	1710.21	6.09	2992.88	357007.38	171021.50
Car	6205	69.30	1.01	1178.18	7643.34	821.76
Passenger car and bus	14,600	328.78	0.77	3874.88	1549.95	1244.66
Truck	5475	290.03	0.80	3272.10	7838.16	1446.42
Total		2398.32	8.68	11318.03	374038.83	174534.33

Table 3. Annual concentration of dust (TSP) in Hanoi at five monitoring stations.

Year	Average annual TSP concentrations (mg/m^3)				
	Thuong Dinh (Thanh Xuan district)	Mai Dong (Hai Ba Trung district)	Ly Quoc Su (Hoan Kiem District)	Van Phuc (Ba Dinh district)	Kim Lien (Dong Da district)
2007	0.28	0.4	0.26	0.19	0.68
2008	0.36	0.29	0.25	0.23	0.48
2009	0.40	0.415	0.26	0.20	0.46

Table 4. Absolute mortality due to PM_{10} in the period 2007-2009.

District	Year	Area	Population	Density	Pop 30+	E_{PM}	Po	D_{PM}	N_{PM}
Ba Dinh	2007	9.25	222,200	24,022	104,656	104.50	0.14	0.46	463
	2008	9.25	223,800	24,195	105,410	126.50	0.12	0.40	507
	2009	9.25	225,000	24,324	105,975	110.00	0.15	0.48	526
Dong Da	2007	9.96	361,100	36,255	170,078	374.00	0.04	0.13	798
	2008	9.96	365,500	36,697	172,151	264.00	0.06	0.19	854
	2009	9.96	371,000	37,249	174,741	253.00	0.06	0.19	832
Hai Ba Trung	2007	10.09	311,200	30,842	146,575	220.00	0.07	0.22	694
	2008	10.09	310,000	30,723	146,010	159.50	0.11	0.35	772
	2009	10.09	297,600	29,495	140,170	228.25	0.08	0.25	776
Hoan Kiem	2007	5.29	150,300	28,412	70,791	143.00	0.15	0.50	484
	2008	5.29	148,600	28,091	69,991	137.50	0.17	0.57	518
	2009	5.29	147,000	27,788	69,237	143.00	0.17	0.56	529
Thanh Xuan	2007	9.08	216,400	23,833	101,924	154.00	0.07	0.23	346
	2008	9.08	221,700	24,416	104,421	198.00	0.05	0.18	356
	2009	9.08	224,900	24,769	105,928	220.00	0.06	0.21	466

The concentrations of PM_{10} are calculated based on the monitored concentration of TSP (**Table 3**). As all monitoring points are located next to streets and are designated as traffic air pollution monitoring stations, the concentrations of TSP are considered 100% traffic-attributable. The average annual concentration of PM_{10} is estimated at 55% of the average annual concentration of TSP (which is the average across all monitoring points) [37].

The results (N_{PM}) indicate the absolute increase in mortality and not the adjusted values reflecting the year-life-lost. At the annual threshold of 7.5 $\mu g/m^3$, the estimated number of people dying by non-external causes that may be associated with traffic air pollution is 2785 in 2007, 3007 in 2008 and 3129 in 2009. **Table 4** shows that the mortality due to PM_{10} has been increased slightly during the period 2007-2009. Amongst the 5 assessed districts, Dong Da has the highest tolls even though it has the lowest number of additional deaths per 1000 people in the age group 30+ for each increment of 10 $\mu g/m^3$ (D_{PM}). This is because it has the worst air quality index. Thanh Xuan has the lowest tolls although it is not the district with the best air quality, thanks mostly to its low crude death rate, which could be the result of younger population overall.

For morbidity, the total number of restricted-activity days due to $PM_{2.5}$ exposure was calculated and presented in **Table 5**. The calculation assumed that $PM_{2.5}$ effect on morbidity has no threshold (or threshold is 0). As there is no measurement for $PM_{2.5}$ concentration, the fraction of 0.7 was applied on the concentration of PM_{10} to calculate the concentration of $PM_{2.5}$, as suggested by Medina *et al*. [22].

Table 5 shows that annually, each person in Hanoi City loses around 14 days in 2007, 12 days in 2008 and 13 days in 2009 due to $PM_{2.5}$ pollution. People in Dong

Da District lost the most days due to air pollution and people in Ba Dinh District lost the least.

4. Discussion

In Hanoi, the economic development and fast urbanisation are associated with a very fast increase in the vehicle fleet, with an annual growth of about 11% in the number of cars and 15% in the number of motorcycles. The average trip per person per day for Hanoi in 2005 was 2.7 with a average daily travelled distance of 20 - 25 km, which is much higher than other countries in the region.

Traffic contributes largely to the air pollution in the city and accounts for about 70% of the total pollution. TSP/PM_{10} in all measurements and monitoring points is higher than the Vietnamese standards and much higher than the level that has impacts on health. Concentrations of other pollutants such as SO_2, NO_x, and CO are in general lower than the annual standard but sometimes exceeds daily standards. Moreover, variations are noticed during the day. Concentrations of pollutants during the morning and afternoon peak hours are always 1.5 to 5 times above the hourly standards. This corresponds with the peak number of vehicles in circulation and frequent congestions on many streets in Hanoi.

Calculation of the health risk caused by traffic shows that mobility in Hanoi causes a high health burden. In 2009, mobility caused more than 3000 extra deaths by traffic related PM_{10}.

Table 6 shows that health impacts due to air pollution is by far larger than the number of fatalities due to traffic accidents. The combination of mortality due to traffic accidents and due to air pollution is called "traffic road toll" [20]. In Hanoi, this "road-tolls" is much higher than other countries [20,23,38].

Table 5. Evolution of the number of restricted-activity days during $PM_{2.5}$ over the period 2008-2009.

District	Year	Population	E_{PM}	$E_{PM2.5}$	N_{RAD}	Average per head
Ba Dinh	2007	222,200	104.50	73.15	1,479,108	6.7
	2008	223,800	126.50	88.55	1,803,392	8.1
	2009	225,000	110.00	77.00	1,576,575	7.0
Dong Da	2007	361,100	374.00	261.80	8,602,774	23.8
	2008	365,500	264.00	184.80	6,146,540	16.8
	2009	371,000	253.00	177.10	5,979,073	16.1
Hai Ba Trung	2007	311,200	220.00	154.00	4,361,157	14.0
	2008	310,000	159.50	111.65	3,149,647	10.2
	2009	297,600	228.25	159.78	4,326,963	14.5
Hoan Kiem	2007	150,300	143.00	100.10	1,369,098	9.1
	2008	148,600	137.50	96.25	1,301,550	8.8
	2009	147,000	143.00	100.10	1,339,038	9.1
Thanh Xuan	2007	216,400	154.00	107.80	2,122,841	9.8
	2008	221,700	198.00	138.60	2,796,213	12.6
	2009	224,900	220.00	154.00	3,151,749	14.0

Table 6. Total mortality due to traffic accident and air pollution.

Country	Population (million)	Mortality due to traffic accidents for all ages (1)	Mortality due to traffic air pollution for adults > 30 (2)	Ratio (1)/(2)
France (1996)[a]	58.3	153 per million	501 per million	1:3.3
Austria (1996)[a]	8.1	119 per million	487 per million	1:4.1
Switzerland (1996)[a]	7.1	84 per million	400 per million	1:4.8
New Zealand (2002)[b]	3.7	137 per million	196 per million	1:1.4
Hai Phong, Vietnam (2007)[c]	0.6	307 per million	1572 per million	1:5.1
Ha Noi, Vietnam (2009)	2.6	174 per million	2473 per million	1:14.2

[a]Kunzli *et al.* [20]; [b]Fisher *et al.* [23]; [c]Vu *et al.* [38].

However, there are several uncertainties in this study. First of all, time-series data on health impacts of air pollution is not available in Vietnam. The use of dose-response functions from other studies worldwide may show discrepancies while applying them to the situation in Hanoi. As dose-response relationships can vary from population to population, specific dose-response relationships for Vietnam are necessitated in future studies.

Also, there is no division between mortality due to fine and ultrafine particulates as the current study on the impact of $PM_{2.5}$ on mortality was not available to be applied in Hanoi. Increasing evidence points to the direction that $PM_{2.5}$ is actually the culprit behind mortality due to particulate matter. Therefore, the higher the proportion of $PM_{2.5}$, the higher mortality will be. From this point of view, current estimate on mortality is conservative.

In addition, exposure to different air pollutants is simplified in this study, due to the lack of monitoring data as well as a reliable air quality modelling methods that allow to accurately calculate the number of people exposed to different levels of pollution by individual air pollutants. This necessitates further studies on air quality and air pollution exposure for Hanoi.

It is worthy to note that the results are indicative, as PM was chosen as the representative pollutant. Other pollutants also contribute to different health effects, such as increase in hospitalisation due to air pollutants (NO_x, SO_2, VOC) or mental health issues due to noise and traffic disturbances. Those health effects cannot be included in the scope of this paper.

5. Acknowledgements

The authors wish to thank the colleagues at the Research Centre for Environmental Monitoring and Modelling (CEMM) for their help with data collection and treatment. This research was performed as part of the PhD research on "Integrating Environmental Modelling and Geographic Information System in Environmental Health Impact Assessment of Transport and Mobility development in Hai Phong, Vietnam", funded by the Flemish Interuniversity Council (VLIR), Belgium.

REFERENCES

[1] Hanoi Statistical Office, "Statistical Year Books of Hanoi 2000-2009," Hanoi, 2010.

[2] Vietnam General Statistical Office, "Vietnam Nationalwide Population Survey 2009," 2011. http://www.gso.gov.vn

[3] Vietnam Register, "Integrated Action Plan to Reduce Vehicle Emissions in Viet Nam," 2002. http://www.adb.org/documents/others/Reduce_Vehicle_E missions_VIE/Reduce_Vehicle_Emissions.pdf

[4] Vietnam Register, "Data on the Vehicle Fleet in Hanoi," Hanoi, 2011.

[5] N. Bruce, R. Perez-Padilla and R. Albalak, "Indoor Air Pollution in Developing Countries: A Major Environmental and Public Health Challenge," *Bulletin of the World Health Organization*, Vol. 78, No. 9, 2000, pp. 1078-1092.

[6] N. Bruce, R. Perez-Padilla and R. Albalak, "The Health Effects of Indoor Air Pollution Exposure in Developing Countries," WHO, Geneva, 2002.

[7] S. Syahril, B. P. Resosudarmo and H. S. Tomo, "Study on Air Quality of Jakarta, Indonesia: Future Trends, Health Impacts, Economic Value and Policy Options," ADB, Jakarta, 2002.

[8] M. Walsh, "Contribution of Vehicles to Overall Air Pollution in Asia," ADB, Manila, 2002.

[9] World Health Organization, "The World Health Report 2002: Reducing Risks, Promoting Healthy Life," Geneva, 2002.

[10] H. Kan and B. Chen "Particulate Air Pollution in Urban Areas of Shanghai, China: Health-Based Economic Assessment," *Science of the Total Environment*, Vol. 322, No. 1-3, 2004, pp. 71-79.

[11] S. Cheng, D. Chen, J. Li, H. Wang and X. Guo, "The Assessment of Emission-Source Contributions to Air Quality by Using a Coupled MM5-ARPS-CMAQ Modeling System: A Case Study in the Beijing Metropolitan Region, China," *Environmental Modelling & Software*, Vol. 22, No. 11, 2007, pp. 1601-1616.

[12] A. Sagar, M. Bhattacharya and V. A. Joon, "Comparative Study of Air Pollution-Related Morbidity among Exposed

Population of Delhi," *Indian Journal of Community Medicine*, Vol. 32, No. 4, 2007, pp. 268-271.

[13] D. W. Dockery, C. A. Pope III, X. Xu, J. D. Spengler, J. H. Ware, M. E. Fay, B. G. Ferris Jr. and F. E. Speizer, "An Association between Air Pollution and Mortality in Six US Cities," *The New England Journal of Medicine*, Vol. 329, No. 24, 1993, pp. 1753-1759.

[14] C. A. Pope III, D. W. Dockery and J. Schwartz, "Review of Epidemiological Evidence of Health Effects of Particulate Air Pollution," *Inhalation Toxicology*, Vol. 7, No. 1, 1995, pp. 1-18.

[15] C. A. Pope III, R. T. Burnett, M. J. Thun, E. E. Calle, D. Krewski, K. Ito and G. D. Thurston, "Lung Cancer, Cardiopulmonary Mortality, and Long-Term Exposure to Fine Particulate Air Pollution," *Journal of the American Medical Association*, Vol. 287, No. 9, 2002, pp. 1132-1141.

[16] World Health Organization, "Health Aspects of Air Pollution with Particulate Matter, Ozone and Nitrogen Dioxide," Bonn, 2003.

[17] H. R. Anderson, R. W. Atkinson, J. L. Peacock, L. Marston and K. Konstantinou, "Meta-Analysis of Time-Series Studies and Panel Studies of Particulate Matter (PM) and Ozone (O_3)," World Health Organization, London, 2004.

[18] G. Fisher, T. Kjellstrom, S. Kingham, S. Hales and R. Shrestha, "Health and Air Pollution in New Zealand: Main Report," Health Research Council of New Zealand, Ministry for the Environment and Ministry of Transport, Wellington, 2007.

[19] F. Ballester, "Air Pollution and Health: An Overview with Some Case Studies," In: P. Nicolopoulou-Stamati, L. Hens and C. V. Howard, Eds., *Environmental Health Impacts of Transport and Mobility*, Springer, Dordrecht, 2005, pp. 53-79.

[20] N. Künzli, R. Kaiser, S. Medina, M. Studnicka, O. Chanel, P. Filliger, M. Herry, F. Horak, V. Puybonnieux-Texier, P. Quénel, J. Schneider, R. Seethaler, J. C. Vergnaud and H. Sommer, "Public-Health Impact of Outdoor and Traffic-Related Air Pollution: A European Assessment," *The Lancet*, Vol. 356, No. 9232, 2000, pp. 795-801.

[21] A. Le Tertre, S. Medina, E. Samoli, B. Forsberg, P. Michelozzi, A. Boumghar, J. M. Vonk, A. Bellini, R. Atkinson, J. G. Ayres, J. Sunyer, J. Schwartz and K. Katsouyianni, "Short-Term Effects of Particulate Air Pollution on Cardiovascular Diseases in Eight European Cities," *Journal of Epidemiology and Community Health*, Vol. 56, No. 10, 2002, pp. 773-779.

[22] S. Medina, E. Boldo, M. Saklad, E. M. Niciu, M. Krzyzanowski, F. Frank, K. Cambra, H. G. Muecke, B. Zorilla, R. Atkinson, A. Le Tertre and B. Forsberg, "APHEIS Health Impact Assement of Air Pollution and Communications Strategy," Institut de Veille Sanitaire, Saint-Maurice, 2005. http://www.apheis.net/vfbisnvsApheis.pdf

[23] G. W. Fisher, K. A. Rolfe, T. Kjellstrom, A. Woodward, S. Hales, A. P. Sturman, S. Kingham, J. Petersen, R. Shrestha and D. King, "Health Effects Due to Motor Vehicle Air Pollution in New Zealand," New Zealand Ministry of Transport, 2002.

[24] T. S. Nguyen, "Assessing Long-Term Effect of Outdoor Dust Pollution on the Population of Hanoi (Lượng giá ảnh hưởng mạn tính của ô nhiễm bụi trong không khí ngoài trời đối với sức khoẻ của dân cư nội thành Hà Nội)," *Journal of Practical Medicine (Y học thực hành)*, Vol. 396, No. 4, 2001, pp. 8-10. (In Vietnamese)

[25] P. Babisch, "Traffic, Noise and Health," In: P. Nicolopoulou-Stamati, L. Hens and C. V. Howard, Eds., *Environmental Health Impacts of Transport and Mobility*, Springer, Dordrecht, 2005, pp. 9-25.

[26] N. Janssen and E. Sanderson, "Air Pollution and the Risks to Human Health—Exposure Assessment," AIRNET Work Group 1: Exposure Assessment, 2004. http://airnet.iras.uu.nl/

[27] K. Katsouyianni, "Ambient Air Pollution and Health," *British Medical Bulletin*, Vol. 68, No. 1, 2003, pp. 143-156.

[28] P. Nicolopoulou-Stamati, L. Hens, P. Lammar and C. V. Howard, "Effects of Mobility on Health—An Overview," In: P. Nicolopoulou-Stamati, L. Hens and C. V. Howard, Eds., *Environmental Health Impacts of Transport and Mobility*, Springer, Dordrecht, 2005, pp. 277-309.

[29] A. Prüss-Ustün, C. Mathers, C. Corvalán and A. Woodward, "Introduction and Methods: Assessing the Environmental Burden of Disease at National and Local Levels," World Health Organization, Geneva, 2003.

[30] T. Roussou and P. Behrakis, "The Respiratory Effects of Air Pollution," In: P. Nicolopoulou-Stamati, L. Hens and C. V. Howard, Eds., *Environmental Health Impacts of Transport and Mobility*, Springer, Dordrecht, 2005, pp. 79-95.

[31] E. Sanderson and F. Hurley, "Air Pollution and the Risks to Human Health—Health Impact Assessment," AIRNET Work Group 4—Risk and Health Impact Assessment, 2003. http://airnet.iras.uu.nl/

[32] World Health Organization, "Transport, Environment and Health," 2007. http://wwweurowhoint/document/e72015pdf

[33] World Health Organization, "Health Aspects of Transport Related Air Pollution," WHO, Copenhagen, 2005.

[34] World Health Organization, "Health Aspects & Risks of Transport Systems: The HEARTS Project," WHO, Copenhagen, 2006.

[35] Center for Environmental Engineering of Towns and Industrial Areas, Hanoi University of Civil Engineering, "Environmental Monitoring Data 2007," Hanoi, 2007.

[36] Minister of Transport of Vietnam, "Report on the Health Impacts of Air Pollution and Related Economical Impacts between 2007-2009 and Future Predictions for Hanoi and Ho Chi Minh Cities," Hanoi, 2010.

[37] J. Dixon, L. Scura, R. Carpenter and P. Sherman, "Economic Analysis of Environmental Impacts," 2nd Edition,

EarthScan, London, 1994.

[38] V. H. Vu, X. Q. Le, P. N. Ho and L. Hens, "Health Impact Assessment for Traffic Pollution (Particulate Matters) for Hai Phong City, Vietnam," In: V. Bhasin and C. Susanne, Eds., *Anthropology Today*: *Trends and Scope of* *Human Ecology*, Anthropologist Special Volume, No. 5, 2010, pp. 67-76.

17

Influence of Grazing Intensity on Soil Properties and Shaping Herbaceous Plant Communities in Semi-Arid Dambo Wetlands of Zimbabwe

E. Dahwa[1*], C. P. Mudzengi[1], T. Hungwe[2], M. D. Shoko[2], X. Poshiwa[3], S. Kativu[4], C. Murungweni[5]

[1]Department of Research and Specialist Services, Makoholi Research Institute, Masvingo, Zimbabwe; [2]Faculty of Agricultural Sciences, Great Zimbabwe University, Masvingo, Zimbabwe; [3]Marondera College of Agricultural Sciences, University of Zimbabwe, Marondera, Zimbabwe; [4]Department of Biological Sciences, University of Zimbabwe, Harare, Zimbabwe; [5]Department of Animal Production and Technology, Chinhoyi University of Technology, Chinhoyi, Zimbabwe.

ABSTRACT

Key issues of concern regarding the environmental impacts of livestock on grazing land are their effects on soil, water quality, and biodiversity. This study was carried out to determine how grazing intensity influences soil physical and chemical properties and occurrence of herbaceous plant species in dambo wetlands. Three categories of grazing intensity were selected from communal, small scale commercial and large scale commercial land. Dambos from the large scale commercial land functioned as the control. Data analysis included ANOVA and multivariate tests from CANOCO. There were significantly negative changes to soil nutrient status in communal dambos though with a higher number of rare taxa. Sodium, phosphorous, pH and infiltration rate were significant determinants of plant species occurrence. Overgrazing is threatening the productivity, stability, and ecological functioning of dambo soils in communal Zimbabwe. These dambos also require special conservation and management priorities as they contain a large number of rare plant species.

Keywords: Dambo Wetlands; Grazing; Soil Nutrients

1. Introduction

Key issues of concern regarding the environmental impacts of livestock on both public and private grazing lands are their effects on soil, water quality, riparian areas, and biodiversity [1]. The direct effects of livestock grazing on ecosystems are well known and include reduction in plant biomass, trampling of plants, including below ground parts and soil, nutrient inputs and bacterial contamination from dung and urine, introduction and dispersal of seeds and other propagules [1-4]. Properly managed grazing lands provide positive environmental benefits, including the provision of clean water supplies, the capacity to sequester atmospheric carbon (C), and the potential to maintain biodiversity [1].

Although many wetlands in the past have been degraded or destroyed as a result of inappropriate land use

or development pressures, more recently they have become the focus of intense conservation interest [5], particularly since the establishment in 1971 of the Convention on Wetlands of International Importance especially as Waterfowl Habitat (the "Ramsar Convention"). This convention promotes sustainable use of wetlands and provides a framework for the conservation of more than 1600 wetlands that have been nominated as internationally important on the basis of their ecological, botanical, zoological, limnological or hydrological values [6].

Dambo wetland is a small-scale environmental resource which is widespread in Africa's tropical plateau savannas [7]. The main area of dambo occurrence is located in Southern and Central Africa, with sporadic occurrence in central West and north Central Africa, south of the Sahara [8,9]. With the exception of a typically narrow (<150 km) humid zone around the southern and eastern margins of the subcontinent, the interior and western

---

*Corresponding author.

Influence of Grazing Intensity on Soil Properties and Shaping Herbaceous Plant Communities in
Semi-Arid Dambo Wetlands of Zimbabwe

181

margin are mostly drylands, with annual potential evaporation greatly exceeding annual precipitation [5]. Across southern Africa, wetlands (including more than 20 Ramsar listed sites) can be found in a variety of coastal and inland settings [5] but the emphasis here is on those wetlands that occur within the dryland interior. In Zimbabwe, wetlands are estimated to cover some 1.28 million hectares of the country's land surface. Some 20% of this wetland area lies in communal areas [10].

Dambo wetlands are highly sensitive to grazing pressure. As such, they are considered as useful environmental indicators of environmental pressures [8,11]. Land pressure has forced communities to concentrate on land and water availability irregardless of the state of these environmental resources [12].

A number of studies have been carried out on the geography and hydrology of dambos. These have focussed on social and agricultural importance of dambos in Zimbabwe [8,13-15]. However, only a few make any reference to important factors shaping the plant communities in wetlands, particularly in the dry ecological Natural Region (NR) of Zimbabwe [16].

Scientific interest in wetlands has also increased rapidly over the last few decades [3,7,14,17,18], largely in response to growing pressures for increased data to inform management decisions regarding conservation, rehabilitation or artificial construction of wetlands, and particularly in view of the potential adjustments that may result from global climate changes [5]. Despite the increased interest in research, there is still lack of scientific understanding especially of wetlands in the world's extensive drylands [5], a collective term that includes subhumid, semiarid, arid and hyperarid regions and incurporates almost 50% of the global land area and nearly 20% of the global population [6].

The objective of this study was to determine the extent at which grazing influences soil nutrient dynamics, interaction with plant community and how these influence occurrence of herbaceous plant species in wetlands. Also to determine if communal dambo wetlands are worthwhile to conserve. In order to make valid comparisons, dambos in the same catchment area under different management were compared. It is hypothesized that grazing in dambos results in changes to important soil physical and chemical properties which influence occurrence of plant species.

2. Materials and Methods

2.1. Description of the Experimental Site

The study area is located to the north of Masvingo town in Zimbabwe. The area falls within Natural Region IV (NR4) of the Zimbabwean ecological classification sys-

tem [19]. Altitude is 1204 m above sea level on latitude 19°50'S and longitude 30°46'E. NR4 is suitable for extensive farming, and receives an annual rainfall of 450 to 650 mm. Mean maximum temperature during summer is about 28°C, and the minimum temperature during winter is about 6°C. It is characterised by sandy soils with low organic matter and humus content, and consequently low fertility. Farming activities in the area are considered risky because of highly variable rainfall [20].

2.2. Dambo Wetland Selection

Three categories of dambos from contrasting land use history were identified based on frequency and severity of defoliation. Two sites per category were selected and were the ungrazed (UG), moderately grazed (MG) and continuously grazed (CG). The three categories were selected from communal, small scale commercial and large scale commercial land respectively. Ungrazed sites functioned as the control treatment. Aerial photographic maps and a Global Positioning Satellite (GPS) altitude-measuring unit were used as aids for selection. Using mileage, a motorbike was run along and across dambos to estimate their varying sizes.

2.3. Vegetation Sampling

At least three transects measuring 100 m or less were laid in each dambo, this depending on size of the dambo sampled (dambos ranged in size from 0.9 - 4.5 ha). All transects were laid perpendicular to the general dambo hydrologic gradient to capture any variations due to moisture gradient [3]. Transects were also laid at least 30 m from roads to minimise border effects. Within each dambo, vegetation was sampled from 0.5 m^2 quadrats systematically placed 10 m intervals along transects [21]. Sampling was started at the center and then transversing to the edge of the dambos. In each dambo, at least 30 subsamples were obtained. Edges of dambos were determined by sampling until the vegetation cover was >90% pasture grass cover [22]. A GPS unit was used to measure altitude and coordinate points within each sampling unit (quadrat). Variables recorded from each quadrat to determine species composition were: name of species (nomenclature followed [23]), erosion estimates (scale of 1 - 10, were 1 is no erosion while 10 is badly eroded).

2.4. Soil Sampling

For soil sampling, subsamples were collected along transects from the center of every third 0.5 m^2 quadrat. A minimum of 10 soil subsamples were collected from each dambo. A soil auger was used to collect soil at a depth of top 15 cm. Upon return from field, subsamples were air dried under shade for a few days and then stored in plastic bags for later analysis. Due to homogeneity of vegetation and soils within dambos, fewer subsamples were collected for analysis [24]. Samples were then analysed for bulk density, Na, P and pH (CaCl$_2$).

2.5. Water Infiltration Rate

A double ring infiltrometer was used to measure water infiltration rate. The instrument was assembled as outlined in the product manual [25]. The instrument was placed at every third quadrat (where a soil sample was taken). This also yielded a minimum of 10 subsamples from each dambo. The rate of infiltration was determined as the amount of water per surface area and time (cm/min) that penetrated the soil. Water percolation is initially fast, but reduces gradually to a constant value, and this is the infiltration rate [25].

2.6. Data Analysis

One way ANOVA from SPSS ver. 13 was used to test for significant differences in soil physical and chemical properties and also in vegetation attributes. Least Significance Difference was used to separate the means. Species data were converted to a presence absence matrix consisting of six dambos by 65 herbaceous plant species. Detrended correspondence analysis (DCA) and hierarchical canonical correspondence analysis from CANOCO ver. 4 were used to assess multivariate relationships between vegetation and environmental data and compared vegetation composition among different dambos. Monte Carlo tests were done to test for significance of soil properties in explaining the observed patterns. Multivariate methods provide a means to structure the data by separating systematic variation from noise [26,27].

3. Results

3.1. Soil Physico-Chemical Properties

Soil physical (**Table 1**) and chemical (**Figures 1** and **2**) properties varied significantly across the three grazing categories. Phosphorous, sodium and water infiltration were significantly higher under moderate grazing when compared to ungrazed and overgrazed sites. Bulk density, erosion and pH were significantly higher under continuous grazing while lower under both moderate and ungrazed.

3.2. Plant Species Relationships with Environmental Variables

The Eigenvalues for axis 1 (horizontally) and axes 2 (vertically) are 0.16 and 0.12 respectively and they explained a total variance of 64.1% of the observed variation (**Table 2**). The first synthetic gradient (axis 1) showed a significantly positive correlation of plants with Na and pH ($P < 0.05$). The second synthetic gradient showed a positive correlation of plants with water infiltration rate and bulk density while negatively with phosphorus and erosion (**Figure 3**). Monte Carlo tests showed significance of all axes in predicting the observed differences in plant species occurrence ($P < 0.05$).

3.3. Plant Species Occurrence across Different Dambos

The ordination diagram (**Figure 4**) is a cluster of species produced from HCCA. Results from the plot shows three clusters based on plant species presence or absence. Results indicate differences in species occurrence across sites. Sites with same species are indicated by forming a cluster. Continuously grazed sites had a significantly higher number of rare plant species ($P < 0.05$).

4. Discussion and Conclusion

4.1. Influence of Grazing on Environmental Variables

While a number of studies have investigated the effects of livestock on soil quality in a range of ecosystem types, very few have been conducted in dambo wetlands [28]. Other studies have also found that in old embanked salty marshes, trampling altered the soil structure (to lamellar structure indicative of compaction) reducing soil infiltration and preventing salt from being leached from the soil [28].

The impact on soils was reflected by differences in soil properties between the differently grazed dambos. Infiltration rates were significantly lower in continuously grazed sites as compared to moderately and ungrazed sites. Continuously grazed sites had significantly higher soil bulk density and higher erosion estimates as compared to other sites. Trampling and compaction of soils by livestock decreases water infiltration and increase runoff hence exposing the soil to erosion hazard [14]. Ex-

Table 1. Mean soil physical properties in dambos subjected to different grazing intensities.

Variable	Ungrazed	Grazing intensity Moderate	Continuous
Basic Infil. rate (cm/min)	0.53[c]	0.78[a]	0.24[b]
Erosion (estimate:1 - 10)	0.85[c]	1.72[a]	1.76[b]
Bulk density (g/cm^3)	1.10[a]	1.10[a]	1.22[b]

Means in rows with different superscripts are significantly different (P < 0.05).

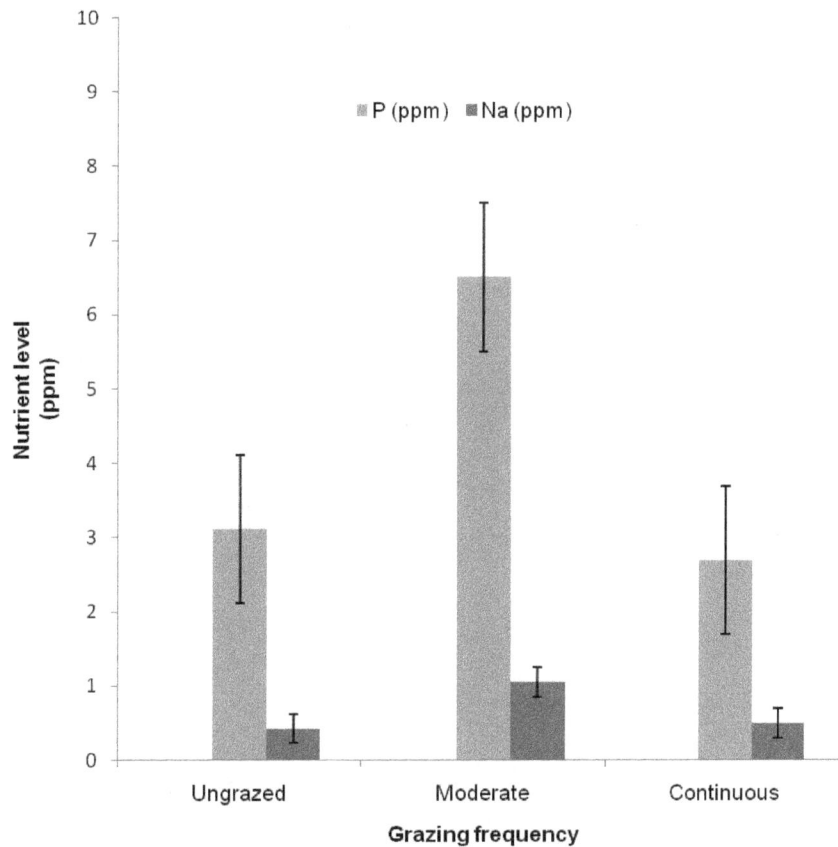

Figure 1. Mean soil phosphorus and sodium levels at sites under different grazing intensities. Values presented are ± standard deviation.

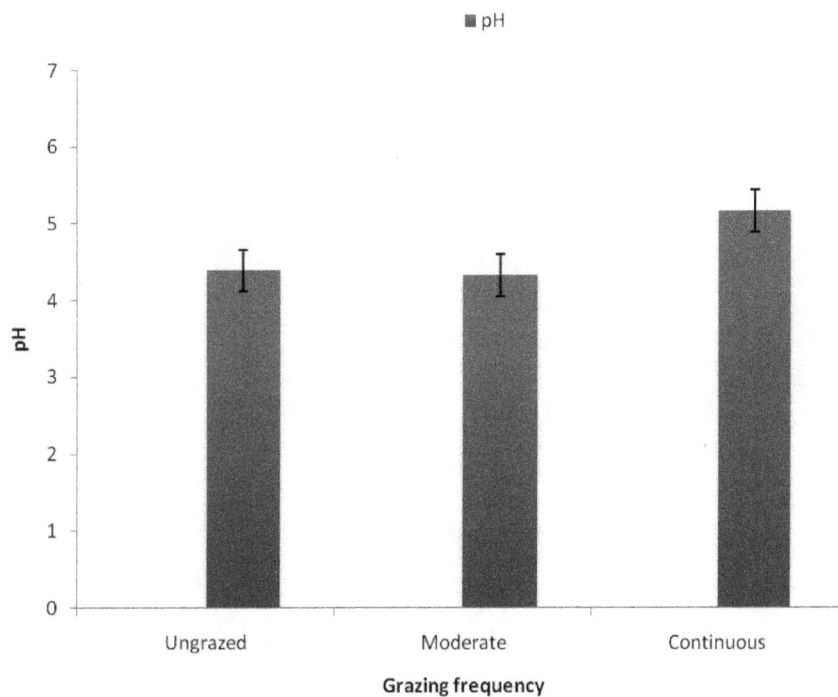

Figure 2. Mean soil pH levels across dambo sites subjected to different grazing management regimes. Values presented are ± standard deviation.

Table 2. Summary of eigenvalues for species environment correlations.

Axes	1	2	3	4	Total inertia
Eigen values	0.16	0.12	0.08	0.56	9.90
Species-environment correlations	0.62	0.54	0.43	0.00	
Cumulative % variance					
of species data:	1.40	2.40	3.00	7.8	
of species-environment relation:	37.0	64.10	81.80	81.90	

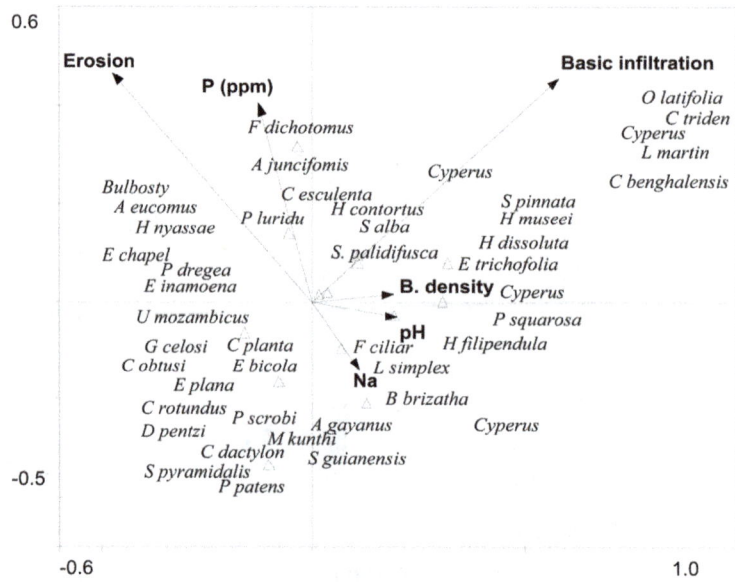

Figure 3. HCCA biplot indicates how environmental variables influence the occurrence of herbaceous plant species in dambos.

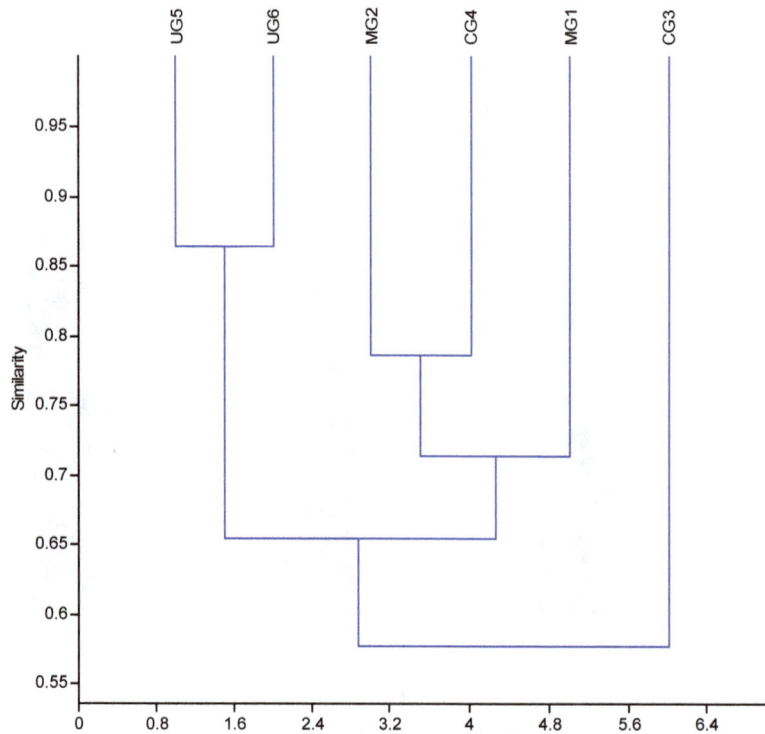

Figure 4. HCCA cluster based on species presence/absence in dambo under different grazing intensities.

Influence of Grazing Intensity on Soil Properties and Shaping Herbaceous Plant Communities in
Semi-Arid Dambo Wetlands of Zimbabwe

185

cessive compaction has negative consequences on plant growth [29]. Grazing increases bulk density or decreases soil porosity [1,30,31]. Even though soil changes can result from excessive compaction during grazing, studies also have shown that natural processes such as soil wetting and drying cycles and livestock can affect soil quality through compaction, erosion, and changes in the plant community [3,4]. Soil compaction by livestock is comparable to farm machinery and is most severe in the top 5 centimetres of the soil but can extend as deep as 30 cm [29]. The degree of compaction depends on soil moisture levels, type of soil and stocking densities. Maximum compaction occurs at soil moisture levels of between 20% to 30% moisture-holding capacity (depending on soil type) and field capacity, as well as at high stocking densities [1].

Dambos in Zimuto have been continuously grazed for over 6 decades with subsequent increase in livestock numbers over the years. Reduced water infiltration rates under continuous grazing suggest soil compaction and reduced pore size of soil. Soil deterioration and erosion is high under continuous grazing. This is evidenced by presence of grass species tolerant to heavy grazing like *Sporobolus pyrimidalis*, *Cynodon dactylon* and several *Eragrostis* species. We suggest that soil deterioration could be limited by reducing stocking intensities during dry periods in the communal areas. Similar results have been reported where water filtration rates on silty clay and silty clay loam soils were 2.5 times greater in an area grazed at 1.35 acres/AUM compared to an area grazed at 3.25 acres/AUM [1,32]. After 22 years of grazing at this intensity, not only had species composition altered but soil properties had been changed as well.

Continuously grazed sites had significantly higher soil pH values (**Figure 2**) as compared to the other sites. This possibly can be explained by the fact that desiccation is higher. As the water evaporates, it leaves the soil with a higher concentration of salts hence a slightly higher pH value as compared to the other sites. Generally dambos have increased agricultural fertility due to higher concentrations of potassium, phosphate and organic matter, and greater cation exchange capacities [7]. Dambo soil characteristics reflect the greater biomass production, lower decomposition rate and the inwash of ions from upslope [7]. In line with our findings, it was reported that calcareous fens had lower pH and higher nitrate (NO_3) levels, with no differences in ammonium between grazed and ungrazed sites [33]. These results suggest that manure inputs were being nitrified and that although the fens were accumulating nitrogen, there may be some resilience to increase in potentially toxic ammoniacal nitrogen levels.

4.2. Species-Environmental Relationships

Our results indicated strong vegetation-explanatory variable relationships in all studied dambo sites. A detrended correspondence analysis (DCA) was first carried out to ascertain on the behaviour of data [27]. Results of DCA indicated that most of the species behaved approximately unimoidal along environmental gradients as indicated by the length of maximum gradient which was greater than 4Sd [34]. Inorder to define the links between species and the environment, hierarchal canonical correspondence analysis (hCCA) followed.

The analysis of species environment relationships and the identification of indicator species are traditional activities in ecology [35]. Knowing how human activity influences the fascinating diversity of biological communities raises much interest in people. However, this very diversity creates problems for the statistical analysis of ecological observations [36]. This implies a large number of species and a large inherent variability. A set of community samples and associated environmental measurements typically yields an enormous amount of noisy data which is difficult to interpret.

By using hCCA, it has been shown that plant structuring and composition in the studied community are considerably influenced by individual effects of ecological factors. Sodium and pH were positively correlated with axis 1 (**Figure 3**) and as such are major gradients influencing the structuring and diversity of the plant communities. Other factors like infiltration rates, bulky density, phosphorous and erosion are linked to the second axis and are important as well. Any factors leading to changes in water filtration, pH and sodium levels in the dambo wetlands will lead to changes in herbaceous species composition and occurrence of these fragile ecosystems.

Weed species like *Oxalis latifolia*, *C. triden* and *C. bangalensis* occurred on well drained soil usually soils close to the top land. Soil moisture gradient could be important in determining or governing the occurrence of these plants. Any factors affecting the drainage conditions of wetlands e.g. infiltration rate will affect what species will occur there. The occurrence or abundance of a species along an environmental gradient often follows shelford's law of Tolerance [37]. Each species thrives best at a particular value (its optimum) and cannot survive when the value is either too low or too high. Each species occurrence is thus confined to a limited range, its niche. Species tend to separate their niche, partly so as to minimise competition. If the separation is strong, successive species replacements occur along environmental gradient. The composition of biotic communities thus changes along environmental gradients according to unimoidal functions [37].

Some individual factors may have a minor influence, for example the effects of erosion and litter cover and total species richness, but their interactions with each other or with other factors, such as litter accumulation, may have a substantial influence on total species diversity and composition. Consequently, knowledge about the role of individual factors only, and ignorance of their interactions, may lead to false predictions about community structure and function.

The cluster (**Figure 4**) indicates that communities of same ecological background are segregating based on species composition and soils. Plant species composition and structure can be altered indirectly through the modified soil environment and directly through trampling [29]. The overgrazed sites had a relatively higher number of rare plant species. This may indicate the need for higher priority for conservation in communal dambo wetlands.

4.3. Importance of Interactions

A number of biotic and abiotic factors have been proposed as determinants shaping plant species structure and diversity [38]. However, when considering the rates of species gain or loss in the community, most attention has been directed to the roles of disturbance, physical resources, species interactions and propagule availability. The failure to identify any single factor as the major determinant of species diversity on a more local scale suggests that interactions between several factors are often more important. There are significant strong correlations between sodium and pH, and also between bulk density and pH. The more factors that are involved in an experiment the more complex the possible interactions, and the more sophisticated the analytical techniques required to interpret them. Although a complexity of interactions between different factors has been assumed in plant community theory [17,37,38], few experiments have combined more than two factors to test such effects.

5. Conclusion

Continuous grazing is threatening the productivity, stability, and ecological functioning of dambo wetlands in communal areas. There are deleterious changes to soil nutrients which are important vegetation determinants in the dambos. There are evident changes in species composition among differently grazed sites. Communal dambos are less fertile and have a higher number of rare taxa, hence requires special conservation and management priority. Debate over grazing needs to move beyond the simple dichotomy of whether it is good or bad. Evaluation of practical alternatives should be done through experimental studies.

6. Acknowledgements

The authors gratefully acknowledge the field assistance of Chiuta T., Mutata L., Muraza I., Madokudya A. Anthony Mapaura from the National Herbarium, Harare, helped with the identification of some plants and Mr. Mutsambiwa from the Soils and Chemistry Laboratory of the University of Zimbabwe carried out the soil analyses. Project went well with financial assistance from the Ministry of Agriculture mechanisation and Irrigation Development. We would like to thank Proffesor CHD Magadza for his input and comments earlier on.

REFERENCES

[1]   CAST (Council for Agricultural Science and Technology), "Environmental Impacts of Livestock on US," *Grazing Lands*, Vol. 6, 2002.

[2]   G. F. Hayes and K. D. Holl, "Cattle Grazing Impacts on Annual Forbs and Vegetation Composition of Mesic Grasslands in California," *Conservation Biology*, Vol. 17, No. 6, 2003, pp. 1694-1702.

[3]   W. M. Jones, L. H. Fraser and J. Curtis, "Plant Community Functional Shifts in Response to Livestock Grazing in Intermountain Depressional Wetlands in British Columbia, Canada," *Biological Conservation*, Vol. 1, No. 144, 2010, pp. 511-517.

[4]   J. T. Marty, "Effects of Cattle Grazing on Diversity in Ephemeral Wetlands," *Conservation Biology*, Vol. 16, 2004, pp. 1626-1632.

[5]   S. Tooth and T. S. McCarthy, "Wetlands in Drylands: Geomorphological and Sedimentological Characteristics, with Emphasis on Examples from Southern Africa," *Progress in Physical Geography*, Vol. 31, No. 1, 2007, pp. 3-41.

[6]   SCBD (Secretariat of the Convention on Biological Diversity), "Global Biodiversity Outlook," SCBD, Montréal, Vol. 3, 2010.

[7]   I. Scoones, "Wetlands in Drylands: Key Resources for Agricultural and Pastoral Production in Africa," *Ambio*, Vol. 8, No. 20, 1991, pp. 366-371.

[8]   N. Roberts, "Dambos in Development of a Fragile Ecological Resource," *Journal of Biogeography*, Vol. 15, No. 1, 1988, pp. 141-148

[9]   J. C. Von der Heyden, "The Hydrology and Hydrogeology of Dambos: A Review," *Progress in Physical Geography*, Vol. 28, No. 4, 2004, pp. 1-21.

[10] M. Mharapara, M. D. Munema and R. Mkwanda, "Wetland Characterization and Classification for Sustainable Agricultural Development. FAO, Zimbabwe Country Paper Experiences on Wetland Characterisation, Classification, Management and Utilization for Agricultural Development in Zimbabwe: A Case for Wetland Research," *FAO*, Vol. 1, No. 132, 1997.

[11] D. Limpitlaw and G. Rudigergens, "Dambo Mapping for Environmental Monitoring Using Landsat TM and SAR

Influence of Grazing Intensity on Soil Properties and Shaping Herbaceous Plant Communities in
Semi-Arid Dambo Wetlands of Zimbabwe

187

Imagery: Case Study in the Zambian Copperbelt," *International Journal of Remote Sensing*, Vol. 27, No. 21, 2006, pp. 4839-4845.

[12] S. Mtetwa and C. F. Schutte, "An Interactive and Participative Approach to Water Quality Management in Agro-Rural Watersheds," Water SA, No. 28, 2002.

[13] M. Bell and N. Roberts, "The Political Ecology of Dambo Soil and Water Resources in Zimbabwe," *Transactions of the Institute of British Geographers, New Series*, Vol. 16, No. 3, 1991, pp. 301-318.

[14] A. Bullock and M. P. McCartney, "Wetland and River Flow Interactions in Zimbabwe. L'hydrologie Tropicale: géOscience et Outil Pour le Development," IAHS, No. 238, 1996.

[15] R. Whitlow, "A Review of Dambo Gullying in South-Central Africa," XVI Department of Geography, University of Zimbabwe, Zambezia, 1989.

[16] J. M. Bullock, J. Franklin, M. J. Stevenson, J. Silvertown, S. Coulson, S. J. Gregory and R. Tofts, "A Plant Trait Analysis of Responses to Grazing in a Long Term Experiment," *Journal of Applied Ecology*, Vol. 38, No. 2, 2001, pp. 253-267.

[17] P. A. Keddy, L. H. Fraser and T. A. Keogh, "Responses of 21 Wetland Species to Shortages of Light, Nitrogen and Phosphorus," *Bulletin of the Geobotanical Institute*, Vol. 67, 20011, pp. 3-25.

[18] P. N. Reeves and P. D. Champion, "Effects of Livestock Grazing on Wetlands," National Institute of Water and Atmospheric Research Ltd., No. EVW042732004, 2004.

[19] V. Vincent and R. G. Thomas, "An Agricultural Survey of Southern Rhodesia," Agro-Ecological Survey, Vol. 1, 1961.

[20] T. Mutimukuru, G. D. Mudimu, S. Siziba, N. Harford and C. Garforth, "Demand Assessment Study of Resource Conserving Technologies in Chivi and Zimuto Communal Areas in Masvingo Province, Zimbabwe," DFID, Natural Resources Systems, Masvingo, 2000.

[21] J. T. Curtis and R. P. McIntosh, "The Interrelations of Certain Analytic and Synthetic Phytosociological Characters," *Ecology*, Vol. 31, No. 3, 1950, pp. 434-455.

[22] E. H. Boughton, A. Quintana, P. J. Bohlen, D. G. Jenkins and R. Pickert, "Land Use and Isolation Interact to Affect Wetland Plant Assemblages," *Ecogeography*, Vol. 33, 2010, pp. 461-470.

[23] H. Wild, "A Rhodesian Botanical Dictionary of African and English Plant Names," National Herbarium, Department of Research and Services, Ministry of Agriculture, Government printer, Salisbury, 1972.

[24] M. L. Amanda, R. G Glenn and T. F. H Allen, "Conceptual Hierarchical Modelling to Describe Wetland Plant Community Organization," *Wetlands*, Vol. 30, No. 1, 2010, pp. 55-65.

[25] Eijkelkamp, "The Double Ring Infiltrometer," Product Manual, 2nd Edition, 1983.

[26] H. G. Gauch, "Multivariate Analysis in Community Ecol-

ogy," Cambridge University Press, Cambridge Cambridgeshire and New York, 1982.

[27] K. McGarigal, S. Cushman and S. Stafford, "Multivariate Statistics for Wildlife and Ecology Research," Springer-Verlag, New York, 2000.

[28] R. R. Blank and T. Morgan, "Influence of Livestock Grazing, Floodplain Position and Time on Soil Nutrient Pools in a Sierra Nevada Montane Meadow," *Soil Science*, Vol. 6, No. 75, 2010, pp. 293-302.

[29] K. L. Greenwood and B. M. McKenzie, "Grazing Effects on Soil Physical Properties and the Consequences for Pastures: A Review," *Australian Journal of Experimental Agriculture*, Vol. 8, No. 41, 20011, pp. 231-250.

[30] V. S. Baron, A. C. Dick, E. Mapfumo and S. S. Malhi, "Grazing Impacts on Soil Nitrogen and Phosphorus under Parkland Pastures," *Journal of Range Management*, Vol. 54, No. 6, 2001, pp. 704-710.

[31] S. D. Warren, W. H. Blackburn and C. A. Jnr. Taylor, "Effects of Season and Stage of Rotation Cycle on Hydrologic Condition of Rangeland under Intensive Rotation Grazing," *Journal of Range Management*, Vol. 6, No. 39, 1986, pp. 486-491.

[32] J. Kauffman and B. Krueger, "Livestock Grazing Effects on Soil Physical Properties," *Journal of Range Management*, Vol. 5, No. 6867, 1984.

[33] D. Van Hoewyk, P. M. Groffman, E. Kiviat, G. Mihocko and G. Stevens, "Soil Nitrogen Dynamics in Organic and Mineral Soil Calcareous Wetlands in Eastern New York," *Soil Science Society of America*, Vol. 64, No. 6, 2000, pp. 2168-2173.

[34] T. C. J. F. Braak, "Canonical Correspondence Analysis: A New Eigenvector Technique for Multivariate Analy- sis," *Ecology*, Vol. 5, No. 67, 1986, pp. 1167-1179.

[35] A. Petraglia and M. Tomaselli, "Ecological Profiles of Wetland Plant Species in the Northern Apennines (N. Italy)," *Journal of Limnology*, Vol. 62, No. 1, 2003, pp. 71-78.

[36] T. C. J. F. Braak and P. F. M. Verdonschot, "Canonical Correspondence Analysis and Related Multivariate Methods in Aquatic Ecology," *Aquatic Sciences*, Vol. 6, No. 57, 1995, pp. 255-289.

[37] P. A. Keddy, "Effects of Competition from Shrubs on Herbaceous Wetland Plants: A Four Years Field Experiment," *Canadian journal of Botany*, Vol. 67, No. 3, 1989, pp. 708-716.

[38] S. Xiong, M. E. Johansson, F. M. R. Hughes, A. Hayes, K. S. Richards and C. Nilsson, "Interactive Effects of Soil Moisture, Vegetation Canopy, Plant Litter and Seed Addition on Plant Diversity in a Wetland Community," *Journal of Ecology*, Vol. 91, No. 6, 2003, pp. 976-986.

18

Anthropogenic Activities and the Degradation of the Environmental Quality in Poor Neighborhoods of Abidjan, Côte d'Ivoire: Abia Koumassi Village

S. Loko[1], K. E. Ahoussi[2], Y. B. Koffi[2], A. M. Kouassi[3], J. Biémi[2]

[1]Laboratoire de chimie et de Microbiologie, Université Technologique et Tertiaire Loko (UTTLOKO), Abidjan, Côte d'Ivoire; [2]Unité de Formation et de Recherche (UFR) des Sciences de la terre et des Ressources Minières (STRM), Université de Cocody, Cocody, Côte d'Ivoire; [3]Département des Sciences de la Terre et des Ressources Minières (STeRMi), Institut National Polytechnique Félix Houphouët Boigny (INP-HB), Yamoussoukro, Côte d'Ivoire.

ABSTRACT

This study deals with the degradation of the quality of the water environment in the village of Abia Koumassi, due to the pollution that has risen in Abidjan. The method used in this study is based on piezometric measurements, the physico-chemical and microbiological analysis. The results were processed using statistical and hydrochemical methods. The groundwater in the village is shallow, with a piezometric average level 0.55 m. The groundwater flows from the north of the village to the south. The Water resources have a neutral pH that varies between 6.8 and 7.43. Water temperature varies from 27.7°C to 29.8°C. The Water is highly mineralized, with electrical conductivity ranging from 585 µS/cm to 1310 µS/cm. The groundwater contains high levels of nitrate (116.81 mg·L^{-1}) greater than the WHO standard for drinking water. High levels of Metallic Trace Elements (Ni, Zn, Co, Cr, Pb, Fe, Cu and Cd) are found in the water. Microbiological analysis shows that the water contains important levels of Escherichia coli, faecal streptococci, Clostridium perfringens and thermo tolerant coliform. These microorganisms create microbiological pollution in the water from the area. The Water resources of the village are facing a recent faecal pollution of human origin. This pollution comes from anthropogenic activities taking place in the area.

Keywords: Anthropogenic Activities; Environment; Pollution; Water Sanitation and Quality

1. Introduction

Abidjan where the village of Abia Koumassi is located is currently experiencing a population growth, rapid Industrialization and urbanization and unplanned growth. This change causes so many problems that the authorities must solve. Among these problems, we can mention the pollution of water resources. Many industrial and domestic activities are probably a real danger to the urban environment.

In this locality, the various districts do not have adequate provisions because all the sewage and sanitation system of Abidjan is insufficient, with many gaps. According to [1], the rate of household connections to the sewerage system is limited to 29%. The uncollected wastes (45%) are found directly in the nature involving the risk of deterioration of groundwater quality. The groundwater in the city of Abidjan is subject to various pollution prob-

lems. Studies by [2-4] showed significant levels of nitrate, fecal coliform and lead in the groundwater from Abidjan. The study of [5] about water and sediments of the Ebrié lagoon showed also high levels of Metallic Trace Elements in the sediments of the same lagoon. According to these authors, this lagoon is subject to anthropogenic pollution due to the discharge of domestic and industrial sewage untreated or inadequately treated. Metallic Trace Elements are factors of important pollution in aquatic ecosystems [6] because of their toxicity and ability to accumulate in biota. According to these authors, one of the important properties of Metallic Trace Elements results from production activities and human consumption. These elements are not biodegradable.

The development of industrial activities notably the sector of oil field in Côte d'Ivoire in the south of Abidjan is not without risk for groundwater pollution. The Ground-

Anthropogenic Activities and the Degradation of the Environmental Quality in Poor Neighborhoods of Abidjan, Côte d'Ivoire: Abia Koumassi Village

189

water from this area is supplied by direct infiltration of rainwater. The stormwater infiltration is favored by a high coefficient of infiltration and a low level of water from the Quaternary [7]. However, a large part of the population, especially in the south of Abidjan in general and in the village of Abia Koumassi in particular uses water of the shallow aquifer for domestic needs. In all the world, water pollution is one of the most crucial problems which every country must solved. To date, many researchers have conducted extensive surveys of water resources contamination [8-10]. The results demonstrated that the pollution of water results from anthropogenic activities of different regions. This study aims to assess environmental risks linked to human activities on the quality of the water in the village of Abia Koumassi.

2. Material and Methods

2.1. Study Area

Abidjan is located in the south of Côte d'Ivoire, between latitudes 5°00' and 5°30' North and longitudes 3°50' and 4°10' West. It consists of ten municipalities, including that of Koumasi where the village of Abia Koumasi is located (**Figure 1**). Abia Koumasi is located in the south of the town of Port-Bouet and has about 4000 inhabitants. People who live in this village are mostly indigenous Ebrié, the Ivorian immigrants from other regions of Côte d'Ivoire. In addition to these populations, there are alien from the Economic Community of West African States and other countries bordering Côte d'Ivoire. The geology of the village of Abia Koumassi is the same with the

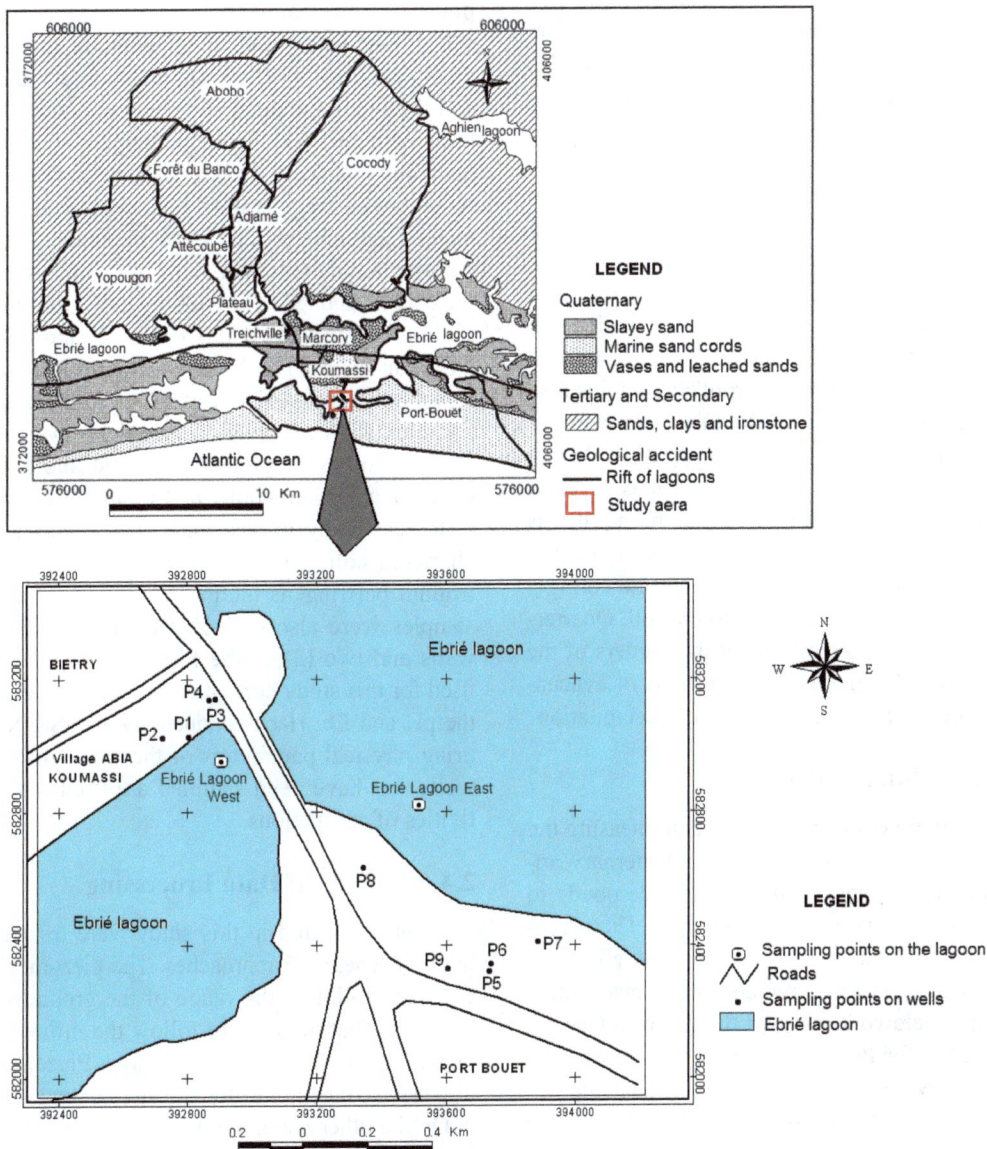

Figure 1. Presentation of the study area and the location of sampling points of water.

Southern District of Abidjan, located on the coastal sedimentary basin of Côte d'Ivoire. The bedrock geological of the District of Abidjan consists of two main types of rocks: Precambrian basement and coastal sedimentary basin. In the Abidjan area, outcrops of Precambrian are rare. It is only found in Anyama and North-East Attiékoi.

The Continental Terminal, formation of Mio-Pliocene age is characterized by a lenticular stratification, coarse sands, clays of variegated ferruginous sandstone and iron ore. On the paleogeograhic, from the Cretaceous to the Quaternary three episodes of transgression are well known: Albo-Aptian: clays and sandstones of the Lower Cretaceous, Maastrichtian-Eocene: glauconitic clays, sands and clays and lower Miocene: black marl relics sharks, colorful clays and lignites. During periods of transgression, marine sediments are clays, marl, sandstone, sand and limestone shells, or lumachelliques. It is also found in these formations Foraminifera, Ammonite and Nautilus features, notably on the beaches of Fresco. However, these episodes of marine sedimentation interspersed with continental phases related to a decline in the sea. From the structural point of view, the ivorian sedimentary basin is crossed by a large fault called the East-West fault lagoons south dipping, with a rejection reaching 5000 m.

In the area, the Quaternary aquifer contains the most vulnerable water in Abidjan. The Quaternary aquifer has from top to bottom four (4) horizons. Formations 1 and 2 are generally marly clay and waterproof. It contains also two types of groundwater. The Oogolien groundwater grows mainly in fine to coarse sand of layer 3, while that of Nouakchottian lays in coarse marine sands layer 4. The piezometric level of groundwater in the aquifer is shallow and varies from 0 to 1 m. The greatest fluctuation is about 1 m depending on the season, indicating a sizable infiltration rate of rainwater into the soil. Quarternary sources are quite numerous along the borders of the Ebrié lagoon, which highlights the conditions of evacuation of part of the excess water provided by precipitation.

2.2. Piezometric Study Method

The aim of the hydrogeological study was to measure the piezometric level of groundwater of the Quaternary aquifer in the month of June 2010, which corresponds to the rainy season in the large city of Abidjan. The piezometric campaign helped in the inventory of all water points in the area. The measurements were made on 9 hand dug wells. Fieldwork required the use of a Garmin GPS Map 60 CSX for the identification of geographical coordinates in UTM different wells taken as piezometers. This study was performed using a probe piezometric OTT light and sound (100 m) to measure the water depth and a digital camera Sony for taking pictures.

2.3. Water Sampling Method

A hydrochemical and microbiological sampling of the water resources of the village Abia Koumassi was conducted in june 2010. This period of water sampling in Abidjan was the rainy season. The samples were focused on well water and water from the Ebrié lagoon. During this campaign, the water samples for chemical analysis were collected in polyethylene bottles of 1 liter capacity, previously washed with nitric acid and then with distilled water. On the field, before filling the bottles, they were washed three times with water to collect.

Bottle filling was done to the brim and the screw cap to prevent gas exchange with the atmosphere. Water samples were then transported in a cooler at 4°C for laboratory for analysis in the hour of sampling. During sampling, the physical parameters of water (the Ebrié lagoon and groundwater) such as temperature (T), pH and electrical conductivity (EC), salinity, dissolved oxygen (O_2) and the redox potential (Eh) were measured. In total, we have collected for chemical analysis 10 samples with 8 hand dug wells and 2 Ebrié lagoon. On the water samples, two types of analysis were performed. It is the analysis of major ions and metallic trace elements. For major ions, elements analyzed are Ca^{2+}, Mg^{2+}, Na^+, K^+, NH_4^+, Fe^{2+} for cations and SO_4^{2-}, HCO_3^-, NO_3^- for anions. Metals such as Mn, Ni, Zn, Cd, Cu, Pb, Cr and Co were determined in water. The Metallic Trace Elements were analyzed using an atomic absorption spectrophotometer (AAS). Samples for bacteriological analysis were collected in 500 ml bottles previously sterilized and kept free from contamination in a cooler at 4°C. In these samples, bacteriological analysis has concerned bacteria such as Escherichia coli, faecal streptococci, coliforms and Clustridium perfringens thermo tolerant. A total of 10 water samples were also collected for this analysis, eight (8) wells and two (2) of the Ebrié lagoon. Field equipment used for this study consists of 4 Star pH meter to measure the pH and Eh, Hach conductivity SENSION 5 for measuring physical parameters of the water. In this study, all wells that have been sampled are used for domestic activities of populations.

2.4. Methods of Data Processing

The data collected in this study were treated from three hydrogeochemical approaches. The piezometric data were collected to know the range of the groundwater level in the area. That helps us to follow the influence of anthropogenic activities in the study area. Piezometric data are also used to determine groundwater flowing main direction.

On the other hand, the data processing involved a combination of statistical methods mutlivariate and hydrochemical method. The hydrochemical approach required

Anthropogenic Activities and the Degradation of the Environmental Quality in Poor Neighborhoods of Abidjan, Côte d'Ivoire: Abia Koumassi Village

191

the use of the triangular diagram of Piper for the classification of the water studied. This method allows a group of water points in key classes that form hydrofacies. The statistical method used is the cluster analysis. It allows the study of phenomena at the origin of the mineralization of the water. Cluster analysis is a powerful tool to analyze data for water chemistry and formulation of geochemical models [11]. This classification system that uses the Euclidean distance for similarity measures and the method of guardianship link that produces the most distinctive classification where each member in a group is more similar to his colleagues that any member outside the group is used in hydrogeology by authors such as [12-14]. This analysis includes 10 descriptors and 18 variables are: salinity (Sal.), electrical conductivity (EC), pH and ions such as Fe^{2+}, Ca^{2+}, Mg^{2+}, Na^+, K^+, NO_3^-, HCO_3^-, SO_4^{2-} and the metallic trace elements Mn, Ni, Zn, Cd, Cu, Pb, Cr and Co. It was conducted using the STATISTICA 6.0 software. Those methods used in this study allowed to know the mechanism of mineralization of water in the village of Abia Koumassi and their relationship with human activities.

3. Results

3.1. Study of Flow Directions of Groundwater

The piezometric measurements carried out in the village

of Abia Koumassi are shown in **Table 1**. In the area, groundwater is shallow. The piezometric level of the groundwater measured varies from 0.13 m to 0.79 m, with an average of 0.55 m. Piezometric study shows that groundwater flow from north to south. Thus the main direction of groundwater flowing during the pollution is from the village of Abia Koumassi to the sea downstream.

3.2. Hydrochemical Study

The results of measurements of physical parameters of groundwater and lagoon water measured are listed in **Table 2**. The temperature of Ebrié lagoon varies from 29°C to 29.3°C, with an average of 29.15°C ± 0.15°C. Groundwater temperature ranging between 27.7°C and 29.8°C, with an average of 28.73°C ± 0.26°C. The pH of Ebrié lagoon water varies from 7.26 to 7.43, with an average of 7.35 ± 0.08. Groundwater sampled from hand dug wells has a pH ranging between 6.8 and 7.62, with an average of 7.15 ± 0.09. That indicates that water of the village has low acidity.

The redox potential of the water varies from −5.06 to −4.08 mV, with an average of −4.57 ± 0.49 mV for water from Ebrié lagoon. For the groundwater, the Eh varies from −22.60 mV to −72.70 mV, with an average of −42.72 ± 5.38 mV. The values of redox potential show that water of the village is in reducer environment. This

Table 1. Piezometric data collected in the field.

Measures points	X (m)	Y (m)	Level piezometric (m)	Water coast (m)	depth (m)	Water film = P-H (m)
P1	392805	583026	0.13	0.87	0.95	0.82
P2	392726	583022	0.695	0.305	1.465	0.77
P3	392890	583139	0.63	0.37	1.19	0.56
P4	392870	583136	0.49	0.51	1.48	0.99
P5	393732	582319	0.64	0.36	1.7	1.06
P6	393738	582342	0.79	0.21	1.68	0.89
P7	393885	582407	0.35	0.65	1.32	0.97
P8	393347	582327	0.76	0.24	1.4	0.64
P9	393606	582327	0.44	0.56	2.09	1.65

Table 2. Summary of *in situ* measurements of the physical parameters of water.

Measures points	Sal ‰	CE (μS/cm)	TDS (mg·L⁻¹)	pH	T°C	Dissolve O₂ (mg·L⁻¹)	Eh (mv)
Lagoon East	3.8	6980	3700	7.26	29	4.57	−4.08
Lagoon West	4.2	7670	4100	7.43	29.3	4.25	−5.06
P1	0.3	585	284	7.17	29.8	2.3	−34.6
P2	0.4	784	383	6.8	29.8	3.72	−22.6
P3	0.3	671	326	7.14	28.7	4.11	−43.6
P4	0.4	762	372	7.62	27.7	3.9	−72.7
P5	0.5	1075	529	7.3	28.4	4.1	−53.2
P6	-	1310	741	7.04	28.7	1.72	−37.8
P7	0.6	1197	590	7.17	28.4	2.13	−45.2
P8	0.4	878	429	6.95	28.3	3.71	−32.1

is also reflected by the low dissolved oxygen values observed in water. Dissolved oxygen values of the Ebrié lagoon range between 4.25 mg·L^{-1} and 4.57 mg·L^{-1}, with an average of 4.41 ± 0.16 mg·L^{-1}. The Groundwater has dissolved oxygen values which vary from 1.72 to 4.11 mg·L^{-1}, with a mean of 3.21 ± 0.35 mg·L^{-1}.

The Water from the village is highly mineralized, with an electrical conductivity average 7325 ± 99.50 µS/cm for the water of the Ebrié lagoon and 907.75 ± 10.26 µS/cm for the groundwater. The average salinity is 4.00‰ ± 0.28‰ for water from Ebrié lagoon and 0.40‰ ± 0.04‰ for groundwater. Total dissolved solids are in average 333.50 ± 49.50 mg·L^{-1} for water from the lagoon and 497.83 ± 62.90 mg·L^{-1} for hand dug wells water.

The results of chemical analysis carried out on groundwater and the lagoon water from Abia village Koumassi are recorded in **Table 3**. The Water sampling contains high concentrations of Fe^{2+}. The values vary from 0.07 mg·L^{-1} to 12.8 mg·L^{-1}, with an average of 17.08 ± 15.86 mg·L^{-1} for the groundwater. These high concentrations (12.8 mg·L^{-1}) occur at the water from a well located in a former iron scrap. In addition, the groundwater contains high levels of nitrate, which vary between 0.19 mg·L^{-1} and 116.81 mg·L^{-1}, with a mean of 32.86 ± 15.66 mg·L^{-1}. These high nitrate levels are a greater danger to the health plan for the population using well water for their domestic needs.

The analysis results of metallic trace elements contained in the water samples are shown in **Table 3**. Analysis shows that manganese is virtually absent in the water. Groundwater and lagoon water contain also Ni (0.072 mg·L^{-1} and 0.085 mg·L^{-1} for water lagoon), Zn (0.077 mg·L^{-1} for well water), Pb (0.004 mg·L^{-1} for well water), Cr (0.649 mg·L^{-1} for lagoon water and 0.613 mg·L^{-1} for well water), Co (0.115 mg·L^{-1} for lagoon water and 0.361 mg·L^{-1} for well water).

3.3. Hydrochemical Classification of Water

Water Classification from Piper diagram identifies two types of water in the area (**Figure 2**). In the area, the most important anion in water is the sulphates and the greatest cation is calcium.

3.4. Multivariate Statistical Study: Cluster Analysis

The analysis from the Ascending Hierarchical Classification gave the graph below. The dendrogram (**Figure 3**) highlights two main groupes of variables. The first consists of Mg^{2+}, Fe^{2+}, K$^+$, NO$_3^-$, SO$_4^{2-}$, pH, EC, HCO$_3^-$, Ca^{2+} and Na$^+$. This group consists of major ions whose dissolution is governed by the contact of water with the substratum. This shows the intervention of the geology in the mineralization of water in the village of Abia Koumassi. However, the presence of nitrate in this group highlights the influence of nitrogen mineralization after domestic activities in water. The second group consists of trace metal elements (Co, Cu, Cd, Ni, Mn, Zn, Cd and Pb) which dissolved in water is related to urban anthropogenic activities. This highlights the influence of human activities in the mineralization of water resources in the study area, therefore the degradation of the physico-chemical quality of water from the village of Abia Koumassi. The classification of water points according to their similarity to part of a follow-up is given in **Figure 4**. This dendrogram shows three clusters of water points. The first group contains the water of the Ebrié lagoon (Lag. E and Lag. W). The second are wells P1, P2, P3, P4, P5, P7 and P8 and the third only represented by the well P6. In the context of monitoring water quality, it is important to choose a track point in each cluster, three monitoring points are needed.

Table 3. Results of chemical analysis of water from Abia Koumassi expressed in mg·L^{-1}.

Sampling points	Fe^{2+}	Ca^{2+}	Mg^{2+}	Na$^+$	K$^+$	NO$_3^-$	HCO$_3^-$	SO$_4^{2-}$	Mn	Ni	Zn	Cd	Cu	Pb	Cr	Co
Lagoon East	0.20	78.40	63.40	200.70	47.80	0.27	31	270	<0.001	0.072	0.011	<0.001	0.026	<0.001	0.591	0.115
Lagoon West	0.25	152	130	202.60	54.30	0.66	52	225	<0.001	0.085	0.012	0.003	0.023	<0.001	0.649	0.109
P1	0.62	80	30	26.50	22.50	0.65	105	25	<0.001	0.074	0.018	0.019	0.025	0.004	0.001	0.126
P2	0.27	102	10.80	72.11	20.30	116.81	23	58	<0.001	0.083	0.077	<0.001	0.026	<0.001	0.094	0.002
P3 Ste Thérèse church	0.38	30	25	70.80	25.60	67.79	103	65	<0.001	0.018	0.03	<0.001	0.021	<0.001	0.380	0.015
P4 Chief	0.07	80	24	79.60	32.01	5.761	140	39	<0.001	0.003	0.013	<0.001	0.020	<0.001	0.34	0.022
P5 Akwuaba	0.19	200	15.30	75.8	45.10	6.99	43	265	<0.001	0.014	0.004	0.005	0.013	0.001	0.474	0.149
P6 old storage of iron	128	450	280	203.40	88.90	0.19	296	350	0.687	0.145	0.021	0.013	0.03	0.002	0.613	0.361
P7 Airport	5.60	32	28.80	56.80	35.99	1.05	180	200	0.07	0.059	0.217	0.027	0.02	0.003	0.277	0.122
P8 Washing	1.50	140	30	65.50	35.10	63.65	113	90	<0.001	0.03	0.054	0.013	0.025	0.001	0.43	0.209

Anthropogenic Activities and the Degradation of the Environmental Quality in Poor Neighborhoods of Abidjan, Côte d'Ivoire: Abia Koumassi Village

193

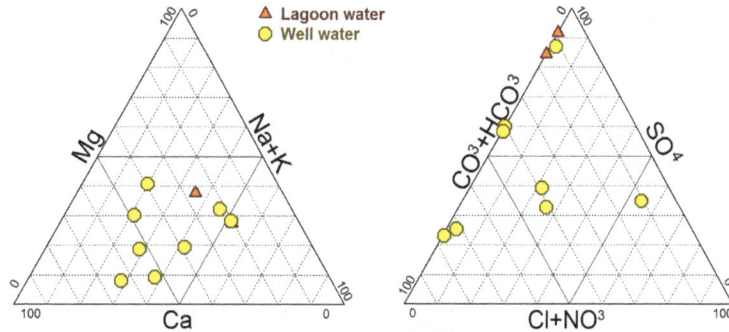

Figure 2. Classification of ions from water of Abia Koumassi village.

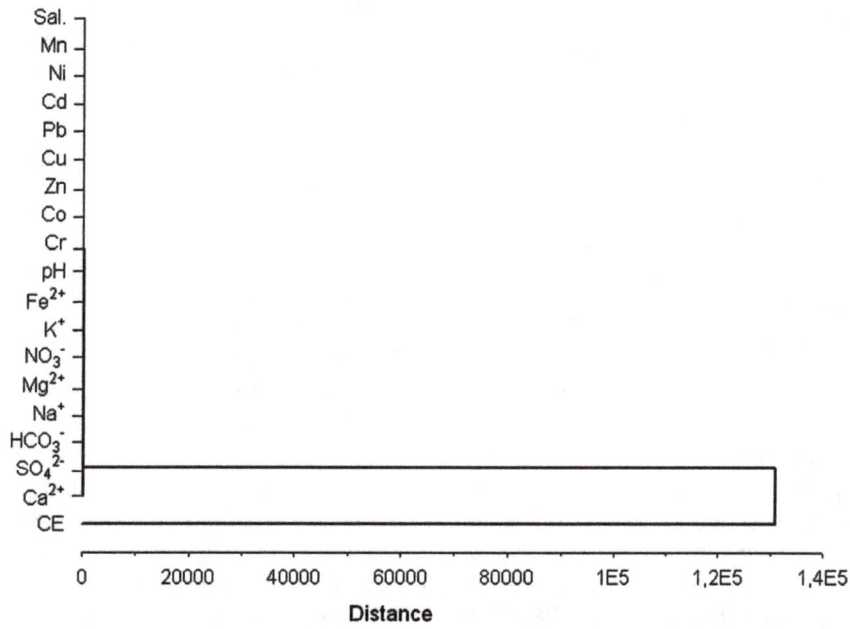

Figure 3. Cluster analysis of physico-chemical parameters of water from Abia Koumassi.

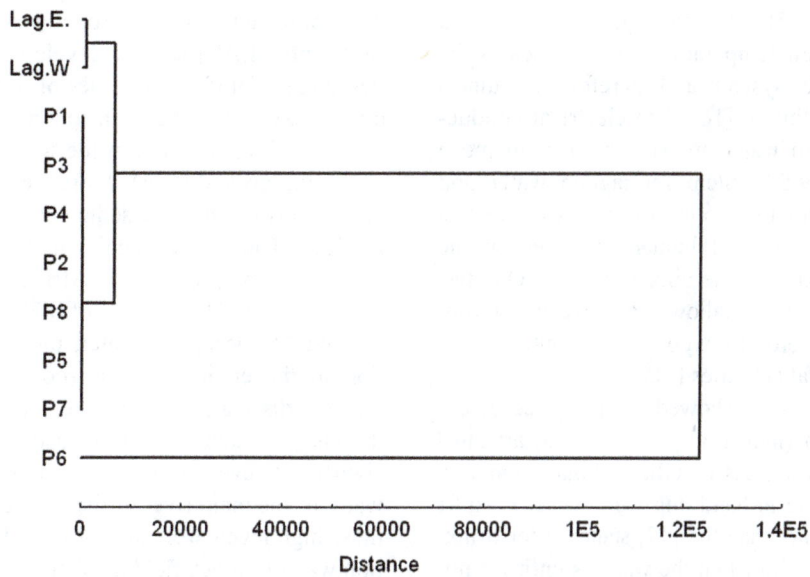

Figure 4. Cluster analysis of water points in the village of Abia Koumassi.

3.5. Study Microbiological Water

The microbiological results of water analysis from Abia Koumassi show the presence of fecal coliforms mainly *Escherichia coli* in the water studied. This is particularly the water of the Ebrié lagoon and hand dug wells P3, P4, P5 and P8. The analysis indicates high rate of faecal streptococci in groundwater. These include wells P1 and P5 respectively (1000 and 330 faecal streptococci per 100 ml of water). The water level of the lagoon is about 240 and 210 faecal streptococci per 100 ml of water. Thermo-tolerant coliforms are the most important in the area, with quantities of 35,000 and 110,000 for lagoon water. For well water, we have more than 5000 - 29,000 thermo tolerant coliform per 100 ml of water in wells P3, P4 and P5.

Clustridium perfringens occur in the water from the Ebrié lagoon the west side (120 Clustridium Perfringens/ 20ml water). In well water, they are present in wells P1 and P2, respectively 120 and 60 Clustridium Perfringens/ 20ml water. Different bacteria in the water of the village of Abia Kumasi show recent water microbiological pollution of human origin, therefore the influence of water by human activities in the study area.

4. Discussion

The Water studied in the village of Abia Koumassi, consist of surface water (water of the Ebrié lagoon) and groundwater (hand dug well water). The physical parameters of water measurement on the field show that the pH is near the neutrality and ranges between 6.8 and 7.43. The Water has a temperature that varies from 28.3°C to 29.8°C. These values fall within the range of values observed temperatures in the water of coastal sedimentary basin by Ivorian [2,14,15]. These temperatures correspond to seasonal ambient air temperatures. This indicates the opening of the aquifer system, and therefore its vulnerability in front of pollution [16]. The electrical conductivities measured from water are very high, with mean values of 7325 ± 99.50 µS/cm for lagoon water and 907.75 ± 10.26 µS/cm for groundwater. These greatest values show the influence of human activities on the physical quality of water in the village. Indeed, Quarternary aquifers are generally shallow groundwater outcropping in places and therefore exposed to pollution from domestic and industrial activities [7].

The multivariate analysis showed the influence of anthropogenic pollution (industrial, domestic and artisanal pollution) of water resources of village Abia Koumassi. The presence of nitrate in hand dug wells water (116.81 mg·L^{-1}), at levels higher than the [17] standard for drinking water (50 mg·L^{-1}) show that the water is unfit for human consumption. Indeed, studies by [13-15] on ground-

water in the city of Abidjan revealed significant levels of nitrate in the water. In fact, this shanty town of Abidjan does not have an adequate sanitation system for the disposal of domestic wastewater.

They, are found in the nature before seeping into the ground and then into groundwater. Surface water is influenced by domestic, industrial and artisanal activities happening in the city. Nitrogen compounds found in water come mainly from the degradation of organic matter by microorganisms in the topsoil. According to [18] in the soil, organic matter is mineralized from the biological oxidation of NH_4^+ to NO_3^-. Changes produced in the surface are caused by leaching, with infiltration of rainwater in depth. The high nitrate levels are also associated with high levels of metalllic trace elements (MTE) in water of the village. According to the statistical studies the high levels of MTE including Co, Cr, Fe, and Cu come from anthropogenic activities related to domestic, Industrial and artisanal activities. Studies of [14] showed excessive lead in well water areas of Koumassi, Marcory and Port-Bouet. These neighborhoods are old and highly urbanized. There is found industrial parks, gas stations and various craft activities that could be the source of this metal pollution of groundwater. To this is added the contribution of runoff from urban storms and leaching of metals from garbage and soluble residues. According to [19], rainwater is universally recognized as being the vector of large amounts of pollution in their path leaching metals (Pb, Ni, Cr, Zn and Cd), salts, hydrocarbons fertilizers, pesticides and other waste or littering roads and parking or suspended in the air. The rainwater seeps into the ground to reach the shallow aquifers later. The shallowness of the water and sewerage failure of the city of Abidjan is the factors that promote the infiltration of these pollutants into the groundwater of the Quaternary. In the city of Abidjan, the development of human activities causes significant releases of domestic and industrial effluents, which is the main source of production of MTE in the environment and in the water of the city (surface water and groundwater). Moreover, [5] have also found high levels of MTE in sediment of the Ebrié lagoon in Abidjan. Today, the metal pollution of water is a real environmental problem in Africa. The studies of [20] also highlighted high levels of MTE in sediments of Lake Fouarat Morocco. For water, the observed metal pollution in the environment is rooted in human activities namely, discharges of garbage, leaching and agricultural discharges associated with industrial activities. In Nigeria, significant levels were observed by MTE and [21] in water of the main river of the state of Ondo. In Morocco, these high levels were also observed by [22] in the irrigation water of a rice field in Gharb.

[20] add to these sources of metal pollution mechanical

garages, workshops, surface treatment by painting vehicles, stations, distribution of petroleum products, and textile industries pollution. According to [23], human activities are causing pollution of groundwater Mzamza community in Morocco. In the plain of El Ma El Abiod Algeria, [24] showed that groundwater is also threatened by many homes anthropogenic pollution such as urbanization, industry, livestock and dumps. Water from Abia Koumassi contains high levels of fecal coliform bacteria (*Escherichia coli*), fecal streptocopes Clustridium perfringens.

The presence of these bacteria in the water shows that the village is facing a recent bacteriological pollution of human origin. [23,25] explain the presence of fecal coliforms and fecal streptococci in wells water by fecal contamination and therefore the possibility that dangerous pathogens are present in the water. Streptococci and fecal coliforms are indicators of fecal pollution and are largely of human origin. The studies of [15,26] showed fecal pollution of human origin and influence of human activities on the deterioration of the quality of water resources in the city.

5. Conclusions

The study on water resources in the village of Abia Koumassi helped highlight the main characteristics of water. In the village, the principal direction of groundwater flow is north-south. This direction could be considered as the direction of the spread of pollutants in the area. The water has a temperature of between 27.7°C and 29.8°C, which corresponds to seasonal variations in ambient temperature. The pH is near neutrality and ranges between 6.98 and 7.62.

The Water resources of the village Abia Koumassi are highly mineralized. The electrical conductivity measured in the water ranges from 585 µS/cm and 1310 µS/cm. In the water from the village Ca^{2+} and SO_4^{2-} are the most important ions. The nitrate level in the water (116.81 $mg \cdot L^{-1}$) is higher than WHO standard for drinking water and has been found in well water in the village. Most of the studied water contains metallic trace elements (MTE) including Ni, Zn, Co, Cr, Pb, Fe, Cu and Cd. The metallic trace elements present in groundwater are of anthropogenic origin. They are mainly from industrial, crafts, urban and domestic discharges. Microbiological analysis showed that the water contains significant levels of fecal coliform type *Escherichia coli*, faecal streptococci, Clustridium perfringens and thermo tolerant coliform. The presence of these microorganisms in the water shows that the water resources of the village are facing a recent faecal pollution of human origin. This study shows that water from Abia Koumassi is confronted with pollution from domestic, industrial and artisanal activities. The water contains recent microbiological pollution. Thus the quality of water resources in Abia Koumassi is controlled by anthropogenic activities in the area.

REFERENCES

[1] K. J. Kouamé, "Contribution à la Gestion Intégrée des Ressources en Eaux (GIRE) du District d'Abidjan (Sud de la Côte d'Ivoire): Outils d'aide à la Décision pour la Prévention et la Protection des Eaux Souterraines Contre la Pollution," Thèse d'Université de Cocody, Abidjan, 2007, p. 227.

[2] K. E. Ahoussi, N. Soro, G. Soro, M. S. Oga and S. Zadé, "Caractérisation de la Qualité Physico-Chimique et Bactériologique des eaux de Puits de la ville d'Abidjan (Côte d'Ivoire)," *Africa Geoscience Review*, Vol. 16, No. 3, 2009, pp. 199-211.

[3] K. E. Ahoussi, S. Loko, Y. B. Koffi, G. Soro, Y. M.-S. Oga and N. Soro, "Evolution Spatio-Temporelle des Teneurs en Nitrates des Eaux Souterraines de la ville D'Abidjan (Côte d'Ivoire)," *International Journal of Pure & Applied Bioscience*, Vol. 1, No. 3, 2013, pp. 45-60.

[4] N. Soro, L. Ouattara, K. Dongo, K. E. Kouadio, K. E. Ahoussi, G. Soro, Y. M.-S. Oga, I. Savane and J. Biémi, "Déchets Municipaux dans le District d'Abidjan en Côte d'Ivoire: Sources Potentielles de Pollution des eaux Souterraines," *International Journal of Biological and Chemical Sciences*, Vol. 4, No. 6, 2010, pp. 364-384.

[5] G. Soro, B. Métongo, N. Soro, E. Ahoussi, F. Kouamé, S. Zadé and T. Soro, "Métaux Lourds (Cu, Cr, Mn et Zn) dans les Sédiments de Surface d'une Lagune Tropicale Africaine: Cas de la Lagune Ebrié (Côte d'Ivoire)," *International Journal of Biological and Chemical Sciences*, Vol. 3, No. 6, 2009, pp. 1408-1427.

[6] A. Coulibaly, S. Mondé, V. A. Wognin and K. Aka, "State of Anthropic Pollution in the Estuary of Ebrié Lagoon (Côte d'Ivoire) by Analysis of the Metal Elements Traces," *European Journal of Scientific Research*, Vol. 19, No. 2, 2008, pp. 372-390.

[7] N. Aghui and J. Biémi, "Géologie et Hydrogéologie des Nappes de la Région D'Abidjan et Risques de Contaminations," *Annales de l'Université de Côte d'Ivoire, Série C (Sciences)*, Vol. 20, 1984, pp. 313-347.

[8] H. Y. Richardson, G. Nichols, C. Lane, I. R. Lake and P. R. Hunter, "Microbiological Surveillance of Private Water Supplies in England. The Impact of Environmental and Climate Factors on Water Quality," *Water Research*, Vol. 43, No. 8, 2009, pp. 2159-2168.

[9] A. Emad, S. Mohammad, A. Z. Tahseen and S. A Ahmed, "Assessment of Heavy Metals Pollution in the Sediments of Euphrates River, Iraq," *Journal of Water Resource and Protection*, Vol. 4, No. 12, 2012, pp. 1009-1023.

[10] K. J. Fatombi, A. T. Ahoyo, O. Nonfodji and T. Aminou, "Physico-Chemical and Bacterial Characteristics of Groundwater and Surface Water Quality in the Lagbe Town: Treatment Essays with *Moringa oleifera* Seeds," *Journal*

of Water Resource and Protection, Vol. 4, No. 12, 2012, pp. 1001-1008.

[11] S. M. Yidana, D. Ophori and B. Banoeng-Yakubo, "A Multivariate Statistical Analysis of Surface Water Chemistry Data—The Ankobra Basin, Ghana," *Journal of Environmental Management*, Vol. 86, No. 1, 2008, pp. 80-87.

[12] C. Güler, G. D. Thyne, J. E. Mccray and A. K. Tuner, "Evaluation of Graphical and Multivariate Statistical Methods for Classification of Water Chemistry Data," *Hydrogeology Journal*, Vol. 10, 2002, pp. 455-474.

[13] L. Matini, J. M. Moutou and M. S. Kongo-Mantono, "Evaluation Hydrochimique des eaux Souterraines en Milieu Urbain au Sud-Ouest de Brazzaville, Congo," *Afrique Science*, Vol. 5, No. 1, 2009, pp. 82-98.

[14] K. E. Ahoussi, N. Soro, A. M. Kouassi, G. Soro, Y. B. Koffi and S. P. Zade, "Application des Méthodes D'analyses Statistiques Multivariées à l'étude de L'origine des Métaux Lourds (Cu^{2+}, Mn^{2+}, Zn^{2+} et Pb^{2+}) dans les eaux des Nappes Phréatiques de la ville D'Abidjan," *International Journal of Biological and Chemical Sciences*, Vol. 4, No. 5, 2010, pp. 1753-1765.

[15] K. E. Ahoussi, N. Soro, G. Soro, T. Lasm, M. S. Oga and S. Zadé, "Groundwater Pollution in Africans Biggest Towns: Case of the Town of Abidjan (Côte d'Ivoire)," *European Journal of Scientific Research*, Vol. 20, No. 2, 2008, pp. 302-316.

[16] A. A. Tandia, E. S. Diop and C. B. Gaye, "Pollution par les Nitrates des Nappes Phréatiques sous Environnement Semi-Urbain non Assaini: Exemple de la Nappe de Yeumbeul, Sénégal," *Journal of African Earth Sciences*, Vol. 29, No. 4, 1999, pp. 809-822.

[17] OMS, "Directives de Qualité pour l'eau de Boisson. Critères D'hygiène et Documentation à L'appui," 2nd Edition, Vol. 2, 2000, p. 1050

[18] L. W. Canter, "Nitrates in Groundwater," Lewis Publishers, New York, 1997, 263 p.

[19] C. Guillemin and J. C. Roux, "La Pollution des eaux Souterraines," *Manuels et Méthodes*, Vol. 23, 1992, 262 p.

[20] H. Ben Bouih, H. Nassali, M. Leblans and A. Srhiri, "Contamination en Métaux Traces des Sédiments du lac Fouarat (Maroc)," *Afrique Science*, Vol. 1, No. 1, 2005, pp. 109-125.

[21] I. A. Ololade and A. O. Ajayi, "Contamination Profile of Major Rivers alongs the Highways in Ondo State, Nigéria," *Journal of Toxicology and Environmental Sciences*, Vol. 1, No. 3, 2009, pp. 38-53.

[22] S. E. Blidi, M. Fekhaoui, A. Serghini, A. El Abidi and L. Drissi, "Comportement des Eléments Traces Métalliques dans l'agrosystème Rizière du Gharb (Maroc) ," *Bulletin de l'Institut Scienctifique, Rabat, section Science de la vie*, Vol. 29, 2007, pp. 63-70.

[23] J. El Asslouij, S. Kholtei, N. El Amira-Paaza and A. Hilali, "Impact des Activités Anthropiques sur la Qualité des eaux Souterraines de la Communauté Mzamza (Chaouia, Maroc)," *Revue des Sciences de l'Eau*, Vol. 20, No. 3, 2007, pp. 309-321.

[24] A. Rouabhia, F. Baali, N. Kherici and L. Djabri, "Vulnérabilité et Risque de Pollution des Eaux Souterraines de la Nappe des Sables Miocènes de la Plaine d'El Ma El Abiod (Algérie)," *Sécheresse*, Vol. 15, No. 4, 2004, pp. 347-352

[25] C. Boutin, "L'eau des Nappes Phréatiques Superficielles, une Richesse Naturelle mais Vulnérable. L'exemple des Zones Rurales du Maroc," *Sciences de l'eau*, Vol. 6, No. 3, 1993, pp. 357-365.

[26] J. S. Claon, "Consommation d'eau de Puits dans Quatre Communes de la ville d'Abidjan Desservies par le Réseau de Distribution d'eau Potable," Thèse de Doctorat en pharmacie, Université d'Abidjan, Côte d'Ivoire, 1997, p. 197.

19

Concealed Environmental Threat in the Coastal Region Requires Persistent Attention: The Panglao Island, Philippines Example

Daniel Edison Husana[1,2*], Tomohiko Kikuchi[2]

[1]Environmental Biology Division, Institute of Biological Sciences, College of Arts and Sciences, University of the Philippines Los Baños, College, Laguna, Philippines; [2]Global Center of Excellence for Environmental Studies, Graduate School of Environment and Information Sciences, Yokohama National University, Yokohama, Japan.

ABSTRACT

Panglao is a small island in the central part of the Philippines and well-known for its world-class beaches and coral reefs. These attract millions of tourists each year thus providing business opportunities and employment, a significant source of revenue for the local economy. Moreover, this island lies in a region with high biodiversity. However, the escalating activity is so alarming that the negative effect to the local environment is very much prevalent but not easily perceivable. Analysis and measurement of physico-chemical parameters of the groundwater revealed high levels of human-induced contaminants. This subterranean pollution was attributable to the leakage of septic tanks, artificial application of disinfectants as well as infiltration of saltwater from the ocean due to over-extraction of groundwater in order to meet the increasing demand for water. The community within the area was oblivious because human impacts to the environment appear to be virtually absent. These findings clearly suggest the concealed vulnerability of the groundwater resources from human activities. Higher standard for the coastal development plan, strong implementation of environmental policy and immediate government action is deemed necessary.

Keywords: Hidden Environmental Threat; Groundwater; Coastal Region; Panglao Island; Philippines

1. Introduction

The coastal areas are known to be one of the world's fastest growing areas for development ever since the ancient times. However, development is usually accompanied by environmental threats if not properly managed and, more often than not, people and authorities react when the problem is already apparent (or the surrounding community is already affected). The case of the underground system that is not readily visible is oftentimes out of people's concern making it the least priority. Humans do not care much about their activities on the surface environment that may impact the hidden subsurface environment. Over extraction of groundwater, for example, could result to the decline of the water table that causes loss of buoyant support to the overlying rocks, which will eventually lead to catastrophic land subsidence [1].

Overexploitation and mismanagement of groundwater resource could lead to the depletion and degradation of water quality [2] as well as saltwater intrusion. Moreover, changes in the physicochemical properties of groundwater are known to affect the physiological development of aquatic biota (e.g. [3-6]). The overall effect will eventually affect the well-being of the human population who are at the losing end. People should therefore be aware of such concealed environmental threats before it becomes too late.

Panglao is a small island located in the central part of the Philippines (**Figure 1**). It is divided into two municipalities, Dauis on the east and Panglao on the west, with a total land area of 95.07 km[2] and bounded by 49.17 kilometers of coastline. The island's population in 2007 is 62,083 with an average growth rate of 1.5% according to the record of the National Statistics Office (NSO) of Dauis and Panglao municipalities. Panglao Island is economically important for the province of Bohol because of

---

*Corresponding author.

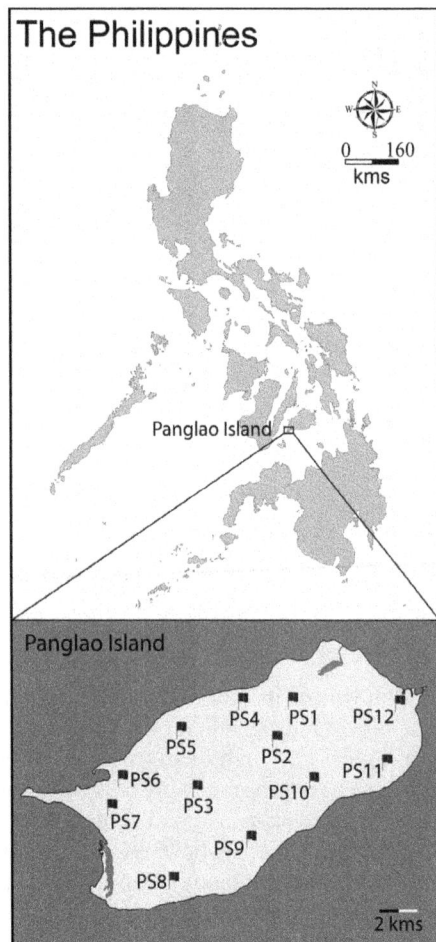

Figure 1. Map of the Philippines showing the sampling sites in Panglao Island.

its tourism industry. This island is known globally for its world-class beach resorts making it frequented by many local and foreign tourists. Bohol provincial tourism office recorded an escalating number of visitors from 128,899 in 2003 to 282,498 in 2008 and was projected by the Bohol Provincial Planning and Development Office (PPDO) to be four times higher in the next ten years and that is on top of its local population that is also increasing every year. Incidentally, [7] detected high concentrations of NO_2^-, NO_3^- and Cl^- in the groundwater of Panglao Island and suggested to have further investigations on what could be the possible cause/s or source/s of the contamination.

On the other hand, Panglao Island is rich in biodiversity. Reference [8] reported that initial estimates of more than 7000 species were sampled during the 2004 marine expedition in the surrounding sea. This does not include the terrestrial species and those species inhabiting the groundwater of this small island (e.g. [9-12]). While unique cave (groundwater) species have been discovered and described in various parts of the Philippines in recent

years (e.g. [13-15]), some of which belong to a higher taxa like new genus [16]) or even a new family [17], it's also possible that there are more undescribed species inhabiting the groundwater of Panglao Island.

This study aimed to assess and identify the problems that exist in a "healthy-looking" coastal area. We intend to provide examples of environmental issues that are frequently overlooked and ignored but could be a serious threat if not properly checked.

2. Materials and Methods

Water samples were collected from 12 sampling points (indicated PS1 to PS12) from deep wells and caves in the island of Panglao, Bohol, Philippines (**Figure 1**) in May 2011. They were analyzed at the University of the Philippines-Natural Sciences Research Institute (UP-NSRI) for microbial content and chemical contaminants.

2.1. Microbial Analyses

Separate analyses were conducted for environmental samples (water samples not ideal for drinking) and those water that is used for drinking.

1) Environmental Water Samples

The samples were serially diluted up to 10^{-2} analyzed by multiple tube fermentation technique (MTFT) using 5 of 1-ml portions of the undiluted sample (10^0) and the 10^{-1} and 10^{-2} dilutions. The presumptive test for coliform bacteria was done on single strength lauryl sulfate tryptose broth (LST). The LST tubes were incubated at 37°C and observed for gas production within 24 - 48 hours. To confirm the presence of coliform bacteria, gassing LST tubes were sub-cultured into brilliant green lactose bile broth (BGLB). BGLB tubes were incubated at 37°C and observed for gas formation within 24 - 48 hours. Gassing LST tubes were likewise sub-cultured into EC broth to confirm the presence of fecal coliforms. EC tubes were incubated at 44°C and observed for gas formation within 24 hours.

The most probable number (MPN) of coliform bacteria per 100 mL sample were determined from the MPN table for 10^0, 10^{-1} and 10^{-2} dilutions.

2) Drinking Water Samples

The samples were analyzed by MTFT using 5 of 20-mL volumes of the sample on triple strength LST, 5 of 10-ml volumes in double strength LST and 5 of 1-ml and 0.1-ml volumes on single strength LST. Procedures to determine presumptive and confirmed fecal coliform levels, as in the environmental water samples, were followed.

The water samples contained high levels of coliform bacteria for drinking water, hence, values obtained using 20-ml volumes were not considered. The MPN levels of coliform bacteria were based on 10-, 1- and 0.1-ml vol-

Concealed Environmental Threat in the Coastal Region Requires Persistent Attention: The Panglao Island, Philippines Example

199

umes of the sample.

2.2. Water Physico-Chemical Analyses

Dissolved oxygen, conductivity, salinity and temperature were measured using YSI 85 while pH was measured using Custom Waterproof pH meter. These parameters were measured in situ during water sampling.

Chloride and nitrate were analyzed using ion chromatography, nitrite using colorimetric N-(1-naphthyl) ethylene diamine dihydrochloride reagent method while Atomic Absorption Spectrophotometry (AAS) Flame method was used to analyze sodium following the standard methods for the examination of water and wastewater [18].

2.3. Survey and Interview

People aged 14 to 84 years old, 90% of them have resided in Panglao Island for 10 years or more, were interviewed during our survey in 2011. One hundred fifty (150) respondents were selected randomly from different areas of the island of which 11 were discarded due to questionable response leaving a total of 139 valid respondents. Questions regarding family background, their environmental and economic points of view on Panglao Island were asked and scored in a sheet.

3. Results and Discussion

3.1. Microbial

The results show high levels of coliform bacteria in all the groundwater sampling sites in Panglao Island (**Figure 2**). It ranged from 2 to 16,000 MPN/100mL. All the sites that were sampled for environmental water quality test were all contaminated with fecal coliform bacteria with

the highest concentration in site PS4 (16,000 MPN/100mL) and with the least concentration in site PS1 (20 MPN/100mL). All the wells that are currently being used for drinking water (PS2, PS3, PS5, PS9, PS10 and PS11) are contaminated with fecal coliform bacteria. The highest concentration was detected in site PS2 (1600 MPN/100mL) and least in PS10 (2 MPN/100mL).

None of the sites passed the standard for drinking water or for recreation use. This is a clear indication that even though there is no apparent sign of human disturbance, deterioration of groundwater system is very evident.

3.2. Physico-Chemical

Table 1 summarizes the environmental parameters of groundwater obtained from Panglao Island. The pH was all within the normal range from 6.89 to 7.93. Dissolved oxygen (DO) ranged from 2.40 to 6.34 mg·L^{-1}, PS3 having the highest concentration and PS6 the lowest. Temperature ranges from 27.7°C to 31.6°C. Water conductivity ranged from 0.002 to 57.4 mS and salinity from 0.1‰ to 6.7‰, PS6 has the highest concentration of both parameters while PS5 has the lowest. Low DO concentration in site PS6 is quite noticeable as well as high EC and salinity.

Sodium ion concentrations (**Figure 3**) ranged from 53.4 to 2120 mg·L^{-1}. Site PS6 has the highest concentration while PS12 has the lowest. Chloride ion concentrations (**Figure 3**) ranged from 33.8 to 3430 mg·L^{-1} with the highest concentration in PS6 and lowest in PS12. Note that PS12 has higher concentration of sodium than chloride unlike the other sites in which chloride ions are almost twice the sodium ions concentrations.

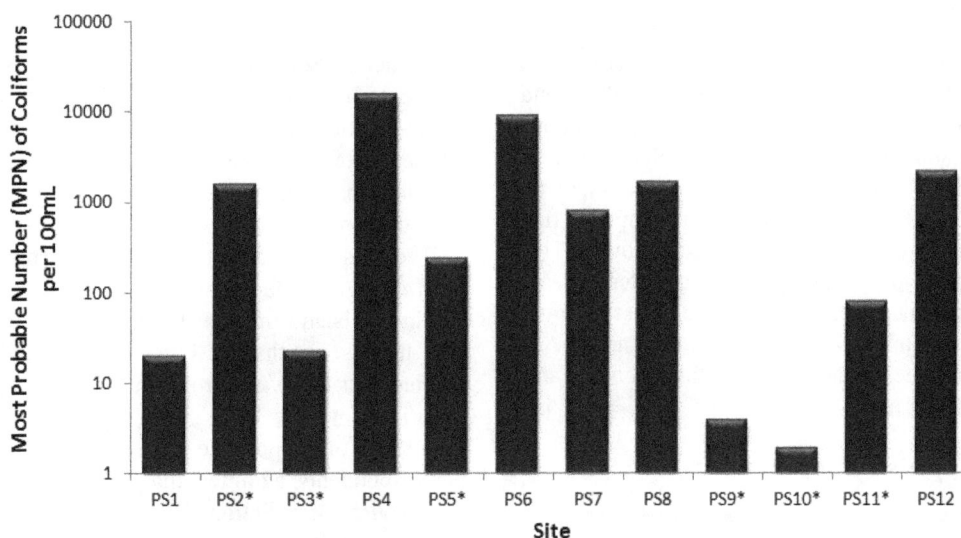

Figure 2. Coliform counts of Panglao Island groundwater samples. Sites with asterisk (*) indicate drinking water samples.

Table 1. Aquatic environmental parameters of groundwater in Panglao Island.

Sampling site	Coordinates	DO (mg·L^{-1})	Conductivity (mS)	Salinity	Temp (°C)	pH
PS1	N9°37.300; E123°49.078	3.20	2.098	1.0	29.2	7.02
PS2*	N9°36.420; E123°48.976	5.12	0.754	0.3	28.7	6.89
PS3*	N9°35.623; E123°47.088	6.34	1.723	0.8	28.5	7.31
PS4	N9°37.523; E123°48.059	5.75	11.340	6.1	27.8	7.25
PS5*	N9°36.703; E123°46.750	5.66	0.002	0.1	28.1	6.97
PS6	N9°35.521; E123°45.034	2.40	12.540	6.7	28.0	7.13
PS7	N9°34.899; E123°44.771	4.38	2.359	1.1	27.8	7.20
PS8	N9°33.316; E123°46.486	4.30	7.540	3.7	28.6	7.19
PS9*	N9°34.207; E123°48.378	5.68	2.517	1.2	27.7	6.93
PS10*	N9°35.442; E123°49.677	5.17	1.608	0.8	28.4	6.90
PS11*	N9°35.855; E123°51.445	5.22	1.133	0.5	28.1	7.00
PS12	N9°37.466; E123°51.928	3.73	0.839	0.4	27.9	7.45

*Drinking water samples.

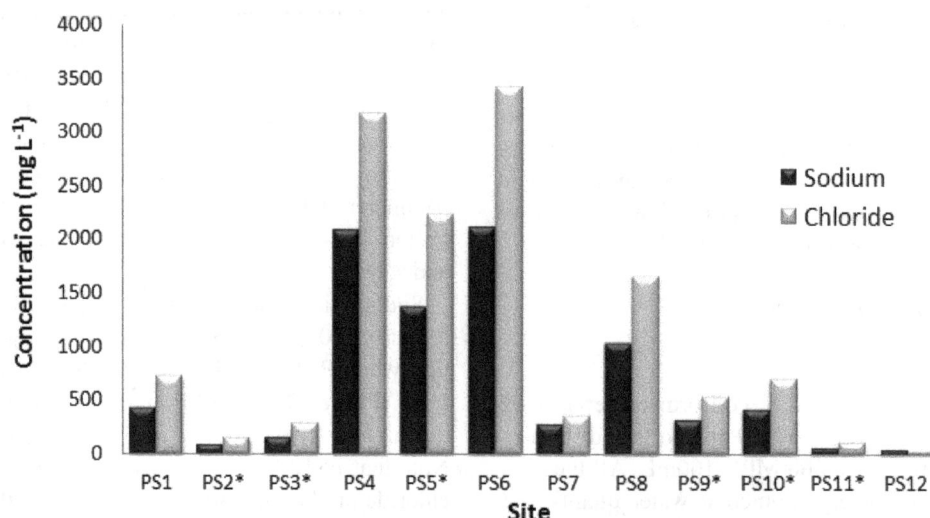

Figure 3. Concentrations of sodium and chloride ions in various parts of Panglao Island. Sites with asterisk (*) indicate drinking water samples.

Nitrate concentrations (**Figure 4**) ranged from 7.2 to 61.9 mg·L^{-1} with the highest concentration in PS12 and lowest in PS3. PS1 and PS12 sites both have above maximum contaminant level (MCL) for NO$_3$-N (11.59 and 14.07 mg·L^{-1} respectively). Nitrite concentrations on the other hand ranged from less than maximum detection limit (MDL) up to 0.025 mg·L^{-1} (data not shown). This high concentration of nitrate in drinking water poses a serious threat to human health especially to babies. Nitrate-nitrogen concentration as low as 10 to 20 mg·L^{-1} is known to cause illnesses and even deaths among infants under six months of age.

3.3. Local People's Perspective

Our random interviews with the local people gave some insights on the situation in Panglao Island as well as some perspective of the residents. Highest number of the

respondents use municipal water supply (38.7%) which is a centralized deep well in the island followed by those using purified (33.8%). About fifteen percent (14.9%) utilizes water from their private deep well for their daily consumption while the rest are using bottled water (9.3%) and other sources not mentioned (3.3%).

Twenty-seven percent (27%) of the respondents noticed changes in the quality of water during their entire life in the island in terms of taste and color while 73% said there's no change, however, some of them commented that the water from the deep well tastes a little brackish.

In terms of the possible sources of pollution in Panglao the respondents admitted that the local people are the main source of pollution (53.7%) while the rest are coming from local industry (19.9%), natural cause (15.4%), tourism (9.6%) and agriculture (1.5%). In this respect,

Concealed Environmental Threat in the Coastal Region Requires Persistent Attention: The Panglao Island, Philippines Example

201

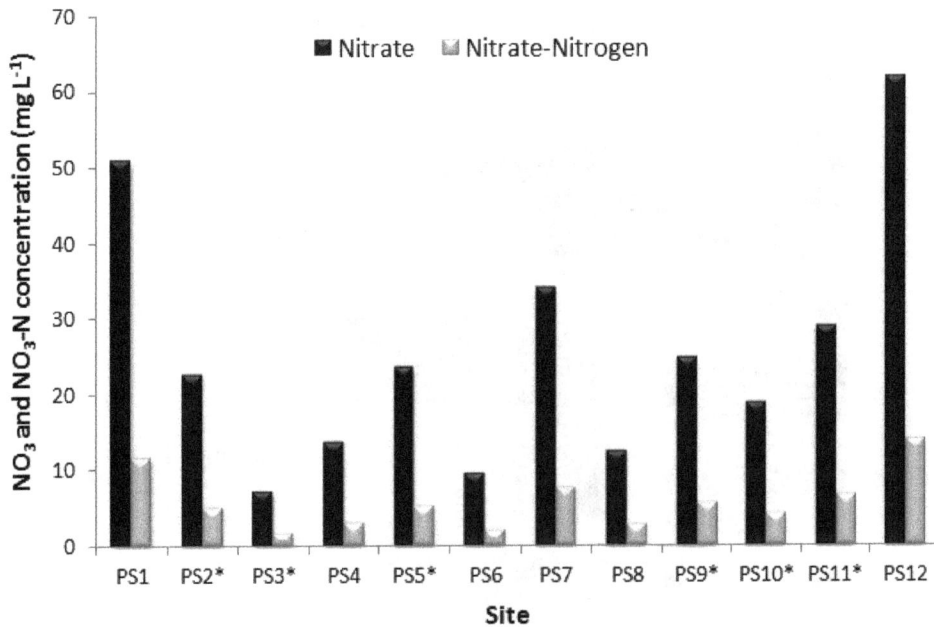

Figure 4. Nitrate and nitrate-nitrogen concentrations of groundwater in Panglao Island. Sites with asterisk (*) indicate drinking water samples.

most people believe that beaches (51%) are the most threatened natural resource in Panglao Island as a result of island development. The other half believed that mangroves (15%), groundwater (13.6%) and caves (12.9%) are also threatened while the smallest percentage of the respondents (6.8%) is not aware of the threat. If pollution will not be controlled, 49% said that beaches would be the most affected followed by groundwater (21%), mangroves (10.2%) and caves (9.5%). While 8% are not aware, 1.5% said that there is no threat at all.

Majority of the respondents (46.6%) believed that tourism is the major source of local employment while 19% said it is the major driver for the development of Panglao Island. Others said that tourism causes the increase in population (17.2%) and pollution (13.2%) while 4% said it causes environmental destruction.

Most of the respondents (46.6%) claimed that Panglao Island in general is a good place for business and employment opportunities. Others know that this island has high biodiversity (19%), overcrowded and polluted (17%), a paradise and tourist destination (13.2%) and 4% said it's a sanctuary for plants and animals.

3.4. Potential Impacts to the Environment

Tourism attracts migration, temporarily and permanently, because of the aesthetic value of the place as well as the employment and business opportunities. These can result to the increase in local population, which if not checked and regulated will skyrocket beyond the local environment's carrying capacity and the ecosystem's resilience

to stressors. This chain reaction could lead to ecological disaster and collapse. The high concentration of nitrate, the presence of coliform bacteria and the increasing groundwater salinity is quite alarming. The presence of these pollutants in coastal groundwater is a clear indication of ignorance and invisible but potentially devastating impacts of human activities to the environment. Local population of the island relies on groundwater (municipal water supply) and the changes in its quality have made the people without a choice but to use purified or filtered water. This issue will become a bigger problem in the long run if the local residents could not afford the (more expensive) potable water for their basic needs.

Contamination of Panglao Island's groundwater with coliform bacteria could primarily be due to the current design of septic tanks as observed personally by the first author (see illustration in **Figure 5**). The lack of water-proof flooring allows the infiltration of waste from the septic tank through the porous limestone base. This contaminated water could easily disperse to the groundwater of the entire island as noticed with the water quality analyses. High chlorine concentration on the other hand could be attributable to the combination of seawater intrusion and application of free chlorine directly to the open wells to disinfect the water for drinking. Sanitation techniques however seem to be ineffective because coliform bacteria in the groundwater were detected.

The groundwater serves as corridor for migrating animals in areas where surface water such as rivers are not present [19,20]. It is also a refuge and breeding ground for some aquatic animals (unpublished report). In this

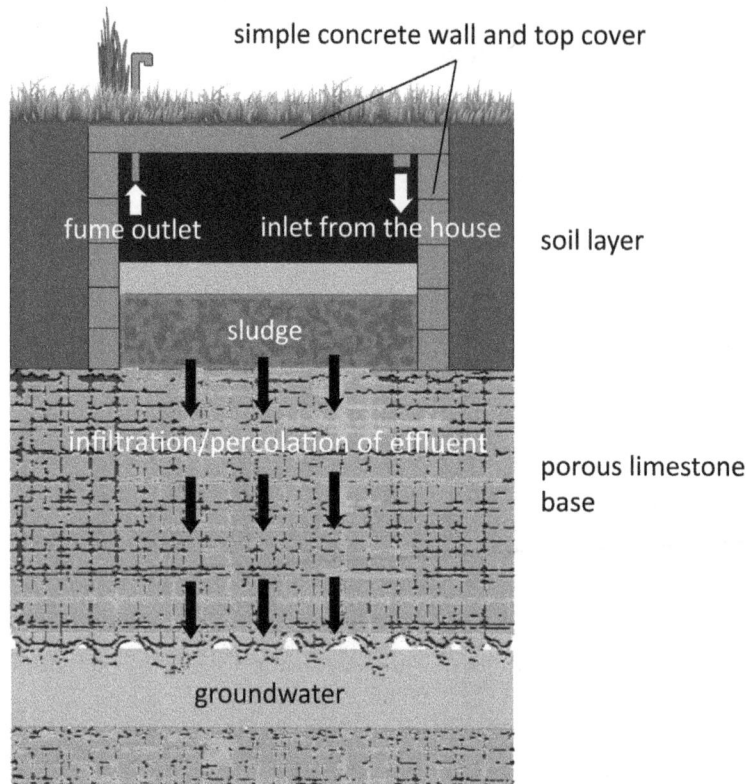

Figure 5. Conceptual model of the common design of septic tanks in Panglao Island illustrating the infiltration of effluent/polluted water through the porous limestone base that causes contamination of the groundwater.

regard, the quality of groundwater will play a vital role in the survival of organisms that depend on it. Aquatic nutrients are important but their excess amounts could be detrimental to the health of the groundwater organisms.

3.5. The Need for a Strategic Management Plan

Panglao Island, though small, is rich in natural resources but these will vanish if not properly managed and taken cared of. Without strong consideration to the biodiversity and natural environment, the island's natural aesthetic quality will collapse in due time. Marine and coastal regions are known to be experiencing the fastest tourism growth among other areas in the world [21] and yet most people are still unaware of its negative impact on the fragile coastal ecosystems. Development of resorts along the coast is economically valuable but ecological consideration and proper management should be implemented in strictest sense. They should not threaten the biodiversity and selfishly exploit the natural resources for the benefit of the few; hence strong political support for sustainable development is deemed essential.

It is important to note that protecting the underground environment does not rely solely on the protection of the groundwater, but also on the protection of the surface environment as well. The rate of percolation to ground-

water is higher in the karst regions thus inhibiting the natural mechanical and biological self-purification of the polluted rainwater that passes through the subterranean cavities [22]. Any pollutants on the surface have the potential to percolate into the groundwater as well. The bigger problem is that once the pollution enters the groundwater, it is difficult to trace its source [23] and by the time it is detected, the pollutant is likely to be widespread and the damage is already irreversible [24].

4. Conclusions and Recommendations

The community within Panglao Island was not aware of the situation because the negative effect of anthropogenic activities to the environment appears to be virtually absent. These findings clearly suggest the vulnerability of the groundwater resources from human activities because of its hidden location (see also [14]). Repercussions of human activities may not be manifested yet while the disturbance is ongoing, but sooner or later their long term effect will be more apparent.

As in the case of Panglao Island, many areas around the world are unfavorably impacted by tourism and threatened by its associated activities, thus, understanding the physiology of underground ecosystem is critically important so that a proper management plan will be implemented. Reference [7] recommended that the surface environment should be protected in order to preserve a good groundwater quality. Therefore, revisiting the current local policy with regards to environmental protection is quite appropriate so that necessary measures will be improved and not be overlooked. Higher standard for the coastal development plan, strict environmental monitoring, especially of groundwater, and public awareness on coastal issues will be beneficial to all concerned especially those people relying on coastal resources.

5. Acknowledgements

We would like to thank the people of Panglao Island who participated in the survey and the local government staffs and mayors of the municipalities of Panglao and Dauis for permitting and supporting this study as well as for providing some local data. Our sincere gratitude is also expressed to Dr. C.G.B. Banaay whose constructive comments and suggestions helped improve this manuscript. The first author is grateful for the assistance provided by the Yokohama National University during his postdoctoral research fellowship. This work was supported in part by Global COE Program, MEXT, Japan.

REFERENCES

[1] P. Lamoreaux and J. Newton, "Catastrophic Subsidence: An Environmental Hazard, Shelby County, Alabama," *Environmental Geology*, Vol. 8, No. 1, 1986, pp. 25-40.

[2] J. De Waele and R. Follesa, "Human Impact on Karst: The Example of Lusaka (Zambia)," *International Journal of Speleology*, Vol. 32, No. 1-4, 2003, pp. 71-83.

[3] J. A. Cuesta and K. Anger, "Larval Morphology and Salinity Tolerance of a Land Crab from West Africa, *Cardisoma armatum* (Brachyura: Grapsoidea: Gecarcinidae)," *Journal of Crustacean Biology*, Vol. 25, No. 4, 2009, pp. 640-654.

[4] B. C. Shock, C. M. Foran and T. A. Stueckle, "Effects of Salinity Stress on Survival, Metabolism, Limb Regeneration, and Ecdysis in *Uca pugnax*," *Journal Crustacean Biology*, Vol. 29, No. 3, 2009, pp. 293-301.

[5] J. S. Mattice, M. B. Burch, S. C. Tsai and W. K. Roy, "A Toxicity Testing System for Exposing Small Invertebrates and Fish to Short Square Wave Concentrations of Chlorine," *Water Research*, Vol. 15, No. 7, 1981, pp. 923-927.

[6] J. M. Capuzzo, "The Effects of Free Chlorine and Chloramines on Growth and Respiration Rates of Larval Lobster (*Homarus americanus*)," *Water Research*, Vol. 11, No. 12, 1977, pp. 1021-1024.

[7] D. E. M. Husana and M. Yamamuro, "Groundwater Quality on Karst Regions in the Philippines," *Limnology*,

Online First, 6 Feb 2013.

[8] P. Bouchet, P. K. L. Ng, D. Largo and S. H. Tan, "Panglao 2004—Investigations of the Marine Species Richness in the Philippines," *The Raffles Bulletin of Zoology*, Supplement No. 20, 2009, pp. 1-19.

[9] P. K. L. Ng, D. Guinot and T. M. Iliffe, "Revision of the Anchialine Varunine Crabs of the Genus *Orcovita* Ng & Tomascik, 1994 (Crustacea: Decapoda: Brachyura: Grapsidae), with Descriptions of Four New Species," *The Raffles Bulletin of Zoology*, Vol. 44, No. 1, 1996, pp. 109-134.

[10] P. K. L. Ng and D. Guinot, "On the Land Crabs of the Genus *Discoplax* A. Milne Edwards, 1867 (Crustacea: Decapoda: Brachyura: Gecarcinidae), with Description of a New Cavernicolous Species from the Philippines," *The Raffles Bulletin of Zoology*, Vol. 49, 2001, pp. 311-338.

[11] P. K. L. Ng, "New Species of Cavernicolous Crabs of the Genus *Sesarmoides* from the Western Pacific, with a Key to the Genus (Crustacea: Decapoda: Brachyura: Sesarmidae)," *The Raffles Bulletin of Zoology*, Vol. 50, 2002, pp. 419-435.

[12] Y. Kano and T. Kase, "Genetic Exchange between Anchialine Cave Populations by Means of Larval Dispersal: The Case of a New Gastropod Species *Neritilia cavernicola*," *Zoological Scripta*, Vol. 33, No. 5, 2004, pp. 423-437.

[13] D. E. M. Husana, T. Naruse and T. Kase, "Two New Cavernicolous Species of the Genus *Sundathelphusa* from Western Samar, Philippines (Decapoda: Brachyura: Parathelphusidae)," *Journal of Crustacean Biology*, Vol. 29, No. 3, 2009, pp. 419-427.

[14] D. E. M. Husana, T. Naruse and T. Kase, "A New Species of the Genus *Karstarma* (Decapoda: Brachyura: Sesarmidae) from Anchialine Caves in the Philippines," *The Raffles Bulletin of Zoology*, Vol. 58, 2010, pp. 65-69.

[15] D. E. M. Husana, "Cave Ecology in the Philippines, a Conservation Perspective: Linking Surface and Subsurface Ecosystems," PhD Thesis, The University of Tokyo, 2010.

[16] D. E. M. Husana, S. H. Tan and T. Kase, "A New Genus and Species of Anchialine Hymenosomatidae (Crustacea, Decapoda, Brachyura) form Samar, Philippines," *Zootaxa*, Vol. 3209, 2011, pp. 49-59.

[17] A. Fosshagen and T. M. Iliffe, "*Boholina*, a New Genus (Copepoda: Calanoida) with Two New Species from an Anchialine Cave in the Philippines," *Sarsia*, Vol. 74, No. 3, 1989, pp. 201-208.

[18] American Public Health Association, "Standard Methods for the Examination of Water and Wastewater," 21st Edition, American Public Health Association (APHA), Washington DC, 20005.

[19] J. Ward and M. Palmer, "Distribution Patterns of Interstitial Freshwater Meiofauna over a Range of Spatial Scales, with Emphasis on Alluvial River—Aquifer Systems," *Hydrobiologia*, Vol. 287, No. 1, 1994, pp. 147-156.

[20] J. Gibert, J. A. Stanford, M. J. Dole-Olivier and J. V.

Ward, "Basic Attributes of Groundwater Ecosystems and Prospects for Research," In: J. Gibert, D. L. Danielopol and J. A. Stanford, Eds., *Groundwater Ecology*, Academic, New York, 1994, pp. 7-40.

[21] C. M. Hall, "Trends in Ocean and Coastal Tourism: The End of the Last Frontier?" *Ocean Coastal Management*, Vol. 44, No. 9-10, 2001, pp. 601-618.

[22] F. A. Assaad and H. Jordan, "Karst Terranes and Environmental Aspects," *Environmental Geology*, Vol. 23, No. 3, 1994, pp. 228-237.

[23] P. J. Wood, J. Gunn and S. D. Rundle, "Response of Benthic Cave Invertebrates to Organic Pollution Events," *Aquatic Conservation: Marine Freshwater Ecosystem*, Vol. 18, No. 6, 2008, pp. 909-922.

[24] P. J. Hancock, A. J. Boulton and W. F. Humphreys, "Aquifers and Hyporheic Zones: Towards an Ecological Understanding of Groundwater," *Hydrogeology Journal*, Vol. 13, No. 1, 2005, pp. 98-111.

Permissions

The contributors of this book come from diverse backgrounds, making this book a truly international effort. This book will bring forth new frontiers with its revolutionizing research information and detailed analysis of the nascent developments around the world.

We would like to thank all the contributing authors for lending their expertise to make the book truly unique. They have played a crucial role in the development of this book. Without their invaluable contributions this book wouldn't have been possible. They have made vital efforts to compile up to date information on the varied aspects of this subject to make this book a valuable addition to the collection of many professionals and students.

This book was conceptualized with the vision of imparting up-to-date information and advanced data in this field. To ensure the same, a matchless editorial board was set up. Every individual on the board went through rigorous rounds of assessment to prove their worth. After which they invested a large part of their time researching and compiling the most relevant data for our readers. Conferences and sessions were held from time to time between the editorial board and the contributing authors to present the data in the most comprehensible form. The editorial team has worked tirelessly to provide valuable and valid information to help people across the globe.

Every chapter published in this book has been scrutinized by our experts. Their significance has been extensively debated. The topics covered herein carry significant findings which will fuel the growth of the discipline. They may even be implemented as practical applications or may be referred to as a beginning point for another development. Chapters in this book were first published by Scientific Research Publishing Inc.; hereby published with permission under the Creative Commons Attribution License or equivalent.

The editorial board has been involved in producing this book since its inception. They have spent rigorous hours researching and exploring the diverse topics which have resulted in the successful publishing of this book. They have passed on their knowledge of decades through this book. To expedite this challenging task, the publisher supported the team at every step. A small team of assistant editors was also appointed to further simplify the editing procedure and attain best results for the readers.

Our editorial team has been hand-picked from every corner of the world. Their multi-ethnicity adds dynamic inputs to the discussions which result in innovative outcomes. These outcomes are then further discussed with the researchers and contributors who give their valuable feedback and opinion regarding the same. The feedback is then collaborated with the researches and they are edited in a comprehensive manner to aid the understanding of the subject.

Apart from the editorial board, the designing team has also invested a significant amount of their time in understanding the subject and creating the most relevant covers. They scrutinized every image to scout for the most suitable representation of the subject and create an appropriate cover for the book.

The publishing team has been involved in this book since its early stages. They were actively engaged in every process, be it collecting the data, connecting with the contributors or procuring relevant information. The team has been an ardent support to the editorial, designing and production team. Their endless efforts to recruit the best for this project, has resulted in the accomplishment of this book. They are a veteran in the field of academics and their pool of knowledge is as vast as their experience in printing. Their expertise and guidance has proved useful at every step. Their uncompromising quality standards have made this book an exceptional effort. Their encouragement from time to time has been an inspiration for everyone.

The publisher and the editorial board hope that this book will prove to be a valuable piece of knowledge for researchers, students, practitioners and scholars across the globe.

List of Contributors

William W. McNeary
Department of Chemical Engineering, University of Missouri-Columbia, Columbia, USA

Larry E. Erickson
Department of Chemical Engineering, Kansas State University, Manhattan, USA

Magdalena Sut and Thomas Raab
Department of Geopedology and Landscape Development, Brandenburg University of Technology Cottbus-Senftenberg, Cottbus, Germany

Frank Repmann
Department of Soil Protection and Recultivation, Brandenburg University of Technology Cottbus Senftenberg, Cottbus, Germany

Thomas Fischer
Central Analytical Laboratory, Brandenburg University of Technology Cottbus-Senftenberg, Cottbus, Germany

Daniel Krewski
Department of Epidemiology and Community Medicine, University of Ottawa, Ottawa, Canada
McLaughlin Center for Population Health Risk Assessment, University of Ottawa, Ottawa, Canada

Michael Jerrett
School of Public Health, University of California at Berkeley, Berkeley, USA

Nawal Farhat and Tim Ramsay
Department of Epidemiology and Community Medicine, University of Ottawa, Ottawa, Canada

S. Julio Friedmann, Lynn Wilder and Lee Neher
Lawrence Livermore National Laboratory, Livermore, USA

Richard A. Beck
Department of Geography, University of Cincinnati, Cincinnati, USA

Yolanda M. Price
International Center for Water Resources Research, Central State University, Wilberforce, USA

Gulnihal Ozbay, Amy Cannon, Amanda Treher, Albert Essel and Dyremple Marsh
College of Agriculture & Related Sciences, Cooperative Extension, Delaware State University, Dover, Delaware, USA

Stephanie Clemens
Master Well Owner Network, University Park, Pennsylvania, USA

John Austin
Office of Sponsored Research, Delaware State University, Dover, Delaware, USA

Wakuru Magigi
Moshi University College of Cooperative and Business Studies (A Constituent College of Sokoine University of Agriculture), Moshi, Tanzania

Francis Rosillon
University of Liège, Water, Environment, Development Unit, Arlon Campus, Arlon, Belgium

Jasim M. Salman and Ahmmed J. Nassar
College of Science, University of Babylon, Hilla, Iraq

Hassan J. Jawad
College of Science, University of Karbala, Holly Karbala, Iraq

Fikrat M. Hassan
College of Science for Woman, University of Baghdad, Baghdad, Iraq

Marilia M. F. de Oliveira, Gilberto C. Pereira and Nelson F. F. Ebecken
Civil Engineering Program, Federal University of Rio de Janeiro-UFRJ, Center of Technology, Fundão Island, Rio de Janeiro, Brazil

Jorge L. F. de Oliveira
Geography Postgraduate Program, Fluminense Federal University-UFF, Geoscience Institute, Niterói, Brazil

Chidsanuphong Chart-asa, Kenneth G. Sexton and Jacqueline MacDonald Gibson
University of North Carolina at Chapel Hill, Chapel Hill, USA

Motoya Hayashi
Department of Life Style and Space Design, Miyagi Gakuin Women's University, Sendai, Japan

Haruki Osawa
Department of Healthy Building and Housing, National Institute of Public Health, Wako, Japan

Peter C. Smiley Jr.
USDA-ARS Soil Drainage Research Unit, Columbus, Ohio, USA

Charles M. Cooper
USDA-ARS National Sedimentation Laboratory, Oxford, Mississippi, USA

Khemiri Sami, Ben Alaya Mohsen and Zargouni Fouad
Department of Geology, University of Sciences of Tunis, Manar, Tunisia

Khnissi Afef
Water Research and Technology Center, Solimane, Tunisia

Juliara Stahl Böhm, Marilia Schuch, Adriana Düpont and Eduardo A. Lobo
Biology and Pharmacy Department, Laboratory of Limnology, University of Santa Cruz do Sul, Santa Cruz do Sul, Brazil

František Hindák and Alica Hindáková
Institute of Botany, Slovak Academy of Sciences, Dúbravská Cesta 9, Bratislava, Slovakia

Ján Machava
Catholic University in Ružomberok, PF, Hrabovská Cesta 1, Ružomberok, Slovakia

Eva Tirjaková
Department of Zoology, Faculty of Natural Sciences, Comenius University, Bratislava, Slovakia

Marta Illyová
Institute of Zoology, Slovak Academy of Sciences, Dúbravská Cesta 9, Bratislava, Slovakia

Vu Van Hieu
Research Centre for Environmental Monitoring and Modelling, Vietnam National University, Hanoi, Vietnam
Department of Hu-man Ecology, Vrije Universiteit Brussel, Brussels, Belgium

Pham Ngoc Ho
Research Centre for Environmental Monitoring and Modelling, Vietnam National University, Hanoi, Vietnam

Le Xuan Quynh
Department of Geography, Vrije Universiteit Brussel, Brussels, Belgium

Luc Hens
Vlaamse Instelling voor Technologisch Onderzoek, Mol, Belgium

E. Dahwa and C. P. Mudzengi
Department of Research and Specialist Services, Makoholi Research Institute, Masvingo, Zimbabwe

C. Murungweni
Department of Animal Production and Technology, Chinhoyi University of Technology, Chinhoyi, Zimbabwe

X. Poshiwa
Marondera College of Agricultural Sciences, University of Zimbabwe, Marondera, Zimbabwe

S. Kativu
Department of Biological Sciences, University of Zimbabwe, Harare, Zimbabwe

T. Hungwe and M. D. Shoko
Faculty of Agricultural Sci-ences, Great Zimbabwe University, Masvingo, Zimbabwe

K. E. Ahoussi, Y. B. Koffi and J. Biémi
Unité de Formation et de Recherche (UFR) des Sciences de la terre et des Ressources Minières (STRM), Université de Cocody, Cocody, Côte d'Ivoire

M. Kouassi
Département des Sciences de la Terre et des Ressources Minières (STeRMi), Institut National Polytechnique Félix Houphouët Boigny (INP-HB), Yamoussoukro, Côte d'Ivoire

S. Loko
Laboratoire de chimie et de Microbiologie, Université Technologique et Tertiaire Loko (UTTLOKO), Abidjan, Côte d'Ivoire

Daniel Edison Husana
Environmental Biology Division, Institute of Biological Sciences, College of Arts and Sciences, University of the Philippines Los Baños, College, Laguna, Philippines
Global Center of Excellence for Environmental Studies, Graduate School of Environment and Information Sciences, Yokohama National University, Yokohama, Japan

Tomohiko Kikuchi
Global Center of Excellence for Environmental Studies, Graduate School of Environment and Information Sciences, Yokohama National University, Yokohama, Japan